计算机技术入门丛书

计算机专业英语

第3版·音频版

卜艳萍　周　伟 ◎ 编著

清华大学出版社
北京

内 容 简 介

本书共包含 25 个单元，涵盖计算机软硬件基础知识、计算机应用技术及前沿学科知识等内容，讲述了计算机硬件基础、计算机系统结构、程序设计语言、操作系统、数据结构、数据库技术、软件工程、计算机图像与动画、计算机网络技术、办公自动化、虚拟现实、人工智能、电子商务基础及其应用以及计算机领域的新技术（如大数据、云计算、物联网和移动商务）等内容。每个单元内容相对独立，单元后均附有重点词汇、课文难点注释、练习、两篇与课文内容相关的阅读材料，以及计算机专业英语相关的语法知识讲解或专业术语介绍。书末提供课文及第一篇阅读材料的参考译文。

本书可作为计算机科学与技术专业及相关电类、信息类专业教材，也可作为相关领域专业技术人员的参考书。

本书封面贴有清华大学出版社防伪标签，无标签者不得销售。
版权所有，侵权必究。举报：010-62782989，beiqinquan@tup.tsinghua.edu.cn。

图书在版编目（CIP）数据

计算机专业英语：音频版 / 卜艳萍，周伟编著. —3 版. —北京：清华大学出版社，2022.9（2025.1重印）
（计算机技术入门丛书）
ISBN 978-7-302-59877-0

Ⅰ. ①计… Ⅱ. ①卜… ②周… Ⅲ. ①电子计算机-英语-教材 Ⅳ. ①TP3

中国版本图书馆 CIP 数据核字（2022）第 003418 号

策划编辑：魏江江
责任编辑：王冰飞
封面设计：刘　键
责任校对：时翠兰
责任印制：曹婉颖

出版发行：清华大学出版社
  网　　址：https://www.tup.com.cn，https://www.wqxuetang.com
  地　　址：北京清华大学学研大厦 A 座　　　　邮　编：100084
  社 总 机：010-83470000　　　　　　　　　　邮　购：010-62786544
  投稿与读者服务：010-62776969，c-service@tup.tsinghua.edu.cn
  质 量 反 馈：010-62772015，zhiliang@tup.tsinghua.edu.cn
  课 件 下 载：https://www.tup.com.cn，010-83470236
印 装 者：三河市少明印务有限公司
经　　销：全国新华书店
开　　本：185mm×260mm　　印　张：22.25　　字　数：566 千字
版　　次：2010 年 8 月第 1 版　2022 年 9 月第 3 版　印　次：2025 年 1 月第 8 次印刷
印　　数：45201～47200
定　　价：58.00 元

产品编号：095268-01

前 言

党的二十大报告指出：教育、科技、人才是全面建设社会主义现代化国家的基础性、战略性支撑。必须坚持科技是第一生产力、人才是第一资源、创新是第一动力，深入实施科教兴国战略、人才强国战略、创新驱动发展战略，开辟发展新领域新赛道，不断塑造发展新动能新优势。高等教育与经济社会发展紧密相连，对促进就业创业、助力经济社会发展、增进人民福祉具有重要意义。

"计算机专业英语"是计算机科学与技术及相关专业的一门重要的专业课。其教学目标在于扩大学生的专业词汇量，提高学生阅读计算机硬件、软件、应用等方面英文资料的能力，培养理解和翻译计算机专业文献的能力。通过本课程的学习，既能掌握计算机软硬件的基础知识、专业技术及其应用系统，也能掌握计算机领域最新发展的比较前沿的新知识、新技术。

本书共包含 25 个单元，涵盖计算机软硬件基础知识、计算机应用技术及前沿学科知识等内容，讲述了计算机硬件基础、计算机系统结构、程序设计语言、操作系统、数据结构、数据库技术、软件工程、多媒体、计算机图像与动画、计算机网络技术、办公自动化、虚拟现实、人工智能、电子商务基础及其应用以及计算机领域的新技术（如大数据、云计算、物联网和移动商务）等内容。

本书每个单元的内容组织为课文、单词、注释、阅读材料 1 及单词、阅读材料 2 及单词、专业英语语法知识或专业术语、单元练习等。每单元练习题的题型有专业词组翻译、单词或短语填空、专业术语匹配等。每个单元的课文、阅读材料 1 和阅读材料 2 都属于同一知识领域范畴，能够覆盖本单元知识点的相关内容。

相对于前两版，本版教材中的单元数量和单元内容都做了修改：去掉了编译原理、远程教育、网格计算、分布式系统、互联网、神经网络、物流与供应链 7 个单元，将其中的部分内容组合到相关的单元中，作为阅读材料；新增加了大数据、云计算、物联网和移动商务 4 个全新的单元，代表计算机发展的新技术；附录中的计算机专业英语缩写词表及计算机专业英语词汇表也做了更新。本版教材依然沿用前两版的单元结构和内容组织方式，每个单元有课文和两篇阅读材料。课文和第一篇阅读材料提供参考译文，第二篇阅读材料不提供参考译文，以便在课堂进行循序渐进的阅读训练，考查学生阅读能力的提高情况。

为便于教学，本书提供丰富的配套资源，包括教学课件、在线作业（包括习题答案）和课文音频。

> **资源下载提示**
>
> **课件等资源**：扫描目录上方的二维码下载。
> **在线作业**：扫描封底的作业系统二维码，登录网站在线做题及查看答案。
> **音频等资源**：扫描封底的文泉云盘防盗码，再扫描书中相应章节中的二维码，可以在线收听。

　　本书由上海交通大学教师卜艳萍和华东理工大学教师周伟编写。卜艳萍编写了 Unit 1～Unit 8、Unit 16～Unit 19，以及完成全书的统稿工作；周伟编写了 Unit 9～Unit 15、Unit 20～Unit 25、附录 A 和附录 B。赵桂钦、王德俊、何飞、邱遥和周烨晴等对本书的选材及内容组织提供了许多宝贵意见和建议，在此一并表示感谢。

　　由于编者水平有限，书中不当之处，敬请读者批评指正。

卜艳萍
2022 年 7 月于上海

配套资源

Unit 1　Introduction of Computers ·· 1
 1.1　Text ··· 1
 1.2　Reading Material 1：Motherboard ·· 4
 1.3　Reading Material 2：Notebook Computer ·· 6
 1.4　专业英语基础知识 ·· 7
 1.5　Exercises ··· 9

Unit 2　CPU and Memory ··· 11
 2.1　Text ·· 11
 2.2　Reading Material 1：Hard Disks ··· 14
 2.3　Reading Material 2：RISC ·· 16
 2.4　专业英语词汇的构成特点 ··· 17
 2.5　Exercises ·· 21

Unit 3　Input and Output Systems ·· 23
 3.1　Text ·· 23
 3.2　Reading Material 1：Mouse ··· 26
 3.3　Reading Material 2：USB ··· 28
 3.4　词汇缩略 ·· 29
 3.5　Exercises ·· 31

Unit 4　C++ Language ·· 33
 4.1　Text ·· 33
 4.2　Reading Material 1：Categories of Computer Software ························· 36
 4.3　Reading Material 2：Visual Basic ··· 38
 4.4　计算机专用术语与命令 ··· 39
 4.5　Exercises ·· 42

Unit 5　Operating System ·· 44
 5.1　Text ·· 44
 5.2　Reading Material 1：Windows Vista ·· 47

5.3 Reading Material 2：Dual-core Computing ·········· 48
5.4 计算机专业英语中长句的运用 ·········· 49
5.5 Exercises ·········· 52

Unit 6 Data Structure ·········· 54

6.1 Text ·········· 54
6.2 Reading Material 1：.NET Framework ·········· 57
6.3 Reading Material 2：Multiprogramming and Multiprocessing ·········· 59
6.4 被动语态的运用 ·········· 60
6.5 Exercises ·········· 62

Unit 7 Database Principle ·········· 64

7.1 Text ·········· 64
7.2 Reading Material 1：Data Warehouse and Data Mining ·········· 67
7.3 Reading Material 2：Library ·········· 69
7.4 专业英语中常用的符号和数学表达式 ·········· 70
7.5 Exercises ·········· 73

Unit 8 Software Engineering ·········· 75

8.1 Text ·········· 75
8.2 Reading Material 1：Enterprise Resource Planning ·········· 78
8.3 Reading Material 2：Software Testing ·········· 80
8.4 专业英语的阅读 ·········· 81
8.5 Exercises ·········· 83

Unit 9 Multimedia ·········· 85

9.1 Text ·········· 85
9.2 Reading Material 1：Computer Aided Design ·········· 88
9.3 Reading Material 2：Creating an Effective Web Presence ·········· 90
9.4 专业术语的翻译及专业英语翻译的基本方法 ·········· 91
9.5 Exercises ·········· 96

Unit 10 Computer Graphics and Images ·········· 98

10.1 Text ·········· 98
10.2 Reading Material 1：Digital Photography ·········· 101
10.3 Reading Material 2：Recent Advances in Computer Vision ·········· 102
10.4 科技论文写作 ·········· 103
10.5 Exercises ·········· 106

Unit 11　Animation ·· 108

11.1　Text ·· 108
11.2　Reading Material 1：Video Compression ························· 111
11.3　Reading Material 2：Lossy Audio Compression ················ 112
11.4　句子成分简介 ··· 113
11.5　Exercises ·· 115

Unit 12　Big Data ··· 117

12.1　Text ·· 117
12.2　Reading Material 1：Using Instant Messaging for Business ··· 120
12.3　Reading Material 2：Applications of Big Data and IoT in
the Health Domain ·· 121
12.4　时态简介 ··· 122
12.5　Exercises ·· 125

Unit 13　Cloud Computing ·· 127

13.1　Text ·· 127
13.2　Reading Material 1：Cloud Computing and Cloud Storage ····· 130
13.3　Reading Material 2：More about the Cloud Computing ········ 132
13.4　分词和动名词 ··· 133
13.5　Exercises ·· 136

Unit 14　Computer Network Basics ··· 138

14.1　Text ·· 138
14.2　Reading Material 1：Communication Media ····················· 141
14.3　Reading Material 2：The Colorful Map of Internet ············· 142
14.4　动词不定式 ·· 143
14.5　Exercises ·· 146

Unit 15　Wireless Network ·· 148

15.1　Text ·· 148
15.2　Reading Material 1：Encryption on the Internet ················· 151
15.3　Reading Material 2：Mobile Networks ··························· 153
15.4　状语从句 ··· 154
15.5　Exercises ·· 157

Unit 16　Internet of Things ·· 159

16.1　Text ·· 159

16.2	Reading Material 1：The Future of IoT	162
16.3	Reading Material 2：RFID Technology	164
16.4	定语从句	164
16.5	Exercises	166

Unit 17　Computer Virus … 168

17.1	Text	168
17.2	Reading Material 1：Computer Virus Timeline	171
17.3	Reading Material 2：Using Antivirus Software	173
17.4	Terms	173
17.5	Exercises	174

Unit 18　Office Automation … 176

18.1	Text	176
18.2	Reading Material 1：Excel	179
18.3	Reading Material 2：Telephone	181
18.4	Terms	182
18.5	Exercises	183

Unit 19　Virtual Reality … 185

19.1	Text	185
19.2	Reading Material 1：Computers and Journalism	188
19.3	Reading Material 2：Virtual Reality in Industry	190
19.4	Terms	191
19.5	Exercises	191

Unit 20　Artificial Intelligence … 193

20.1	Text	193
20.2	Reading Material 1：Expert System	196
20.3	Reading Material 2：Genetic Algorithm	198
20.4	Terms	199
20.5	Exercises	200

Unit 21　Electronic Commerce … 202

21.1	Text	202
21.2	Reading Material 1：The Introduction of E-Commerce	205
21.3	Reading Material 2：E-Government	207
21.4	Terms	208

21.5　Exercises 209

Unit 22　Electronic Payment System 211

22.1　Text 211
22.2　Reading Material 1：E-Commerce Security 214
22.3　Reading Material 2：Using Credit Cards for Online Purchases 216
22.4　Terms 217
22.5　Exercises 218

Unit 23　Electronic Marketing 220

23.1　Text 220
23.2　Reading Material 1：Logistics and Supply-Chain Management 223
23.3　Reading Material 2：Customer Relationship Management 225
23.4　Terms 226
23.5　Exercises 227

Unit 24　Mobile Commerce 229

24.1　Text 229
24.2　Reading Material 1：O2O 232
24.3　Reading Material 2：A Mobile Web 233
24.4　Terms 234
24.5　Exercises 235

Unit 25　Network Security 237

25.1　Text 237
25.2　Reading Material 1：Network Firewall 240
25.3　Reading Material 2：Digital Signature and Certificates 242
25.4　Terms 243
25.5　Exercises 244

参考译文 246

附录 A　计算机专业英语缩写词表 319

附录 B　计算机专业英语词汇表 329

参考文献 342

Unit 1　Introduction of Computers

1.1　Text

A computer is a digital electronic data processing system. It is now an acknowledged definition. The digital computer is a digital system performs various computation tasks. The word digital implies that the information in the computer is represented by variables that take a limited number of discrete values. These values are processed internally by components that can maintain a limited number of states. Digital computer uses the binary number system, which has two digits: 0 and 1. A binary is called a bit. Information represented in digital computers is in groups of bits. By using various coding techniques, group of bits can be made to represent not only binary number but also other discrete symbols, such as decimal digits or letter of alphabet.[1] By judicious use of binary arrangements and by using various coding techniques, the groups of bits are used to develop complete set of instructions for performing various types of computation.

Computer Development

- The First Generation of Computers (1946 through 1959)

The first generation of computers was characterized by the most prominent feature of the ENIAC-vacuum tubes. Through 1950, several other notable computers were built, each contributing significant advancements, such as binary arithmetic, random access, and the concept of stored programs. These computer concepts are common in today's computers.

- The Second Generation of Computers (1959 through 1964)

To most people, the invention of the transistor meant small portable radios. To those in the data processing business, it signaled the start of the second generation of computers. The transistor meant more powerful, more reliable, and less expensive computers that would occupy less space and give off less heat than did vacuum-tube-powered computers.[2]

- The Third Generation of Computers (1964 through 1971)

Integrated circuits did for the third-generation what transistors did for the second generation. The compatibility problems of second-generation computers were almost eliminated in third-generation computers. However, third-generation computers differed radically from second-generation computers. The change was revolutionary, not evolutionary, and caused conversion nightmares for thousands of computer users. In time, the conversion of information systems from second-generation to third-generation hardware was written off as the price of progress.

◆ The Fourth Generation of Computers (1971 through now)

The fourth generation of computers is more difficult to define than the other three generations. This generation is characterized by more and more transistors being contained on a silicon chip. First there was large scale integration (LSI), with hundreds and thousands of transistors per chip, then came very large scale integration (VLSI), with tens of thousands and hundreds of thousands of transistors. The trend continues today. One of the most significant contributions to the emergence of the fourth generation of computers is the microprocessor.

Most computer vendors classify their computers as being in the fourth generation of computers, and a few call theirs "the fifth generation". The first three generations were characterized by significant technological breakthroughs in electronics — the use of vacuum tubes, then transistors, and then integrated circuits. Some people prefer to pinpoint the start of the fourth generation as 1971, with introduction of large-scale integration (more circuits per unit space) of electronic circuitry. However, other computer designers argue that if we accept this premise, then there would probably have been a fifth, a sixth, and maybe a seventh generation since 1971.

Computer Types

Computers can be general classified by size and power as follows:

Microcomputer: Microcomputer is generally a synonym for the more common term, personal computer, or PC, which is a small single-user computer based on a microcomputer. In addition to the microprocessor, a personal computer has a keyboard for entering data, a monitor for displaying information, and a storage device for saving data.[3]

Workstation: Workstation is a powerful single-user computer. It is like a personal computer, but it has a more powerful microprocessor and a higher-quality monitor.

Mainframe: Mainframe or mainframe computer is a powerful multi-user computer. It is capable of supporting many hundreds or thousands of users at the same time. It is now usually referred to a "large server".

Minicomputer: Minicomputer (a term no longer much used) is a multi-user computer of a size between a microcomputer and a mainframe.

Supercomputer: Supercomputer is an extremely fast computer that can perform hundreds of millions of instructions per second, but now it refers to a "very large server" and sometimes includes a system of computers using parallel processing.

Desktop Computers

Desktop computers are the natural choice when a computer remains in the same place for all of its working life.[4] The modular design of a desktop system makes it relatively easy to configure it with exactly the right set of features and functions and for your specific needs.

The parts inside a desktop computer usually follow one or more design standards, so it is often possible to replace a component that fails with a new one from a different manufacturer. And when you want to add more memory, a large hard drive or monitor to your system, you can be confident that you won't have to limit yourself to products from a single manufacturer. Just because the label on the case says Compaq or Gateway, you can still go to a big box-retailer and choose from among many different brands.

This combination of modular design and competition is one reason that the prices of most desktop computer components are lower than the comparable, non-standard parts in a laptop. In addition, the common parts specifications allow a repair shop to maintain a smaller inventory because they can use the same parts in many different desktop computer makes and models.[5]

Desktop computers are modular systems that make it easy to add or replace individual parts to meet each user's particular requirements. A computer intended for an illustrator or a computer-aided designer might have a higher-quality graphics controller and video display, where a purchasing agent may not use anything more demanding than a word processor and a spreadsheet. Most computer manufacturers let you order exactly the set of features and specifications that you want.

When you need change, it is usually easy to open up a desktop case and reconfigure the system, unless your computer uses proprietary parts. You can be confident that the sockets on the motherboard and the mounting holes in the drive bays fit the new expansion card or disk drive, and the main printed circuit board that controls the rest of the system works with the new parts.

Modular design also means that you can transfer some old parts to your new computer when your old faithful machine that has finally become obsolescent. Of course, there are some limits to this flexible design. You can not use a brand-new memory module or the latest disk drives with a 10-year-old motherboard because the designs have changed to accommodate newer and better processors and other devices.

You can improve the computer's performance by adding new components and replacing existing parts with new ones that have faster speed, greater capacity, or more features. Once again, the desktop computer's modular design makes it easy to work inside the case. Of course, there is a point of diminishing returns where it is better and less costly to buy a new system, but just about every desktop computer has room for economical improvement. The most common and effective motherboard has one or more sockets for memory modules, so you can increase the total amount of memory by adding one or more new modules to the memory that is already in place. You can also remove the existing memory and replace it with the same number of modules with more memory on each module. Adding memory is easier in a desktop system because there is plenty of space inside the case.

The CPU chip in a desktop system—the central processing unit that controls everything else—is also relatively easy to remove and replace with a faster CPU with similar architecture, and that fits in the same socket. A new CPU can offer faster processing and better performance than the one that was originally supplied with the computer. Unlike most of the other integrated circuits on the motherboard, the CPU mounts in a special socket that uses a latching mechanism to hole it in place.

Keywords

accommodate	v.	供应，适应，调节	discrete	adj.	离散的，分立的
alphabet	n.	字母表	eliminate	v.	消去，淘汰
arrangement	n.	安排，整理，约定	emergence	n.	出现，发生
compatibility	n.	兼容性，一致性	expansion	n.	扩展，扩大
configure	v.	配置	evolutionary	adj.	发展的，进化的
conversion	n.	变化，转换	illustrator	n.	说明者

Introduction of Computers

instruction	n.	指令		proprietary	adj.	专有的，独占的	
latch	v.	获得，抓住		reliable	adj.	可靠的，确实的	
mainframe	n.	大型机		represent	v.	描述，表示	
manufacturer	n.	制造商		revolutionary	adj.	革命性的	
notable	adj.	显著的，著名的		silicon	n.	硅	
obsolescent	adj.	废弃的，淘汰的		socket	n.	插座，插槽	
occupy	v.	占据，占用		transistor	n.	三极管	
portable	adj.	便携式的		vacuum	n.	真空	
prominent	adj.	突出的，重要的					

Notes

[1] By using various coding techniques, group of bits can be made to represent not only binary number but also other discrete symbols, such as decimal digits or letter of alphabet.

译文：通过应用各种编码技术，字位的组合不但可以用来表示二进制数，还可以表示其他离散符号，如十进制数字和字母表上的字母等。

说明：本句的 By using various coding techniques 作状语，主句采用被动语态。

[2] The transistor meant more powerful, more reliable, and less expensive computers that would occupy less space and give off less heat than did vacuum-tube-powered computers.

译文：晶体管计算机意味着功能更强、更可靠、更价廉，它与真空管计算机相比占地面积小，功耗小。

说明：本句的 that 引导定语从句，修饰宾语 computers。

[3] In addition to the microprocessor, a personal computer has a keyboard for entering data, a monitor for displaying information, and a storage device for saving data.

译文：除了微型处理器外，个人计算机还有输入数据的键盘、显示信息的显示器和存储数据的存储装置。

说明：本句中的 In addition to the microprocessor 是状语，宾语是 keyboard、monitor 和 storage device，宾语后的 for 结构是宾语补足语。

[4] Desktop computers are the natural choice when a computer remains in the same place for all of its working life.

译文：当一台计算机在它整个工作期间一直都被放在同一个地方时，台式计算机是一个理所当然的选择。

说明：本句由 when 引导时间状语从句。

[5] In addition, the common parts specifications allow a repair shop to maintain a smaller inventory because they can use the same parts in many different desktop computer makes and models.

译文：此外，一般的配件说明书中都允许维修车间维持较少的库存量，这是因为在不同的品牌和模式的台式计算机上可以使用相同的配件。

说明：本句由 because 引导原因状语从句，to maintain a smaller inventory 作宾语补足语。

1.2　Reading Material 1：Motherboard

　　Motherboard is also known as main board or system board. It is installed in the chassis. It is also one of the most important components of a computer. Motherboard is generally rectangular circuit board. If the

CPU is considered as the brain of a computer, then the motherboard is the body of a computer. When the computer has a high performance brain (CPU), it also needs a healthy strong body (motherboard) to operate. Motherboard affects the performance of the entire computer system.

Motherboard Structure

Motherboard is usually composed of a CPU slot, an AGP slot, a CNR slot, five PCI slots, three DIMM slots, two IDE interfaces, a floppy drive interface, two serial ports, one parallel port, a PS/2 keyboard interface, a PS/2 mouse interface, two USB interfaces, erasable BIOS, control chip sets and other components.

CPU slot consists of Socket series and Slot series. Socket series CPU slot uses ZIF (zero insert force) standards. Socket is next to a lever. Pull the lever, each stitch of CPU can be easily inserted into each orifice of the socket, then press lever to its original position, CPU can be fixed. Slot series adopt the form of slot, looking like a common expansion slot on the motherboard. Any external interface card, such as graphics, sound card, NIC, etc., would have to be planted on the expansion slot on the motherboard so that it can communicate with the motherboard and export images and sound.

AGP slot is devoted to the high-speed image processing. AGP slot is similar to PCI expansion slot on shape, and the color is brown. AGP can only be inserted graphics into, therefore on the motherboard there is only an AGP interface. Now most of the motherboards adopt AGP $8\times$ interfaces with AGP $8\times$ display cards, which greatly enhances the computer's ability to handle 3D.

The role of memory lots is to install memory. Memory slots have 30 lines, 72 lines, 168 lines and 184 lines, now 168 lines.

BIOS (Basic Input/Output System) is a ROM chip installed on the motherboard, with the most important basic input/output procedures of a computer system, the setup about CMOS, the self-inspection procedures and so on.

Floppy drive interface is generally a socket with 34 double row needles, with the label of Floppy, FDC or FDD. IDE interface is a socket with 40 double row needles, which is used to connect to IDE hard drives or CD-ROM drives. Now in order to improve the reliability of data transmissions, the needles are up to 80.

The motherboard provides a standard AT keyboard DIN connector for attaching a keyboard. You can directly plug a keyboard cable to this connector.

The motherboard also provides a 5-pin connector for PS/2 mouse cable (optional). You can directly plug a PS/2 mouse cable to it.

Power socket has two specifications, AT and ATX. ATX is the power socket specification used widely now, with the use of ATX power supply. ATX power socket is a 20-double socket, with anti-inserted wrong structure, if the plug is in wrong way, it could not get on. Hence there is no need to worry about burning the motherboard.

Computer system is closely related to the clock, and when the computer is turned off, the battery provides the power needed by the system clock. The battery, usually a button battery, is inserted into the battery slot of the motherboard.

Sound Card and Graphics

Sound card is audio sound card. Its basic function is to produce and handle voice signals, and then transfer the signals to the speakers. These signals exist in the computer in the form of digit. The digital signals will be processed by sound card chip, and then can be converted into analog signals which can be identified by people's ears by using the digital-to-analog (D/A) converter on the sound card. If it has to deal with external input sound signals, at the first step, sound card should convert these signals into the digital signal which can be recognized by the computer by using the analog-to-digital (A/D) converter, and then transfer the digitals to the sound card chip for processing.

Graphics is short for graphics display card. The early graphics was not took too much attention to, its role was merely to convert the data from the processor and then display them on the screen strictly. With the continuous increase of the graphics software, particularly the increasing popularity of the graphics system, just relying on the CPU to handle all the data would greatly affect the overall performance of the system, so graphics appeared.

Graphics works more as a coprocessor. It is right for accelerating graphics processing operation, and the CPU can handle other more important tasks. In fact, the graphics is already graphics accelerator card now, which can perform a number of graphic processing functions. The accelerator capability is to provide graphics function computing power by the chip set on the accelerator card, which is also known as accelerator or graphics processor. Generally speaking, there is a clock generator, VGA core and the hardware accelerator in the chip set.

Keywords

accelerate	v.	加速，促进	needle	n.	针，磁针
attention	n.	注意，留意	orifice	n.	口，孔
chassis	n.	机箱	popularity	n.	普及，流行
lever	n.	杠杆	rectangular	adj.	矩形的
merely	adv.	只，纯粹	strictly	adv.	精密地，严格地

1.3 Reading Material 2：Notebook Computer

Notebook computers are the further development of PCs. The primary innovation is to ensure the performance and speed of the new generation of notebook computers, so that users who are used to their powerful PCs on the desktops need not compromise on performance and speed when they switch to notebook computers while they are on the road, for example, traveling on duty, doing business outside office, interviewing guests or audience in public places, or even accessing to Internet anywhere they go. Today, notebook computers are no longer the "luxury toy" of executives as they are now more affordable than ever before.

Higher performance of CPUs, multimedia technology, as well as bigger memory and large-capacity silicon chips allow manufacturers to put in fancier things into notebook computers. For notebook computers to be part of the enterprise computing system, manufacturers are including built-in networking, modem and other forms of communications capabilities in notebook computers.

To make the Compaq notebook easy to use, the company preloads all models with the Tabwords, software for launching, organizing and accessing applications and documents. This graphical interface looks like a notebook-diary to encourage intuitive working. Versions of the ThinkPad 750 series use both pen and keyboard input. Voice input will soon be available for basic commands on all its 750 series units. When configured with speech recognition software, IBM promises full dictation in near future.

In summary, the trend of notebook computers is as follows:

- Faster speed. Processors used are now usually fast CPUs.
- Large storage capacity.
- Multimedia capabilities. To support microphones, headphones, video-performance, and CD-ROM.
- Easier input methods. Pen, voice, and speech recognition in future.
- Built-in data communication support. Networking, built-in modem, even wireless support.

Although the battery life of notebook computers is longer than what it used to be, manufacturers continue to work to extend the battery life of notebook computers.

Weight is an important consideration and notebook computers are getting lighter although more features have been packed into the system.

Keywords

affordable	adj.	负担得起的	innovation	n.	创新，革新
compromise	n.	折中方案，调和	interview	v.	协商，会谈
fancier	n.	行家，空想家	intuitive	adj.	直觉的，直观的
headphone	n.	头戴受话器，耳机			

1.4 专业英语基础知识

计算机专业是融科学性与技术性于一体的学科，在专业文献的表达中应当遵循科技文体的规范。科学著作、学术论文、实验报告、设计报告、科技产品说明书、科技产品操作指导等都属于科技文体。科技文体讲究逻辑的条理清楚和叙述的准确、严密。计算机专业英语以表达计算机专业知识和技术的概念、理论和事实为主要目的，专业英语的主要特点是具有很强的专业性。因此，在表达中要注重客观事实和真相，要求逻辑性强、条理规范、精练及正式。

因为专业科技文献所涉及的内容（如科学定义、定理、方程式或公式、图表等）一般没有特定的时间关系，所以在专业文献中大部分都使用一般现在时。至于一般过去时、一般完成时也在专业英语中经常出现，如科技报告、科技新闻、科技史料等。用尽可能少的单词来清晰地表达原意，这就导致了非限定动词、名词化单词或词组及其他简化形式的广泛使用。

为了准确精细地描述事物过程，有时句子很长，会出现一段就是一个句子的情况。长句反映了客观事物中复杂的关系，它与前述精练的要求并不矛盾，句子虽长，但结构仍是精练的，只是包含的信息量大，准确性较高。

计算机专业文章一般重在客观地叙述事实，力求严谨和清楚，避免主观成分和感情色彩，这就决定了专业文体具有以下语法特点。

（1）叙述方式常避免用第一人称单数，而用第一人称复数 we，或者用 the author 等第三人称形式。

（2）被动语态使用频繁，而且多为没有行为的被动语态。

（3）时态形式使用比较单一，最常用的有五种时态，即一般现在时、现在进行时、一般过去时、一般将来时和现在完成时。

（4）谓语经常使用静态结构，用来表示状态或情况。

（5）定语经常使用名词作定语，以取得简洁的效果。

（6）经常使用动词非限定形式来扩展句子，如动词不定式短语、动名词短语、分词短语及独立分词结构。

（7）名词性词组多，即以名词为中心词构成短语以取代句子。

（8）多重复合句较多，句子中又嵌入句子。

（9）在说明书、手册中广泛使用祈使句。

（10）逻辑词语使用很频繁，明确表示出内容的内在联系，有助于清楚地叙述、归纳、推理、论证和概括，如 as a result、consequently、on the contrary、as mentioned above 等。

（11）插图、插画、表格、公式、数字所占比例大。

以下句子能充分体现计算机专业英语的语法特点。

（1）The procedure by which a computer is told how to work is called programming.

句子的主要结构为 The procedure is called programming。用一般现在时和被动语态。by which 为"介词+关系代词"引导定语从句，从句的谓语也为被动语态，which 指代 procedure。

译文： 告诉计算机如何工作的过程称为程序设计。

（2）Written language uses a small number of symbols which are easily encoded in digital form and can be combined in innumerable ways to convey meaning.

句中 are encoded 和 can be combined 是并列谓语，用被动语态，in digital form 和 in innumerable ways 中的介词 in 表示以什么形式、用什么方式。

译文： 书面语言只使用少数符号，它们很容易用数字形式编码，并且可以用数不清的方法进行组合以便表达意义。

（3）Having developed the capacity to store vast quantities of data, and compress it into a small space, and now having the capability to accurately and quickly retrieve information via adaptive pattern recognition make up the core which is taking multimedia solutions from a fringe technology to something which is at the beginning of pervading every aspect of our lives, and finally, living up to the promises that the hype has generated during the last five years.

这是长句，翻译时要注意准确和精练。本句中的主语是动名词短语 Having developed the capacity to store vast quantities of data, and compress it into a small space, and now having the capability to accurately and quickly retrieve information via adaptive pattern recognition，谓语是 make up，宾语是 the core，which is at the beginning of pervading every aspect of our lives, and finally, living up to the promises 是定语从句，修饰 something。that the hype has generated during the last five years 也是定语从句，修饰 promises。

译文： 已经能够存储大量数据并把它压缩到一个小的空间，能够迅速而准确地通过自适应模式识别来恢复信息，这些构成了这个核心。该核心正在使多媒体从一项边缘技术变成一项渗透我们生活的每一方面，并且最后将实现广告在近五年里所做出许诺的技术。

（4）Technology for development will allow construction of larger projects, artificial intelligence (intelligent agents, knowledge based systems, data mining and intelligent filtering, and so on) will be increasingly feasible as costs decrease, performance improves and widespread networking are available.

这也是一个长句。本句的主语是 Technology for development，谓语是 will allow，其后是一宾语从句，作 allow 的宾语。在该宾语从句中，construction of larger projects, artificial intelligence (intelligent agents, knowledge based systems, data mining and intelligent filtering, and so on)作主语，will be increasingly feasible 是系表结构作谓语，as 引导的原因状语从句作状语，修饰谓语。

译文：技术发展带来的成本的降低、性能的提高，以及网络的广泛应用使得建造更大的工程、人工智能（智能代理、知识库系统、数据挖掘及智能过滤等）将更加可行。

计算机专业术语多，而且派生和新出现的专业用语还在不断地增加。这些术语的出现是和计算机技术的高速发展分不开的。在词汇方面，体现为专业术语多、缩略词多、半技术词汇多、合成新词多、希腊词根和拉丁词根的词多等特点。例如，Internet、Intranet、Extranet 等都是随着网络技术的发展而出现的。CPU(central processing unit)、WPS(word processing system)、NT(network technology)、IT(information technology)等都是典型的缩略语。这些词汇的掌握首先要有一定的英语词汇量，还要有对新技术的了解。

科学技术本身的性质要求专业英语与专业内容相互配合，相互一致，这就决定了专业英语与普通英语有很大的差异。

1.5　Exercises

1. **Translate the following phrases.**

   data processing system

   digital computer

   integrated circuit

   very large scale integration

   single-user computer

   video display

   desktop computers

   printed circuit board

   central processing unit

   decimal digit

   sound card

   notebook computer

2. **Fill in the blanks with appropriate words or phrases.**

   a. second　　　　b. microprocessor　　　c. multi-user　　　d. desktop computers

   e. in place　　　f. change　　　　　　　g. computation　　　h. instructions

(1) The digital computer is a digital system performs various_____tasks.

(2) Integrated circuits did for the third-generation what transistors did for the _____ generation.

(3) One of the most significant contributions to the emergence of the fourth generation of computers is the_____.

(4) Mainframe computer is a powerful_____computer.

(5) Supercomputer is an extremely fast computer that can perform hundreds of millions of _____per second.

(6) _____ are modular systems that make it easy to add or replace individual parts to meet each user's particular requirements.

(7) When you need _____, it is usually easy to open up a desktop case and reconfigure the system.

(8) The CPU mounts in a special socket that uses a latching mechanism to hole it _____.

3. Match the following terms to the appropriate definition.

    a. agent   b. applet   c. analog-to-digital

    d. authentication   e. application   f. broadcasting

(1) The process of identifying an individual, usually based on a username and password.

(2) Client or robot programs, often able to act autonomously.

(3) A small Java program that can be embedded in an HTML page.

(4) The dissemination of any form of radio electric communications by means of Hertzian waves intended to be received by the public.

(5) A device that converts a signal whose input is information in the analog form and whose output is the same information in digital form.

(6) Also called end-user programs, includes database programs, word processors, and spreadsheets.

Unit 2　CPU and Memory

2.1　Text

音频

CPU

The CPU means the central processing unit. It is the heart of a computer system. A processor is a functional unit that interprets and carries out instructions. Every processor comes with a unique set of operations such as ADD, STORE, or LOAD that represent the processor's instruction set.[1] Computer designers are fond of calling their computer machines, so the instruction set is sometimes referred to as machine instructions and the binary language in which they are written is called machine language.

A CPU can be a single microprocessor chip, a set of chips, or a box of boards of transistors, chips, wires, and connectors. Differences in CPUs distinguish mainframes, minicomputers and microcomputers. A processor is composed of two functional units: a control unit and an arithmetic/logic unit, and a set of special workspaces called registers.

Two main performance factors of CPU are as follow:

Clock rate: A computer contains a system clock that emits pluses to establish the timing for system operations. The system is not the same as a "real-time clock" that keep track of the time of day. Instead, the system clock sets the speed of "frequency" for data transport and instruction execution. The clock rate set by the system clock determines the speed at which the computer can execute an instruction and, therefore, limits the number of instructions that a computer can complete within a specific amount of time.[2] The time to complete an instruction cycle is measured in megahertz (MHz), or millions of cycles per second.

The microprocessor in the original IBM PC performed at 4.77MHz. Today's processors perform at speeds exceeding 2GHz. If all other specifications are identical, higher megahertz rating mean faster processing.

Word size: It refers to the number of bits that the central processing unit can manipulate at one time. Word size is based on the size of the registers in the CPU and the number of data lines in the bus. For example, a CPU with an 8-bit word size is referred to as an 8-bit processor, it has 8-bit registers and manipulates 8-bit at a time.

A computer with a large word size can process more data in each instruction cycle than a computer with a small word size. Processing more data in each cycle contributes to increased performance. For example, the first microcomputers contained 8-bit microprocessors, but today's faster computers contain

32-bit or 64-bit microprocessor.

64-bit Processors

The newest thing in processor design is 64-bit ALUs, and people are expected to have these processors in their home PCs in the next decade. There has also been a tendency toward special instructions that make certain operations particularly efficient, and the addition of hardware virtual memory support and L1 caching on the processor chip. All of these trends push up the transistor count, leading to the multi-million transistor powerhouses available today. These processors can execute about one billion instructions per second.

Sixty-four-bit processors have been with us since 1992, and in the 21st century they have started to become mainstream. Both Intel and AMD have introduced 64-bit chips, and the Mac G5 supports a 64-bit processor. Sixty-four-bit processors have 64-bit ALUs, 64-bit registers, 64-bit buses and so on.

One reason why the world needs 64-bit processors is because of their enlarged address spaces. Thirty-two-bit chips are often constrained to a maximum of 2GB or 4GB of RAM access. However, a 4GB limit can be a severe problem for server machines and machines running large databases. And even home machines will start bumping up against the 2GB or 4GB limit pretty soon if current trends continue. A 64-bit chip has none of these constraints because a 64-bit RAM address space is essentially infinite for the foreseeable future—2^{64} bytes of RAM is something on the order of quadrillion gigabytes of RAM.

With a 64-bit address bus and wide, high-speed data buses on the motherboard, 64-bit machines also offer faster I/O speeds to things like hard disk drives and video cards.[3] These features can greatly increase system performance.

Servers can definitely benefit from 64 bits, but what about normal users? Beyond the RAM solution, it is not clear that a 64-bit chip offers "normal users" any real, tangible benefits at the moment. They can process data (very complex data features lots of real numbers) faster. People doing video editing and people doing photographic editing on very large images can benefit from this kind of computing power.[4] High-end games will also benefit, once they are re-coded to take advantage of 64-bit features.

Multiprocessing

Multiprocessing refers to the organizational technique in which a number of processor units are employed in a single computer system to increase the performance of the system in its application environment above the performance of a single processor.

Multiprocessor systems may be classified into four types: single instruction stream, single data stream (SISD); single instruction steam, multiple data stream (SIMD); multiple instruction stream, single data stream (MISD); and multiple instruction stream, multiple data stream (MIMD). Systems in the MISD category are rarely built. The other three architectures may be distinguished simply by the differences in their respective instruction cycles:

In an SISD architecture there is a single instruction cycle; operands are fetched serially into a single processing unit before execution.

An SIMD architecture also has a single instruction cycle, but multiple sets of operands may be fetched to multiple processing units and may be operated upon simultaneously within a single instruction cycle. Multiple-functional-unit, array, vector, and pipeline processors are in this category.

In an MIMD architecture, several instruction cycles may be active at any given time, each independently fetching instructions and operands into multiple processing units and operating on them in a concurrent fashion. This category includes multiple processor systems in which each processor has its own program control, rather that sharing a single control unit.

Memory

The memory unit is an essential component in any digit computer since it is needed for storing the programs that are executed by the CPU.[5] RAM and ROM play important roles of storage devices. RAM is an acronym of random access memory. RAM bars are used for main memory bars. ROM means Read Only Memory. The control and specialized programs are stored in ROM chips and these programs can not be changed by users. The basic operating instructions are stored in ROM and are not erased when the computer is turned off. The computer has a module, called the flash EEPROM (electrically erasable programmable read-only memory) that can be updated. The BIOS instructions and the configuration utility program are stored in the flash EEPROM in the computer.

The instructions that the computer gets and the information the computer processes remain in RAM during your work sessions. RAM is not a permanent storage place for information. When you turn your computer off, the information you entered during the work session does not remain in memory. Since RAM is only active when the computer is on, the computer uses disk drives to store information even when the computer is off.

Computer memory is measured in kilobytes or megabytes of information. (A byte is the amount of storage needed to hold one character, such as a letter or a numeric digit.) One kilobyte (KB) equals 1024 bytes, and one megabyte (MB) is about 1 million bytes. Software requires the correct amount of RAM to work properly. If you want to add new software to your computer, you can usually find the exact memory requirements on the software packaging.

Cache is another factor affecting CPU performance. It is special high speed memory that gives the CPU more rapid access to data. A very fast CPU can execute an instruction so quickly that it often must wait for data to be delivered from RAM, which slow processing. The cache ensures that data is immediately available whenever the CPU requests it.

Keywords

constraint	n.	约束，强制力	permanent	adj.	永久的，不变的
contribute	v.	贡献	quadrillion	num.	千的五次幂，千万亿
distinguish	v.	区别，辨别	register	n.	寄存器
emit	v.	发射，发行	respective	adj.	分别的，各自的
independently	adv.	独立地，任意地	serially	adv.	连续地，顺次
kilobyte	n.	千字节	server	n.	服务器
megahertz	n.	兆赫	simultaneously	adv.	同时发生地，同时
mainstream	n.	主流	specification	n.	规范，规格，说明
manipulate	v.	处理，控制	tendency	n.	倾向，趋势
performance	n.	性能			

Notes

[1] Every processor comes with a unique set of operations such as ADD, STORE, or LOAD that represent the processor's instruction set.

译文：每个处理器都有一个独特的诸如 ADD、STORE 或 LOAD 这样的操作集，这个操作集就是该处理器的指令系统。

说明：本句的宾语是"operations such as ADD, STORE, or LOAD", that represent the processor's instruction set 作定语，修饰宾语。

[2] The clock rate set by the system clock determines the speed at which the computer can execute an instruction and, therefore, limits the number of instructions that a computer can complete within a specific amount of time.

译文：系统时钟的频率决定了计算机执行指令的速度，因此限制了计算机在一定时间内所能执行的指令数。

说明：这是一个并列句，主语只有一个，set by the system clock 作定语，修饰主语 The clock rate，谓语有 determines 和 limits 两个，at which the computer can execute an instruction 作定语，修饰宾语 speed，而 that a computer can complete within a specific amount of time 也是定语，修饰 instructions。

[3] With a 64-bit address bus and wide, high-speed data buses on the motherboard, 64-bit machines also offer faster I/O speeds to things like hard disk drives and video cards.

译文：利用主板上的 64 位地址总线与宽的高速数据总线，64 位机器也为硬盘驱动器和显示卡等提供了较快的 I/O 速度。

说明：本句的 With a 64-bit address bus and wide, high-speed data buses on the motherboard 作状语，to things like hard disk drives and video cards 是宾语补足语。

[4] People doing video editing and people doing photographic editing on very large images can benefit from this kind of computing power.

译文：这类计算功能对进行视频编码和超大图像照片编辑的人有好处。

说明：本句的主语是并列的两个 people，doing video editing 修饰第一个 people，doing photographic editing on very large images 修饰第二个 people。

[5] The memory unit is an essential component in any digit computer since it is needed for storing the programs that are executed by the CPU.

译文：任何一台数字计算机都需要存储 CPU 所执行的程序，因此，存储器是计算机最重要的部件之一。

说明：本句由 since 引导原因状语从句。

2.2　Reading Material 1：Hard Disks

　　Nearly every desktop computer and server in use today contains one or more hard-disk drivers. Every mainframe and supercomputer is normally connected to hundreds of them. These billions of hard disks do one thing well—they store changing digital information in a relatively permanent form. They give computers the ability to remember things when the power goes out.

　　Hard disks were invented in the 1950s. They started as large disks up to 20 inches in diameter holding just a few megabytes. A modern hard disk is able to store an amazing amount of information in a small space. A hard disk can also access any of its information in a fraction of a second. A typical desktop machine will have a hard disk with a capacity of between 10 and 40 gigabytes. Data is stored onto the disk in the form of files. A file is simply named collection of bytes. The bytes might be the ASCII codes for the characters of a text file, or they could be the instructions of a software application for the computer to execute, or they could be the records of a database, or they could be the pixel colors for a GIF image. No matter what it contains, however, a file is simply a string of bytes.

There are two ways to measure the performance of a hard disk:

♦ Data Rate

The data rate is the number of bytes per second that the drive can deliver to the CPU. Rates between 5 and 40 megabytes per second are common.

♦ Seek Time

The seek time is the amount of time between when the CPU requires a file and when the first byte of the file is sent to the CPU. Times between 10 and 20 milliseconds are common.

The other important parameter is the capacity of the drive, which is the number of bytes it can hold.

A typical hard-disk drive is a sealed aluminum box with controller electronics attached to one side. The electronics control the read/write mechanism and the motor that spins the platters. The electronics also assemble the magnetic domains on the drive into data (reading) and turn data into magnetic domains (writing). The electronics are all contained on a small board that detaches from the rest of the drive.

Underneath the board are the connections for the motor that spins the platters, as well as a highly-filtered vent hole that lets internal and external air pressures equalize. Removing the cover from the drive reveals an extremely simple but very precise interior.

♦ The Platters

The platters typically spin at 3600 or 7200 rpm when the drive is operating. These platters are manufactured to amazing tolerances and are mirror-smooth.

♦ The Arm

The arm holds the read/write heads and is controlled by the mechanism in the upper-left corner. The arm is able to move the heads from the hub to the edge of the drive. The arm and its movement mechanism are extremely light and fast. The arm on a typical hard-disk drive can move from hub to edge and back up to 50 times per second.

In order to increase the amount of information the drive can store, most hard disks have multiple platters. The mechanism that moves the arms on a hard disk has to be incredibly fast and precise. It can be constructed using a high-speed linear motor. Many drives use a "voice coil" approach—the same technique used to move the cone of a speaker on your stereo is used to move the arm.

Data is stored on the surface of a platter in sectors and tracks. Tracks are concentric circles, and sectors are pie-shaped wedges on a track. A sector contains a fixed number of bytes—for example, 256 or 512. Either at the drive or the operating system level, sectors are often grouped together into clusters. The process of low-level formatting a drive establishes the tracks and sectors on the platter. The starting and ending points of each sector are written onto the platter. This process prepares the drive to hold blocks of bytes high-level formatting then writes the file-storage structures, like the file-allocation table, into the sectors. This process prepares the drive to hold files.

Keywords

aluminum	n.	铝	platter	n.	圆盘
coil	n. v.	一圈，盘绕	seal	v.	密封，隔离
detach	v.	分开，脱离	tolerance	n.	忍受，耐性
equalize	v.	均衡，调整	underneath	prep. adv.	在下面，在底部
incredibly	adv.	惊人地，不可思议地	wedge	n.	楔，楔形

2.3 Reading Material 2：RISC

The simplest way to examine the advantages and disadvantages of RISC (reduced instruction set computer) architecture is by contrasting it with its predecessor: CISC (complex instruction set computer) architecture.

The primary goal of CISC architecture is to complete a task in as few lines of assembly as possible. This is achieved by building processor hardware that is capable of understanding and executing a series of operations. For this particular task, a CISC processor would come prepared with a specific instruction ("MULT"). "MULT" is what is known as a "complex instruction". When executed, this instruction loads the two values into separate registers, multiples the operands in the execution unit, and then stores the product in the appropriate register. Thus, the entire task of multiplying two numbers can be completed with one instruction.

RISC processors only use simple instructions that can be executed within one clock cycle. Thus, the "MULT" command could be divided into three separate commands: "LOAD", which moves data from the memory bank to a register, "PROD", which finds the product of two operands located within the registers, and "STORE", which moves data from a register to the memory banks.

At first, this may seem like a much less efficient way of completing the operation. Because there are more lines of code, more RAM is needed to store the assembly level instructions. The compiler must also perform more work to convert a high-level language statement into code of this form.

However, this RISC strategy also brings some very important advantages. Because each instruction requires only one clock cycle to execute, the entire program will execute in approximately the same amount of time as the multi-cycle "MULT" command. These RISC "reduced instructions" require less transistors of hardware space than the complex instructions, leaving more room for general purpose registers. Because all of the instructions execute in a uniform amount of time (i.e. one clock), pipelining is possible.

Despite the advantages of RISC based processing, RISC chips took over a decade to gain a foothold in the commercial world. This was largely due to a lack of software support. Without commercial interest, processor developers were unable to manufacture RISC chips in large enough volumes to make their price competitive.

The CISC approach attempts to minimize the number of instructions per program, sacrificing the number of cycles per instruction. RISC does the opposite, reducing the cycles per instruction at the cost of the number of instructions per program.

Keywords

appropriate	adj.	合适的，适当的	operand	n.	操作数
approximately	adv.	接近地，近似地	pipeline	n.	管线，管道
commercial	adj.	商业上的，贸易的	predecessor	n.	祖先，前任，前辈
competitive	adj.	竞争的，竞赛的	primary	adj.	主要的，最初的
foothold	n.	立足点	sacrifice	n.	牺牲，损失

2.4 专业英语词汇的构成特点

词汇是阅读、翻译和写作的基础。随着社会和信息传递的发展，新的词汇层出不穷，尤其在计算机科学与技术领域，这种现象更为突出。专业英语的构词有两个显著特点：一是大部分专业词汇来自希腊语和拉丁语；二是前缀和后缀的出现频率非常高。希腊语和拉丁语是现代专业英语词汇的基础。各行各业都有一些自己领域的专业词汇，有的是随着本专业的发展应运而生的，有的是借用公共英语中的词汇，有的是借用外来语言词汇，有的则是人为构造的词汇。

计算机专业英语中常见的词汇类型有以下几种。

1. 技术词汇和次技术词汇

技术词汇的意义狭窄、单一，一般只使用在各自的专业范围内，因而专业性很强。这类词一般较长并且越长词义越狭窄，出现的频率也不高，如 superconductivity（超导性）、hexadecimal（十六进制）、amplifier（放大器）、bandwidth（带宽）、flip-flop（触发器）等。

次技术词汇是指不受上下文限制的各专业中出现频率都很高的词。这类词往往在不同的专业中具有不同的含义。例如，register 在计算机系统中表示寄存器，在电学中表示计数器、记录器，在乐器中表示音区，而在日常生活中则表示登记簿、名册、挂号信等。

2. 合成词

合成词是专业英语中另一大类词汇，其组成面广，多数以短画线-连接单词构成，或者采用短语构成。合成方法有名词+名词、形容词+名词、动词+副词、名词+动词、介词+名词、形容词+动词等。但是合成词并非随意可以构造，否则会形成一种非正常的英语句子结构，虽然可由多个单词构成合成词，但这种合成方式太冗长，应尽量避免。

下面这些是由短画线-连接而成的合成词。

file + based → file-based 基于文件的
Windows + based → Windows-based 以 Windows 为基础的
object + oriented → object-oriented 面向对象的
thread + oriented → thread-oriented 面向线程的
point + to + point → point-to-point 点到点
plug + and + play → plug-and-play 即插即用
pear + to + pear → pear-to-pear 对等的
front + user → front-user 前端用户
push + up → push-up 上拉
pull + down → pull-down 下拉
line + by + line → line-by-line 逐行

CPU and Memory

paper + free → paper-free 无纸的
jumper + free → jumper-free 无跳线的
user + centric → user-centric 以用户为中心的
power + plant → power-plant 发电站
cast + iron → cast-iron 铸铁
conveyer + belt → conveyer-belt 传送带
machine + made → machine-made 机制的
reading + room → reading-room 阅览室

随着词汇的专用化，合成词中间的连接符被省略掉，形成了一个独立的单词。例如：

in + put → input 输入
out + put → output 输出
feed + back → feedback 反馈
work + shop → workshop 车间
fan + in → fanin 扇入
fan + out → fanout 扇出
on + line → online 在线
air + craft → aircraft 飞机
metal + work → metalwork 金属制品

英语中有很多专业术语由两个或更多的词组成，称为复合术语。它们的构成成分虽然看起来是独立的，但实际上合起来构成一个完整的概念。

liquid crystal　　　　　液晶
computer language　　　计算机语言
machine building　　　　机器制造
linear measure　　　　　长度单位
civil engineering　　　　土木工程

3. 派生词

派生方法也称缀合，这类词汇非常多，专业英语词汇大部分都是用派生法构成的，它是根据已有的词，通过对词根加上各种前缀和后缀来构成新词。这些词缀有名词词缀，如 inter-、sub-、in-、tele-、micro- 等；形容词词缀，如 im-、un-、-able、-al、-ing、-ed 等；动词词缀，如 re-、under-、de-、-en、con- 等。其中，采用前缀构成的单词在计算机专业英语中占了很大比例。

英语的前缀是有固定意义的，记住其中的一些常用前缀对于记忆生词和猜测词义很有帮助。加前缀构成新词只改变词义，不改变词性。加后缀构成新词可能改变也可能不改变词义，但一般改变词性。有的派生词加后缀的时候，语音或拼写可能发生变化。从一个词的后缀可以判别它的词类，这是它的语法意义。它们的词汇意义往往并不明显。

下面是一些典型的派生词。

1）hyper- 超级
hypertext　超文本　　　　　hypermedia　超媒体
hyperswitch　超级交换机　　hypersonic　超音速的

2）inter- 相互，在……之间
interface　接口　　　　　　intercommunication　通信
interlace　隔行扫描　　　　interactive　交互的

3）micro- 微型
microprocessor 微处理器　　　　microelectronics 微电子
microcomputer 微型计算机　　　　microwave 微波

4）multi- 多，多的
multimedia 多媒体　　　　　　　　multiprocessor 多处理器
multiprogram 多道程序　　　　　　multicast 多点传送

5）poly- 多，复，聚
polycrystal 多晶体　　　　　　　　polytechnical 多工艺的
polyatomic 多原子的　　　　　　　polymorphism 多态性

6）re- 再，重新
rerun 重新运行　　　　　　　　　　rewrite 改写
resetup 重新设置　　　　　　　　　reexchange 再交换

7）semi- 半
semiconductor 半导体　　　　　　　semiautomatic 半自动的
semidiameter 半径　　　　　　　　　semicircular 半圆的

8）super- 超级
supercomputer 超级计算机　　　　　superclass 超类
superstructure 上层建筑　　　　　　superuser 超级用户

9）tele- 远程的，电的
telephone 电话　　　　　　　　　　teleconference 远程会议
telescope 望远镜　　　　　　　　　telegraph 电报

10）ultra- 超过，极端
ultrashort 超短（波）的　　　　　　ultrared 红外线的
ultraspeed 超高速的　　　　　　　　ultramicroscope 超显微镜

11）un- 反，不，非
unformat 未格式化的　　　　　　　　undelete 恢复
uninstall 卸载　　　　　　　　　　　unimportant 不重要的

12）-able 可能的
programmable 可编程的　　　　　　　portable 便携的
adjustable 可调整的　　　　　　　　considerable 值得重视的

13）-ate 成为……，处理
eliminate 消除　　　　　　　　　　　circulate 循环，流通
terminate 终止　　　　　　　　　　　estimate 估计，估算

14）-ic 有……特性的，属于……的
academic 学术的　　　　　　　　　　elastic 灵活的
atomic 原子的　　　　　　　　　　　periodic 周期的

15）-ive 有……性质的，与……有关的
productive 生产的　　　　　　　　　expensive 昂贵的
active 主动的　　　　　　　　　　　attractive 有吸引力的

16）-ize ……化，变成

characterize 表示……的特性　　industrialize 使工业化
optimize 完善　　realize 实现

17）-lity ……性能

reliability 可靠性　　confidentiality 保密性

18）-ment 行为，状态

development 发展　　agreement 同意，协议
equipment 设备　　adjustment 调整

19）-meter 计量仪器

barometer 气压表　　telemeter 测距仪

20）-ware 件，部件

hardware 硬件　　software 软件

4．借用词

借用词是指借用公共英语及日常生活用语中的词汇来表达专业含义。借用词一般来自厂商名、商标名、产品代号名、发明者名、地名等，也可将普通公共英语词汇演变成专业词意而实现。也有对原来词汇赋予新的意义的。例如：

cache 高速缓存　　firewall 防火墙
fitfall 子程序入口　　flag 标志，状态
mailbomb 邮件炸弹　　semaphore 信号量

英语科技问题中有很多词汇并不是专业术语，但在日常口语中用得不是很多，它们多见于书面语中。掌握这类词对阅读科技文献或写科技论文十分重要。例如：

accordance 按照　　acknowledge 承认
alternative 交替的　　application 应用
appropriate 恰当的　　circumstance 情况
compensation 补偿　　imply 隐含
confirm 证实　　modification 修改
inclusion 包括　　indicate 指示
induce 归纳　　initial 初始的
nonetheless 然而　　nevertheless 然而

5．通过词类转化构成新词

通过词类转化构成新词是指一个词不变化词形，而由一种词类转化为另一种或几种词类，有时发生重音或尾音的变化。英语中名词、形容词、副词、介词可以转化成动词，而动词、形容词、副词、介词可以转化成名词，但最活跃的是名词转化成动词和动词转化成名词。例如：

break（动词）打破 → break（名词）间歇
but（连词）但是 → but（介词）除了
by（介词）在……旁边 → by（副词）在一旁
center（名词）中心 → center（动词）集中
clear（形容词）明确的 → clear（动词）清除
close（动词）关上 → close（副词）靠近

coordinate（动词）协调→ coordinate（名词）坐标

hard（形容词）坚硬的 → hard（副词）努力地

island（名词）小岛→ island（动词）隔离

名词化是英语科技文体中一种常见的现象。所谓名词化，就是把动词变成有动作含义的名词。如果是动词短语或是句子，则把这个动词短语或句子变成名词短语。

air moves 可以转化成 the motion of air，其含义为"空气运动"，air 是 motion 的行为主体。

to apply force 可以转化成 the application of force，其含义为"应用力"，force 是 application 的宾语。

analytical chemists develop the equipment 可以转化成 the development of the equipment by analytical chemists，其含义为"由分析化学家开发的设备"，analytical chemists 是 the development 的行为主体。

2.5　Exercises

1. **Translate the following phrases.**

   random access memory

   high-speed data bus

   binary language

   machine language

   system clock

   word size

   multiprocessor

   photographic editing

   single instruction stream

   multiple data stream

   disk drive

   reduced instruction set computer

2. **Fill in the blanks with appropriate words or phrases.**

   a. single computer system　　b. instructions　　c. 64-bit processors　　d. storage

   e. RAM　　　　　　　　　　f. SISD　　　　　g. performance　　　　h. cache

(1) A processor is a functional unit that interprets and carries out _____.

(2) Thirty-two-bit chips are often constrained to a maximum of 2GB or 4GB of _____ access.

(3) Processing more data in each cycle contributes to increased _____.

(4) The _____ ensures that data is immediately available whenever the CPU requests it.

(5) One reason why the world needs _____ is because of their enlarged address spaces.

(6) In an _____ architecture there is a single instruction cycle.

(7) Multiprocessing refers to the organizational technique in which a number of processor units are employed in a _____.

(8) RAM and ROM play important roles of _____ devices.

3. **Match the following terms to the appropriate definition.**

   a. broadband   b. bookmark   c. BBS
   d. benchmark   e. binary   f. bit

   (1) Short for binary digit, the smallest unit of information on a machine. It can hold only one of two values: 0 or 1.

   (2) Any system able to deliver multiple channels and services to its users or subscribers.

   (3) An electronic message center. Most bulletin boards serve specific interest groups.

   (4) It lets you save the address of a Web page so that you can easily revisit the page at a later time.

   (5) A test used to compare performance of hardware or software.

   (6) Information consisting entirely of ones and zeros. Also, commonly used to refer to files that are not simply text files, e.g. images.

Unit 3　Input and Output Systems

3.1　Text

音频

A computer is a powerful machine that can do everything people assign it. However the computer can not communicate with people directly. By the aid of input and output devices, a computer and people can "know" each other. Using input devices, people "tell" the computer what it should do and the computer feedback the result through output devices. Input and output devices are the interfaces of man and machine. They usually include keyboard, mouse, monitor, printer, disk (hard disk or floppy disk), input pen, scanner, and microphone, etc. Let's consider some input and output devices that are in common use.

Keyboard

The keyboard is used to type information into the computer or input information. There are many different keyboard layouts and sizes with the most common for Latin based languages being the QWERTY layout (named for the first six keys). The standard keyboard has 101 keys. Notebooks have embedded keys accessible by special keys or by pressing key combinations. Some of the keys on a standard keyboard have a special use. They are referred to as command keys. The three most common are the Control or Ctrl, Alternate or Alt and the Shift keys. Each key on a standard keyboard has one or two characters. Press the key to get the lower character and hold Shift to get the upper.

Mouse

A mouse is a small device that a computer user pushes across a desk surface in order to point to a place on a display screen and to select one or more actions to take from that position.[1] The most conventional kind of mouse has two buttons on the top: the left one is used most frequently. A mouse consists of a metal or plastic housing or casing, a ball that sticks out of the bottom of the casing and rolls on a flat surface, one or more buttons on the top of the casing, and a cable that connects the mouse to the computer. As the ball is moved over the surface in any direction, a sensor sends impulses to the computer that causes a mouse-responsive program to re-position a visible indicator (called a cursor) on the display screen. The positioning is relative to some starting place. Viewing the cursor's present position, the user readjusts the position by moving the mouse.[2]

CRT Monitor

CRT monitor means a cathode ray tube monitor. Cathode ray tubes have five main components: electron gun, deflection coil, shadow mask, phosphor layer and the glass casing. CRT monitor is one of the most widely used monitors now. The core component of CRT monitor is cathode ray tube. Its working principle is similar to our TV's tube, and we can see it as a TV with more precise images. Classic CRT uses electron gun to fire high-speed electrons, and controls the deflection angle of high-speed electrons through vertical and horizontal deflection coils. At last, the high-speed electrons hit phosphorescence material on the screen to make it radiate. Regulating the power of electron beam through voltage will form light points of different chiaroscuro on the screen to form various images and text.

The flat-screen CRT monitor has the following advantages: a large viewing angle, no dead pixel, high color reproduction, color uniformity, adjustable multi-resolution mode and short response time. Another important advantage is that the CRT monitor is cheaper than many LCD monitors.

LCD Monitor

LCD (liquid crystal display) uses a liquid crystal control transmittance technology to display color. In view of the structure of LCD monitors, whether a laptop or desktop system, the adoptive LCD screens are hierarchical structure composed of different parts. LCD is constituted by the two glass panels, 1mm thick, which are separated by 5μm even spacing containing liquid crystal material. Because the liquid crystal material itself is not light, on both sides of the screen there are not lamps as light sources, but on the back of the LCD screen there is a backlit panel (or optical absorption panel) and reflector membrane. Backlit panel is composed of fluorescent material and can emit light. It is mainly used to provide even background light.

The light fired by the backlit panel passes through the first layer's polarization filter into liquid crystal layer which contains thousands of liquid crystal droplets. Transparent electrodes are between the glass and liquid crystal material. The electrodes are divided into rows and columns, and on their intersections, the liquid crystal's optical state is changed by changing the voltage. The role of liquid crystal material is similar to many small light valves. Around the liquid crystal material, there is control circuit and driver circuit. When the electrodes in LCD generate electric field, the liquid crystal molecules have been distorted, thus the light through them is regularly refracted, and then through the second layer's filter, it will be displayed on screen.[3]

Compared to the CRT monitor, the LCD monitor's advantages are obvious. On account of its controlling bright and dark by controlling the transmission, when the color doesn't change, LCD doesn't do any change at all either. Thus we have no need to consider the refresh rate problem. For the LCD monitor with picture stability and flicker-free sense, the refresh rate is not high, but the image is also very stable. The LCD monitor also adopts LCD control transmittance technological principles for the overall luminous floor, so it is realty the complete plane.[4] Some slap-up digital LCD monitors use digital technology to transfer data and show images, this will not make the color error or loss because of the graphics card. It also has the advantage of no radiation, even watching LCD screen for a long time will not cause great harm to the eyes. The small size and low power consumption are other advantages that CRT monitors can not have.

Compared to the present CRT monitor, the image quality of the LCD monitor is still not perfect enough. LCD monitors are defeated by CRT monitors in different degrees in color display and saturation, and LCD monitors' response time is also longer than CRT monitors'.

I/O Subsystem Organization and Interfacing

Each I/O device is connected to the computer system's address, data, and control buses. Each I/O device includes I/O interface circuitry. It is actually this circuitry that interacts with the buses. The circuitry also interacts with the actual I/O device to transfer data.

As for the generic interface circuitry for an input device, the data from the input device goes to the tri-state buffers. When the values on the address and control buses are correct, the buffers are enabled and data passes on to the data bus. The CPU can then read in this data. When the conditions are not right, the logic block does not enable the buffers. They are tri-stated and do not place data onto the bus. The key to this design is the enable logic. Just as every memory location has a unique address, each I/O device also has a unique address. The enable logic must not enable the buffers unless it receives the correct address from the address bus. It must also get the correct control signals from the control bus.

The design of the interface circuitry for an output device is somewhat different from that for the input device. Tri-state buffers are replaced by the register. The tri-state buffers are used in input device interfaces to make sure than no more than one device writes data to the bus at any time. Since the output devices read data from the bus, rather than write data to it, they do not need the buffers. The data can be made available to all output devices. Only the device with the correct address will read it in.

Some devices are used for both input and output. A personal computer's hard disk drive falls into this category. Such a device requires a combined interface that is essentially two interfaces, one for input and the other for output.[5] Some logic elements, such as the gates that check the address on the address bus, can be used to generate both the buffer enable and register load signals.

I/O devices are much slower than CPUs and memory. For this reason, they can have timing problems when interacting with the CPU. To illustrate this, consider what happens when a CPU wants to read data from a disk. It may take the disk drive several milliseconds to position its heads properly to read the desired value. In this time, the CPU could have read in invalid data and fetched, decoded, and executed thousands of instructions.

Most CPUs have a control input signal called READY (or sometimes similar). Normally this input is high. When the CPU outputs the address of the I/O device and the correct control signals, enabling the tri-state buffers of the I/O device interface, the I/O device sets READY low. The CPU reads this signal and continues to output the same address and control signals, which cause the buffers to remain enabled. In the hard disk drive example, the drive rotates the disk and positions its read heads until it reads the desired data. The CPU then reads the data from the bus and continues its normal operation. The extra clock cycles generated by having READY set low are called wait states.

Keywords

absorption	n.	吸收，合并	electron	n.	电子
accessible	adj.	容易理解的，能接近的	embed	v.	嵌入，嵌进
chiaroscuro	n.	明暗对比法	feedback	n.	反馈
conventional	adj.	常规的，约定的	hierarchical	adj.	等级的
cursor	n.	光标	invalid	adj.	无效的，无用的
deflection	n.	偏转，偏离	interface	n.	接口，界面

layout	n.	布局		phosphorescence	n.	荧光，磷光
millisecond	n.	毫秒		radiate	v.	发光，辐射
phosphor	n.	荧光粉，磷光体		shadow	n.	阴影，影像

Notes

[1] A mouse is a small device that a computer user pushes across a desk surface in order to point to a place on a display screen and to select one or more actions to take from that position.

译文：鼠标是一种小型的设备，用户可以在桌面上移动，以指向屏幕上的某个位置并从该位置选择一项或多项操作。

说明：本句的 that a computer user pushes across a desk surface 作定语，修饰宾语，in order to 引导目的状语。

[2] Viewing the cursor's present position, the user readjusts the position by moving the mouse.

译文：看到光标的当前位置后，用户可以移动鼠标再次进行调整。

说明：本句的 Viewing the cursor's present position 是状语，by moving the mouse 也是状语。

[3] When the electrodes in LCD generate electric field, the liquid crystal molecules have been distorted, thus the light through them is regularly refracted, and then through the second layer's filter, it will be displayed on screen.

译文：当 LCD 中的电极产生电场时，液晶分子就会产生扭曲，从而将穿越其中的光线进行有规则的折射，然后经过第二层过滤层的过滤在屏幕上显示出来。

说明：本句是并列句，由 When 引导时间状语从句，thus 引导的是结果状语，and then through the second layer's filter 作状语，修饰并列句的第二部分。

[4] The LCD monitor also adopts LCD control transmittance technological principles for the overall luminous floor, so it is realty the complete plane.

译文：LCD 显示器还采用液晶控制透光度技术的原理让底板整体发光，所以它是真正的完全平面。

说明：本句的 for the overall luminous floor 是宾语补足语，so it is realty the complete plane 是结果状语。

[5] Such a device requires a combined interface that is essentially two interfaces, one for input and the other for output.

译文：这样的设备需要一个组合接口，本质上是两个接口，一个用于输入，另一个用于输出。

说明：本句的 that 引导定语从句。

3.2 Reading Material 1：Mouse

The mouse first broke onto the public stage with the introduction of the Apple Macintosh in 1984, and since then it has helped to completely redefine the way we use computers. It is amazing how simple and effective a mouse is, and it is also amazing how long it took a mouse to become a part of everyday life.

When the mouse hit the scene, it was an immediate success. There is something about it that is completely natural. Compared to a graphics tablet, mice are extremely inexpensive and take up very little desk space. In the PC world, mice took longer to gain ground, mainly because of a lack of support in the operating system. Once Windows 3.1 made graphical user interfaces (GUIs) a standard, the mouse became the PC-human interface of choice very quickly. The main goal of any mouse is to translate the motion of your hand into signals that the computer can use. Almost all mice today do the translation using five components:

1. A ball inside the mouse touches the desktop and rolls when the mouse moves.

2. Two rollers inside the mouse touch the ball. One of the rollers is oriented so that it detects motion in the x direction, and the other is oriented 90 degrees to the first roller so it detects motion in the y direction. When the ball rotates, one or both of these rollers rotate as well.

3. The rollers each connect to a shaft, and the shaft spins a disk with holes in it. When a roller rolls, its shaft and disk spin.

4. On either side of the disk there is an infrared LED and an infrared sensor. The holes in the disk break the beam of light coming from the LED so that the infrared sensor sees pulses of light. The rate of the pulsing is directly related to the speed of the mouse and the distance it travels.

5. An on-board processor chip reads the pulses from the infrared sensors and turns them into binary data that the computer can understand. The chip sends the binary data to the computer through the mouse's cord.

With advances it mouse technology, it appears that the venerable wheeled mouse is in danger of extinction. The now-preferred device for pointing and clicking is the optical mouse. Able to work on almost any surface, the mouse has a small, red light-emitting diode (LED) that bounces light off that surface onto a complimentary metal-oxide semiconductor (CMOS) sensor.

The CMOS sensor sends each image to a digital signal processor (DSP) for analysis. The DSP, operating at 18 MIPS, is able to detect patterns in the images and see how those patterns have moved since the previous image. Based on the change in patterns over a sequence of images, the DSP determines how far the mouse has moved and sends the corresponding coordinates to the computer. The computer moves the cursor on the screen based on the coordinates received from the mouse. This happens hundreds of times each second, making the cursor appear to move very smoothly.

The optical mouse has several benefits over the wheeled mouse:
- No moving part means less wear and a lower chance of failure.
- Increased tracking resolution means smoother response.
- They don't require a special surface.
- There's no way for dirt to get inside the mouse and interfere with the tracking sensors.

Most mice in today use the standard PS/2 type connector. Whenever the mouse moves or the user clicks a button, the mouse sends 3 bytes of data to the computer. The next 2 bytes contain the x and y movement values, respectively. These 2 bytes contain the number of pulses that have been detected in the x and y direction since the last packet was sent. The data is sent from the mouse to the computer serially on the data line, with the clock line pulsing to tell the computer where each bit starts and stops.

Keywords

amazing	adj.	令人惊异的	rotate	v.	轮流，旋转
complimentary	adj.	互补型的	scene	n.	场面，场景，景色
motion	n.	移动，动作	semiconductor	n.	半导体
optical	adj.	光学上的	wheel	n.	轮，轮状物

3.3 Reading Material 2: USB

USB (universal serial bus) is a serial bus standard to interface devices. A major component in the legacy-free PC, USB was designed to allow peripherals to be connected using a single standardized interface socket, to improve plug-and-play capabilities by allowing devices to be connected and disconnected without rebooting the computer. Other convenient features include powering low-consumption devices without the need for an external power supply and allowing some devices to be used without requiring individual devices to be installed.

A USB system has an asymmetric design, consisting of a host controller and multiple daisy-chained peripheral devices. Additional USB hubs may be included in the chain, allowing branching into a tree structure, subject to a limit of 5 levels of branching per controller. No more than 127 devices, including the bus devices, may be connected to a single host controller. Modern computers often have several host controllers, allowing a very large number of USB devices to be connected. USB cables do not need to be terminated.

So-called "sharing hubs" also exist, allowing multiple computers to access the same peripheral devices, either switching access between PCs automatically or manually. They are popular in small-office environments. In network terms they converge rather than diverge branches.

In USB terminology, individual devices are referred to as functions, because each individual physical device may actually host several functions, such as a webcam with a built-in microphone. Functions are linked in series through hubs. The hubs are special-purpose devices that are not considered functions. There always exists one hub known as the root hub, which is attached directly to the host controller.

When a device is first connected, the host enumerates and recognizes it, and loads the device driver it needs. When a function or hub is attached to the host controller through any hub on the bus, it is given a unique 7 bit address on the bus by the host controller. The host controller then polls the bus for traffic, usually in a round-robin fashion, so no function can transfer any data on the bus without explicit request from the host controller.

The connectors which the USB committee specified were designed to support a number of USB's underlying goals, and to reflect lessons learned from the varied menagerie of connectors then in service.

The connectors are designed to be robust. The electrical contacts in a USB connector are protected by an adjacent plastic tongue, and the entire connecting assembly is further protected by an enclosing metal sheath. As a result USB connectors can safely be handled, inserted, and removed.

Keywords

asymmetric	adj.	不对称的	explicit	adj.	明确的,显而易见的
automatically	adv.	自动地,机械地	legacy	adj.	传统的,老化的
branching	n.	分支,分流	peripheral	adj. n.	外围的,外设
diverge	v.	分开,分岔	sheath	n.	护套,外壳,包装
enumerate	v.	枚举,列举,数	terminate	v.	结束,终止

terminology	n.	术语，专门名词	webcam	n.	网络摄像机
tongue	n.	舌状物			

3.4 词汇缩略

词汇缩略是指将较长的单词取其首部或主干构成与原词同义的短单词，或者将组成词汇短语的各个单词的首字母拼接为一个大写字母的字符串。通常词汇缩略在文章索引、前序、摘要、文摘、电报、说明书、商标等科技文章中频繁采用。对计算机专业来说，在程序语句、程序注释、软件文档、互联网信息、文件描述中也采用了大量的缩略词汇作为标识符、名称等。缩略词汇的出现方便了印刷、书写、速记及口语交流等，但也同时增加了阅读和理解的困难。现在的趋势是缩略词的数目不断增大，使用面不断扩大。

1．词汇缩略的形式

词汇缩略有以下 3 种形式。

1）节略词

某些词汇在发展过程中为方便使用逐渐用它们的前几个字母来表示，这就是节略词；或者在一个词组中取各词的一部分，重新组合成一个新词，表达的意思与原词组相同。

例如：maths——mathematics　　数学
　　　　ad——advertisement　　广告
　　　　kilo——kilogram　　千克
　　　　dir——directory　　目录
　　　　lab——laboratory　　实验室
　　　　radar——radio detection and ranging　　雷达
　　　　transceiver——transistor receiver　　收发机
　　　　telesat——telecommunications satellite　　通信卫星

2）缩略词和首字词

缩略词是指由某些词组的首字母所组成的新词。例如：ROM——read only memory，RAM——random access memory，RISC——reduced instruction set computer（精简指令集计算机），CISC——complex instruction set computer（复杂指令集计算机），COBOL——common business oriented language（面向商务的通用语言）等。

首字词与缩略词基本相同，区别在于首字词必须逐字母念出。例如：CAD——computer aided design，CPU——central process unit，DBMS——database management system（数据库管理系统），UFO——unidentified flying object（不明飞行物），CGA——color graphics adapter（彩色图形适配器）等。

3）缩写词（abbreviation）

缩写词并不一定由某个词组的首字母组成，有些缩写词仅由一个单词变化而来，而且大多数缩写词每个字母后都附有一个句点，例如 e.g.——for example、Ltd.——limited、sq.——square 等。

2．Internet 的相关缩写词

在 Internet 技术描述中，通常会用到大量的专用缩写术语，下面是常用的与 Internet 技术相关的缩写术语。

1）ARP（address resolution protocol）

地址解析协议，用于从 IP 地址找出对应的以太网地址。

2）BBS（bulletin board system）

电子公告板系统，又称网上论坛或 Usenet 新闻组，是人们在一起交流观点、公布公共注意事项、寻求帮助的地方。

3）BOOTP（bootstap protocol）

自举协议，用于无盘工作站的启动。

4）E-mail（electronic-mail）

电子邮件，是指通过计算机网络收发信息的服务。电子邮件是 Internet 上最普遍的应用，它是加强人与人之间沟通的渠道。

5）FTP（file transfer protocol）

文件传输协议，利用 FTP 服务可以直接将远程系统上任何类型的文件下载到本地计算机上，或者将本地文件上传到远程系统。它是实现 Internet 上软件共享的基本方式。

6）HTTP（hypertext transfer protocol）

超文本传输协议，用于 World Wide Web 服务。

7）ICMP（Internet control message protocol）

Internet 控制信息协议。

8）IGMP（Internet group multicast protocol）

Internet 成组广播协议。

9）NFS（network file system）

网络文件系统，用于实现计算机间共享文件系统。

10）RARP（reverse address resolution protocol）

逆向地址解析协议，用于从以太网地址找出对应的 IP 地址。

11）SMTP（simple mail transfer protocol）

简单邮件传输协议，用于电子邮件传输。

12）SNMP（simple network management protocol）

简单网络管理协议，用于网络管理。

13）TCP/IP

Internet 使用的一组网络协议，其中 IP 是 Internet Protocol，即网际协议；TCP 是 Transmission Control Protocol，即传输控制协议，这是最核心的两个协议。IP 协议提供基本的通信，TCP 协议提供应用程序所需要的其他功能。

14）TELNET

远程登录协议，可以看成是 Internet 的一种特殊通信方式，它的功能是把用户正在使用的终端或主机变成它要在其上登录的某一远程主机的仿真远程终端，该系统根据用户账号判断用户对本系统的使用权限。

15）UDP（user datagram protocol）

用户数据报协议，用于可靠性要求不高的场合。

16）WWW（world wide web）

万维网，当前 Internet 上最重要的服务方式。WWW 是由欧洲粒子物理研究中心（CERN）研制的，它将位于全球 Internet 上不同地点的相关多媒体信息有机地组织在一起，称为 Web 页的集合。

3．命令和指令

每一个处理器都具有很多指令，每一台机器也具有很多系统命令，不同的操作系统也定义了不同的操作命令，它们通常是缩写的，牢记这些指令，就熟悉了计算机的操作；了解缩写的含义，也就了解了所用的操作的含义。例如：

创建目录　MD（make directory）
改变目录　CD（change directory）
删除目录　RD（remove directory）
列表目录　DIR（directory）
取偏移地址指令　LEA（load effective address offset）
取数据段地址指令　LDS（load doubleword pointer）
总线封锁命令　LOCK（assert bus lock signal）
串装入操作指令　LODS（load string operand）
中断请求　INT（call to interrupt procedure）
中断返回　IRET（interrupt return）
重命名　REN（rename）

3.5　Exercises

1. **Translate the following phrases.**

   tri-state buffer
   cathode ray tube
   electron beam
   standard keyboard
   liquid crystal display
   control circuit
   input and output device
   graphical user interface
   light-emitting diode
   digital signal processor
   universal serial bus
   electronic gun

2. **Fill in the blanks with appropriate words or phrases.**

   a. obvious　　b. Shift　　c. interface circuitry　d. monitor
   e. keyboard　　f. control　　g. CRT monitor　　h. characters

(1) Each I/O device includes I/O _____.
(2) The _____ is used to type information into the computer or input information.
(3) The core component of _____ is cathode ray tube.
(4) Compared to the CRT monitor, the LCD monitor's advantages are _____.

(5) Each I/O device is connected to the computer system's address, data, and _____ buses.

(6) CRT monitor means a cathode ray tube _____.

(7) Each key on a standard keyboard has one or two _____.

(8) Press the key to get the lower character and hold _____ to get the upper.

3. **Match the following terms to the appropriate definition.**

   a. banner advertising   b. chatroom   c. cyberspace
   d. cybercash   e. connection speed   f. coaxial cable

(1) It combines features from cash and checks. It offers Internet payment options, including credit card, micro-payment, and check payment services.

(2) A graphic advertising image on a Web site.

(3) A place in cyberspace where you and your friends hang out to talk and share ideas.

(4) Actual line of transmission for carrying television signals. Its principal conductor is either copper or copper-coated wire surrounded by insulation and then incased in aluminum.

(5) The world that exists only on the computer. You can't physically go there. It only exists on your screen and in your head.

(6) Tells you how fast the modem is able to talk to other computers and get information from them. The higher the number, the faster it talks.

Unit 4   C++ Language

4.1 Text

音频

A programming language represents a special vocabulary and a set of grammatical rules for instructing a computer to perform specific tasks. Broadly speaking, it consists of a set of statements or expressions understandable to both people and computers. People understand these instructions because they use human expressions. Computers, on the other hand, process these instructions through using special programs, which, known as translators, can decode the instructions from people and create machine-language code.[1] These instructions are represented by long strings of ones and zeros. And this language used 0 and 1 is called machine language.

High-level programming languages, while simply compared to human languages, are more complex than the languages the computer actually understands, which are called machine languages.[2] Each different type of CPU has its own unique machine language. Lying between machine languages and high-level languages are assembly languages, which are directly related to a computer's machine language. In other words, it takes one assembly command to generate one machine-language command. Machine languages consist entirely of numbers and are almost impossible for humans to read and write. Assembly languages have the same structure and set of commands as machine languages, but they enable a programmer to use characters instead of numbers.

Lying above high-level languages are those called fourth-generation languages (usually abbreviated 4GL). 4GLs are far removed from machine languages and represent the class of computer languages closest to human languages. Most 4GLs are used to access databases.

C++ is a general-purpose programming language with high-level and low-level capabilities. It is a statically typed, free-form, multi-paradigm, usually compiled language supporting procedural programming, data abstraction, object-oriented programming, and generic programming.

C++ is often considered to be a superset of C, but this is not strictly true. Most C code can easily be made to compile correctly in C++, but there are a few differences that cause some valid C code to be invalid in C++, or to behave differently in C++.

C++ is an enhanced version of the C language. C++ includes everything that is part of C and adds support for object-oriented programming. C is a procedure-oriented programming language. C++ contains many improvements and features.

The basic syntax and semantics of C and C++ are the same. If you are familiar with C, you can program in C++ immediately. C++ has the same types, operators, and other facilities defined in C that usually correspond directly to computer architecture.

Object-oriented programming is a programming technique that allows you to view concepts as a variety of objects. By using objects, you can represent the tasks that are to be performed, their interaction, and any given conditions that must be observed. A data structure often forms the basis of an object; thus, in C or C++, the struct type can form an elementary object. Communicating with objects can be done through the use of messages. Using messages is similar to calling a function in a procedure-oriented program. When an object receives a message, methods contained within the object respond. Methods are similar to the functions of procedure-oriented programming. However, methods are part of an object.

C++ allows the programmer to create classes, which are somewhat similar to C structures.[3] In C++, it can be assigned methods, functions associated to it, of various prototypes, which can access and operate within the class, somewhat like C functions often operate on a supplied handler pointer. The C++ class is an extension of the C language structure. Because the only difference between a structure and a class is that structure members have public access by default and a class member has private access by default, you can use the keywords class or struct to define equivalent classes.

The C++ class is an extension of the C and C++ struct type and forms the required abstract data type for object-oriented programming. The class can contain closely related items that share attributes. Stated more formally, an object is simply an instance of a class.

Ultimately, there should emerge class libraries containing many object types. You could use instances of those object types to piece together program code.

Typically, an object's description is part of a C++ class and includes a description of the object's internal structure, how the object relates with other objects, and some form of protection that isolates the functional details of the object from outside the class. The C++ class structure does all of this.

In a C++ class, you control functional details of the object by using private, public, or protected descriptors. In object-oriented programming, the public section is typically used for the interface information (methods) that makes the class reusable across applications. If data or methods are contained in the public section, they are available outside the class.[4] The private section of a class limits the availability of data or methods to the class itself. A protected section containing data or methods is limited to the class and any derived subclasses.

The C++ class actually serves as a template or pattern for creating objects. The objects formed from the class description are instances of the class. It is possible to develop a class hierarchy where there is a parent class and several child classes. In C++, the basis for doing this revolves around derived classes. Parent classes represent more generalized tasks, while derived child classes are given specific tasks to perform.

Inheritance in object-oriented programming allows a class to inherit properties from a class of objects. The parent class serves as a pattern for the derived class and can be altered in several ways. If an object inherits its attributes from multiple parents, it is called multiple inheritance. Inheritance is an important concept since it allows reuse of a class definition without requiring major code changes.[5] Inheritance encourages the reuse of code since child classes are extensions of parent classes.

Another important object-oriented concept that relates to the class hierarchy is that common messages can be sent to the parent class objects and all derived subclass objects. In formal terms, this is called polymorphism.

Polymorphism allows each subclass object to respond to the message format in a manner appropriate to its definition. Imagine a class hierarchy for gathering data. The parent class might be responsible for gathering the name, social security number, occupation, and number of years of employment for an individual. You could then use child classes to decide what additional information would be added based on occupation. In one case a supervisory position might include yearly salary, while in another case a sales position might include an hourly rate and commission information. Thus, the parent class gathers general information common to all child classes while the child classes gather additional information relating to specific job descriptions. Polymorphism allows a common data-gathering message to be sent to each class. Both the parent and child classes respond in an appropriate manner to the message.

Polymorphism gives objects the ability to responds to messages from routines when the object's exact type is not known. In C++ this ability is a result of late binding. With late binding, the addresses are determined dynamically at run time, rather than statically at compile time, as in traditional compiled languages. This static method is often called early binding. Function names are replaced with memory addresses. You accomplish late binding by using virtual functions. Virtual functions are defined in the parent class when subsequent derived classes will overload the function by redefining the function's implementation. When you use virtual functions, messages are passed as a pointer that points to the object instead of directly to the object.

Virtual functions utilize a table for address information. The table is initialized at run time by using a constructor. A constructor is invokes whenever an object of its class is created. The job of the constructor here is to link the virtual function with the table of address information. During the compile operation, the address of the virtual function is not known; rather, it is given the position in the table of addresses that will contain the address for the function.

Keywords

abbreviate	v.	缩写，省略	prototype	n.	原型，样板，标准
constructor	n.	建造者，建设者	semantics	n.	语义学
facility	n.	方便，简便，机构	statically	adv.	静止地，静态地
hierarchy	n.	等级，层次	superset	n.	超集
inheritance	n.	继承，承受	syntax	n.	语法
interaction	n.	互动，交互，相互作用	template	n.	模板
invoke	v.	引起，产生	ultimately	adv.	最终的，极限的
paradigm	n.	范例，示例	vocabulary	n.	词汇，单词表
polymorphism	n.	多态性			

Notes

[1] Computers, on the other hand, process these instructions through using special programs, which, known as translators, can decode the instructions from people and create machine-language code.

译文：另一方面，计算机通过使用专门的程序来处理这些指令，这些专门的程序就是人们所熟知的翻译程序，它能对用户发出的指令进行解码并生成机器语言代码。

说明：本句的 through using special programs 是宾语补足语，which 引导的是非限定性定语从句。

[2] High-level programming languages, while simply compared to human languages, are more complex than the languages the computer actually understands, which are called machine languages.

译文：简单地与人类语言相比，高级程序设计语言比计算机实际识别的语言——机器语言，复杂得多。

说明：本句的 while simply compared to human languages 作状语，the computer actually understands 作定语，修饰 languages，which are called machine languages 是非限定定语从句。

[3] C++ allows the programmer to create classes, which are somewhat similar to C structures.

译文：C++允许程序员创建类，这些类类似于 C 中的结构。

说明：本句中的 which 引导的是非限定性定语从句，修饰 classes。

[4] If data or methods are contained in the public section, they are available outside the class.

译文：如果数据或方法被包含在公共部分，它们在该类外部也可用。

说明：本句由 If 引导条件状语从句。

[5] Inheritance is an important concept since it allows reuse of a class definition without requiring major code changes.

译文：继承是一个重要概念，因为它使得无须对代码做大的改变就能重用类定义。

说明：本句由 since 引导原因状语从句。

4.2 Reading Material 1：Categories of Computer Software

Today, seven broad categories of computer software present continuing challenges for software engineers.

♦ System Software

System software refers to a collection of programs written to service other programs. Some system software (e.g., compilers, editors, and file management utilities) processes complex, but determinate, information structures. Other systems applications (e.g., operating system components, drivers, networking software, telecommunications processors) process largely indeterminate data. In either case, the system software area is characterized by heavy interaction with computer hardware; heavy usage by multiple users; concurrent operation that requires scheduling, resource sharing, and sophisticated process management; complex data structures; and multiple external interfaces.

♦ Application Software

Application software refers to stand-alone programs that solve a specific business need. Applications in this area process business or technical data in a way that facilitates business operations or management / technical decision making. In addition to conventional data processing applications, application software is used to control business functions in real time (e.g., point-of-sale transaction processing, real-time manufacturing process control).

♦ Engineering / Scientific Software

Applications of engineering/scientific software range from astronomy to volcanology, from automotive stress analysis to space shuttle orbital dynamics, and from molecular biology to automotive manufacturing. However, modern applications within the engineering / scientific area are moving away from conventional numerical algorithms. Computer-aided design, system simulation, and other interactive applications have begun to take on real-time and even system software characteristics.

- Embedded Software

Embedded software resides within a product or system and is used to implement and control features for the end user and for the system itself. Embedded software can perform limited and esoteric functions (e.g., key pad control for a microwave oven) or provide significant function and control capability (e.g., digital functions in an automobile such as fuel control, dashboard displays, and braking systems).

- Product-line Software

Product-line software can focus on a limited and esoteric marketplace (e.g., inventory control products) or address mass consumer markets (e.g., word processing, spreadsheets, computer graphics, multimedia, entertainment, database management, and personal and business financial applications).

- Web Applications

Called "WebApps", this network-centric software category spans a wide array of applications. In their simplest form, WebApps can be little more than a set of linked hypertext files that present information using text and limited graphics. However, as Web 2.0 emerges, WebApps are evolving into sophisticated computing environments that not only provide stand-alone features, computing functions, and content to the end user, but also are integrated with corporate databases and business applications.

- Artificial Intelligence Software

Artificial intelligence software makes use of non-numerical algorithms to solve complex problems that are not amenable to computation or straightforward analysis, applications within this area include robotics, expert systems, pattern recognition (image and voice), artificial neural networks, theorem proving, and game playing.

Millions of software engineers worldwide are hard at work on software projects in one or more of these categories. In some cases, new systems are being built, but in many others, existing applications are being corrected, adapted, and enhanced.

- Open Source

Open source is a growing trend that results in distribution of source code for systems applications (e.g., operating systems, databases, and development environments) so that many people can contribute to its development. The challenge for software engineers is to build source code that is self-descriptive, but more importantly, to develop techniques that will enable both customers and developers to know what changes have been made and how those changes manifest themselves within the software.

- Open-world Computing

The rapid growth of wireless networking may soon lead to true pervasive, distributed computing. The challenge for software engineers will be to develop systems and application software that will allow mobile devices, personal computers, and enterprise systems to communicate across vast networks.

The WWW (world wide web) is rapidly becoming a computing engine as well as a content provider. The challenge for software engineers is to architect simple (e.g., personal financial planning) and sophisticated applications that provide a benefit to targeted end-user markets worldwide.

Each of these new challenges will undoubtedly obey the law of unintended consequences and have effects (for businesspeople, software engineers, and end users) that cannot be predicted today. However, software engineers can prepare by instantiating a process that is agile and adaptable enough to accommodate dramatic changes in technology and to business rules that are sure to come over the next decade.

Keywords

agile	adj.	敏捷的，灵活的	instantiate	v.	用具体例子说明，示例
amenable	adj.	顺从的，易作出响应的	inventory	n.	存货（清单），库存
astronomy	n.	天文学	manifest	v.	显示，使显现
automated	adj.	自动化的	pattern recognition		模式识别
automotive	adj.	汽车的	point-of-sale	adj.	销售点的
concurrent	adj.	同时发生的，并行的	robotics	n.	机器人学
dashboard	n.	仪表盘	simulation	n.	模拟，仿真
editor	n.	编辑程序，编辑器	space shuttle	n.	航天飞机
esoteric	adj.	只有内行才懂的，难理解的	telecommunication	n.	电信
			theorem	n.	定理
facilitate	v.	使变得(更)容易，促进	volcanology	n.	火山学
indeterminate	adj.	不确定的，未限定的			

4.3　Reading Material 2：Visual Basic

Visual Basic is a great starter programming language. Not only is it easy to learn, but many business applications use it extensively for their applications. If you are just starting to learn to develop applications—this is a great language to start with. This tutorial and the many other Visual Basic tutorials will give you a solid foundation in your Visual Basic knowledge.

Visual Basic allows you to do event driven programming. This is a concept that is very powerful and easy to use. Event driven programming works as follows: Visual Basic has many different events defined that occur when a specified thing happens. We as the programmer can link into these events and write our custom code to do what we want. One vary useful event is the Click event. This event occurs any time the user clicks on the specified object.

MsgBox is a built-in Visual Basic function that causes a message box to be displayed to the user. The first parameter this function takes is the text string you wish to have displayed. We choose the text string "Hello, World!". MsgBox also takes other parameters to specify things such as what buttons to display and what caption to use.

Controls are reusable, predefined components used for visual programming. Controls serve as building blocks that can be quickly combined to create a working application. Visual Basic controls exist in two varieties-intrinsic controls and ActiveX controls. Intrinsic controls (also called standard controls) are the default controls provided in the Visual Basic toolbox.

A variable is a word or letter used to reference data used in a program. At the most basic level, all variables used in a program are held on the computer as a sequence of 1s and 0s which represent numbers.

In a sane programmer's code the variable names are easy to understand because they clearly state what the variable is used for inside of the variable name.

The following are the requirements when naming the variables in Visual Basic.

- It must be less than 255 characters.
- No spacing is allowed.
- It must not begin with a number.
- Period is not permitted.

For the sake of making sure other people can look at your code and know what you were thinking.

- Prefix your variable name with the appropriate prefix for your variable's data type.
- Make sure the body of your variable name makes it easy to tell what is used for.
- Don't use an ambiguous name like "intAAaa" or "strXY" unless its use is within a very small scope of the program.

Keywords

ambiguous	*adj.*	含糊的，不明确的	reusable	*adj.*	可重用的
caption	*n.*	标题，题目	sane	*adj.*	合乎情理的，稳健的
control	*n.*	控件	tutorial	*adj.*	指导的，辅导的
foundation	*n.*	基础，根本	variable	*adj.*	变量的，变换的
intrinsic	*adj.*	本来的，内在的			

4.4 计算机专用术语与命令

在计算机语言、程序语句、程序文本注释、系统调用、命令字、保留字、指令字及网络操作中广泛使用专业术语进行信息描述。随着计算机技术的发展，这样的专业术语还会进一步增加。

1. Internet 地址

域名（domain name）是 Internet 中主机地址的一种表示方式。域名采用层次结构，每一层构成一个子域名，子域名之间用点号隔开并且从右到左逐渐具体化。域名的一般表示形式为计算机名、网络名、机构名、一级域名。一级域名有一些规定，用于区分机构和组织的性质，如 edu（教育机构）、com（商业单位）、mil（军事部门）、gov（政府机关）、org（其他组织）。

用于区分地域的一级域名采用标准化的两个字母的代码。例如：

cn	中国	ca	加拿大
us	美国	au	澳大利亚
gb	英国（官方）	uk	英国（通用）
tw	中国台湾	hk	中国香港
fr	法国	un	联合国
nz	新西兰	dk	丹麦
ch	瑞士	de	德国
jp	日本	sg	新加坡
aq	南极洲	it	意大利

在 Internet 上，电子邮件（E-mail）地址具有如下统一的标准格式：用户名@主机域名。例如，wang@online.sh.cn 是一个电子邮件的地址，其中 wang 是用户名，@是连接符，online.sh.cn 是"上海热线"的主机域名，这是注册"上海热线"后得到的一个 E-mail 地址。

2．专用的软件名称

人类相互交流信息所用的语言称为自然语言，但是当前的计算机还不能理解自然语言，它能理解的是计算机语言，即软件。软件分为系统软件和用户软件。近几年来，随着计算机技术的发展，新的软件不断推出。下面是一些常用软件的名称。

NextStep	面向对象操作系统（Next）
Netware	局域网络操作系统（Novell）
Cairo	面向对象操作系统（MS）
Daytona	视窗型操作系统（MS）
Java	网络编程语言（Sun）
Excel	电子表格软件（MS）
Delphi	视窗系统开发工具（Borland）
Informix	关系数据库系统（Informix）
Navigator	互联网浏览器（Netscape）

3．专用计算机厂商及商标名

下面给出的是一些著名计算机公司的译名。

Microsoft	微软	Philip	飞利浦
Compaq	康柏	DELL	戴尔
Panasonic	松下	ASUS	华硕
Acer	宏碁	Intel	英特尔
Hewlett-Packard (HP)	惠普	Samsung	三星
Hisense	海信	Epson	爱普生

4．DOS 系统

DOS（Disk Operating System）是个人计算机磁盘操作系统，DOS 是一组非常重要的程序，它帮助用户建立、管理程序和数据，也管理计算机系统的设备。DOS 是一种层次结构，包括 DOS BIOS（基本输入输出系统）、DOS 核心部分和 DOS COMMAND（命令处理程序）。

一般情况下，在 DOS 启动盘上有配置系统文件 CONFIG.SYS，在该文件内给出有关系统配置命令，能确定系统的环境。配置系统包括以下 9 个方面的内容。

1）设置 Ctrl-Break（BREAK）检查

格式：BREAK = [ON] | [OFF]

隐含是 BREAK = OFF，这时 DOS 只是对标准输出操作、标准输入操作、标准打印操作和标准辅助操作过程检查 Ctrl-Break。

如果设置 BREAK = ON，则 DOS 在它被调用的任何时候都检查 Ctrl-Break，如编译程序，即使没有标准设备操作，在编译过程中遇到错误，也能使编译停止。

2）指定磁盘缓冲区的数目（BUFFERS）

格式：BUFFERS = X，X 为 1~99 的数。

3）指定国家码及日期时间格式（COUNTRY）

格式：COUNTRY = XXX，XXX 是电话系统使用的三数字的国际通用国家码。

4）建立能由文件控制块打开的文件数（FCBS）

格式：FCBS = m，n

m 为 1～255。

n 指定由 FCBS 打开但不能由 DOS 自动关闭的文件数，n 为 0～255，约定值为 0。

5）指定能一次打开的最大文件数（FILES）

格式：FILES = X

X 为 8～255，约定值为 8。

6）指定能访问的最大驱动器字母（LASTDRIVE）

格式：LASTDRIVE = X

X 可以是 A 到 Z 之间的字母，它表示 DOS 能接受的最后一个有效驱动器字母，约定值为 E。

7）指定高层命令处理程序的文件名（SHELL）

格式：SHELL = [d：] [path] filename [.ext] [parm1] [parm2]

8）安装驱动程序（DEVICE）

格式：DEVICE = [d：] [path] filename [.ext]

除标准设备外，如果用户增加了别的设备，就要由用户自己提供相应的驱动程序。在 CONFIG.SYS 中加上命令 DEVICE = 驱动程序名称。

9）指定堆栈空间（STACKS)

格式：STACKS = n，s

n 是堆栈的框架个数，为 8～64。

s 是每层堆栈框架的字节数，为 32～512。

5．计算机专用命令和指令

程序设计语言同任何一门自然语言一样，有它自己的一套词法和语法规则，只是它的规则很少，也很死板，每一条语句的规定都很严格，一旦违反就可能导致整个系统瘫痪。到目前为止，大部分的计算机语言的词汇都是取自英语词汇中一个很小的子集和最常用的数学符号。由于各个计算机指令系统所具有的功能大致相同，各个程序设计语言也大体包含了函数、过程、子程序、条件、循环，以及输入和输出等部分，因此它们必然存在一些共同的词汇特点和语法特点。

系统命令与程序无关，而且语法结构简单。主要的系统命令有系统连接命令、初始化命令、程序调试命令和文件操作命令。即：

系统命令<CR>

其中<CR>为回车换行符，一般而言，系统命令总是立即被执行，但某些系统命令也可以用于程序执行。至于词汇特点，同一功能的命令在不同机器中一般以相同的单词表示，如删除文件命令 DELETE，列磁盘文件命令 DIR，复制文件命令 COPY。但有相当一部分系统命令名是由各厂家或公司自己定义的，如清屏命令就有 CLEAR、CLS、CLR、HOME 等几种。

6．屏幕信息

随着软件技术的发展，特别是微机商品软件的大量涌现，软件质量的标准发生了深刻的变化。目前大部分软件都采用了菜单技术及其他人机对话技术。能否正确地阅读和理解这些屏幕信息，关系到我们能否正确使用这些软件，以及能否充分发挥软件所提供的全部功能。大部分软件都备有详细的解释或说明，以供用户阅读理解之用。常见的文件有 Readme、Message、Help、Assist 等。很多软件的求助命令是可供随时调用的，使用起来非常方便。

当一个程序具有若干项供用户选择的功能时，通常使用交互技术进行分支处理。实现的过程是：

屏幕首先显示出提供的功能名称，用户根据需要指出希望完成的功能，然后由程序分析用户的选择并调用不同的功能块进行处理。此过程称为"菜单技术"（menu technique）。

为了使程序具有较好的通用性、坚固性及人机交互性，在设计屏幕"菜单"和其他输入输出内容时，应符合下述主要原则。

- 输入方式要简单，并尽可能每一步给出屏幕提示。
- 屏幕显示信息应简洁、易懂，避免二义性。
- 尽可能在一屏中包含更多的信息。
- 检测所有输入数据的合法性。
- 尽量减少用户处理出错的工作量。
- 如果用户选择的功能可能会产生严重的后果（如删除文件、格式化磁盘等），应再次予以确认，以提醒用户不致误操作。

屏幕菜单的语言特色为：菜单中所显示的备选功能一般不用英语句子来表示，而是用表示该功能的单词、词组或动词短语来表示，个别菜单内容也用不完全句。所以，菜单语言的主要特点是精练且一般不存在时态和语态问题。

为了方便用户使用，尽量减少击键次数，菜单的功能选择通常用编号或单词和词组首字母来表示。有些则通过移动标记块来选择功能。

屏幕输出信息涉及的语法现象较多，输出的内容也较广泛，下面是常见的几类。

- 引导用户按一定步骤进行某些操作。这种类型通常使用祈使语句。

例 1：Insert new diskette for drive A: and strike any key when ready.

译文：将新盘插入 A 驱动器，准备好后按任意键。

例 2：Press Esc key to return to main menu.

译文：按 Esc 键返回主菜单。

- 要求用户回答某一个问题。所用句型通常是 Yes/No 型疑问句，也有按填空的方式向用户提出要求的。

例 3：Do you want to make another copy? [Y / N]

译文：还要复制吗？[Y / N]

例 4：Is this list correct [Y / N]?

译文：列表正确吗[Y / N]？

4.5　Exercises

1. **Translate the following phrases.**

    assembly language

    click event

    standard control

    resource file

    high-level language

    object-oriented program

    structure member

    derived classes

late binding

virtual function

machine-language command

procedure-oriented programming language

2. **Fill in the blanks with appropriate words or phrases.**

   a. compile   b. messages   c. machine language   d. C language

   e. instance   f. properties   g. functions   h. C++

(1) Inheritance in object-oriented programming allows a class to inherit _____ from a class of objects.

(2) The language used 0 and 1 is called _____.

(3) _____ is often considered to be a superset of C, but this is not strictly true.

(4) Most C code can easily be made to _____ correctly in C++.

(5) C++ is an enhanced version of the _____.

(6) Methods are similar to the _____ of procedure-oriented programming.

(7) An object is simply an _____ of a class.

(8) Polymorphism gives objects the ability to responds to _____ from routines when the object's exact type is not known.

3. **Match the following terms to the appropriate definition.**

   a. control unit   b. compiler   c. custom controls

   d. cookie   e. communication server   f. collision

(1) An Internet site's way of keeping track of you. It is a small program built into pages you may visit. It can identify you, track sites you visit, and topics you search.

(2) It is the most complex block of computer hardware from a designer's point of view. Its function is to generate control signals needed by the other blocks of the machine in a predetermined sequence to bring about the sequence of actions called for by each instruction.

(3) It is a computer program that translates a series of instructions written in one computer language (called the source language) into a resulting output in another computer language.

(4) The result of two network nodes transmitting on the same channel at the same time. The transmitted data is not usable.

(5) It means an object in a window or dialog box. Examples of them include push-buttons, scroll bars, radio buttons, and pull-down menus.

(6) A dedicated, standalone system that manages communications activities for other computers.

Unit 5　　Operating System

音频

5.1　Text

The most important program on any computer is the operating system or OS. The OS is a large program made up of many smaller programs. It controls how the CPU communicates with other hardware components. It also makes computers easier to operate by people who don't understand programming languages. In other words, operating systems make computer users friendly.

Operating systems are normally unique to their manufacturers and the hardware in which they are run. Generally, when a new computer system is installed, operational software suitable to that hardware is purchased. Users want reliable operational software that can effectively support their processing activities. Though operational software varies with manufacturers, it has similar characteristics.[1] Modern hardware, because of its sophistication, requires that operating systems meet certain specific standards. For example, considering the present state of the field, an operating system must support some form of online processing.

The operating system is the system software that manages and controls the activities of the computer. It supervises the operation of the CPU, controls input, output, and storage activities; and provides various support services. It can be visualized as the chief manager of the computer system. The operating system determines which computer resources will be used for solving which problems and the order in which they will be used.[2]

The operating system has some principal functions:

♦ Allocating and assigning system resources.

A master control program called a supervisor, executive or monitor oversees computer operations and coordinates all of the computer's work. The supervisor, which remains in primary storage, brings other programs from secondary storage to primary storage when they are needed. As each program is activated, the supervisor transfers control to that program. Once the program ends, control returns to the supervisor. A command language translator controls the assignment of system resources. The command language translator reads special instructions to the operating system that contain specifications for retrieving, saving, deleting, copying, or moving files; selecting input/output devices; selecting programming languages and applications programs; and performing other processing requirements for a particular application. These instructions are called command language.

♦ Scheduling the use of resources and computer jobs.

A very important responsibility of any operational software is the scheduling of jobs to be handled by a computer system. This is one of the main tasks of the job management function. Thousands of pieces of work can be going on in a computer at the same time. The operating system decides when to schedule them as computer jobs are unnecessarily performed in the order in which they are submitted. For example, payroll or online order processing may have a higher priority than other kinds of work. Other processing jobs, such as software program testing, would have to wait until these jobs were finished or left enough computer resources free to accommodate them. The operating system coordinates scheduling in various areas of the computer so that different parts of different jobs can be worked on simultaneously. For example, while some programs are executing, the operating system is also scheduling the use of input and output devices.

♦ Monitoring computer system activities.

The operating system is also responsible for keeping track of the activities in the computer system. It maintains logs of job operations, notifying end users or computer operators of any abnormal terminations or error conditions. It also terminates programs that run longer than the maximum time allowed. Operating system software may also contain security monitoring features, such as recording who has logged on and off the system, what programs they have run, and unauthorized attempts to access the system.

♦ Control of I/O operations.

Allocation of a system's resources is closely tied to the operational software's control of I/O operations. As access is often necessary to a particular device before I/O operations may begin, the operating system must coordinate I/O operations and the devices on which they are performed. In effect, it sets up a directory of programs undergoing execution and the devices they must use in completing I/O operations.

To facilitate execution of I/O operations, most operating systems have a standard set of control instructions to handle the processing of all input and output instructions.[3] These standard instructions, referred to as the input/output control system (IOCS), are an integral part of most operating systems. They simplify the means by which all programs being processed may undertake I/O operations. The controlling software calls on the IOCS software to actually complete the I/O operation. Considering the level of I/O activity in most programs, the IOCS instructions are extremely vital.

Within the board family of operating systems, there are generally four types, categorized on the types of computers they control and the sort of applications they support.

♦ Real-time operating system.

Real-time operating systems are used to control machinery, scientific instruments and industrial systems. A very important part of a real-time operating system is managing the resources of the computer so that a particular operation executes in precisely the same amount of time every time it occurs.

♦ Single-user, single-task operating system.

As the name implies, this operating system is designed for one user to effectively do one thing at a time. The Palm OS for Palm handheld computers is a good example of a modern single-user, single-task operating system.

♦ Single-user, multi-tasking operating system.

This is the type of operating system most people use on their desktop and laptop computers today.

Microsoft's Windows and Apple's Mac OS platforms are both examples of single-user, multi-tasking operating systems that will let a single user for a Windows to edit a text in a word processor while downloading a file from the Internet while printing the text of an E-mail message.[4]

♦ Multi-user operating system.

A multi-user operating system allows many different users to take advantage of one computer's resources simultaneously. The operating system must make sure that the requirements of the various users are balanced, and that each of the programs they are using has sufficient and separate resources so that a problem with one user doesn't affect the entire community of users.[5] UNIX, VMS and mainframe operating systems, such as MVS, are examples of multi-user operating systems.

Operating systems are constantly being improved or upgraded as technology advances. Upgrading an operating system can have several advantages, such as simplifying tasks and navigation. However, there can be disadvantages, too. Many older programs would no longer run within the new operating system. By far the most popular microcomputer operating system is Microsoft's Windows. Windows is custom-made to run with Intel CPU and Intel-compatible CPUs, such as the Pentium IV.

Keywords

activate	v.	使活动，起动		precisely	adv.	准确地，清楚地
assignment	n.	分配，指定		principal	adj.	主要的，重要的
categorize	v.	把……分类		responsible	adj.	有责任的
community	n.	团体，共同组织		retrieve	v.	检索，恢复
compatible	adj.	相容的，兼容的		schedule	n.	程序表，进度，计划
effectively	adv.	有效地，实质上地		separate	v.	分离，离开，隔开
extremely	adv.	极端地，非常地		termination	n.	末端，终点
navigation	n.	导航		undergo	v.	经受，经历
notify	v.	通知，通告		undertake	v.	承担，承办
oversee	v.	监督，监视		vital	adj.	重大的，紧要的

Notes

[1] Though operational software varies with manufacturers, it has similar characteristics.

译文：尽管各厂家的操作系统软件各不相同，但特性都是相似的。

说明：本句的 Though 引导让步状语从句。

[2] The operating system determines which computer resources will be used for solving which problems and the order in which they will be used.

译文：操作系统决定调用解决某个问题所需要的计算机资源及资源的使用顺序。

说明：本句的 which 引导宾语从句。

[3] To facilitate execution of I/O operations, most operating systems have a standard set of control instructions to handle the processing of all input and output instructions.

译文：为便于实施 I/O 的操作，大多数操作系统都有一套标准的控制指令集来处理所有输入输出指令。

说明：本句的 To facilitate execution of I/O operations 是目的状语，to handle the processing of all input and output instructions 作宾语补足语。

[4] Microsoft's Windows and Apple's Mac OS platforms are both examples of single-user, multi-tasking operating systems that will let a single user for a Windows to edit a text in a word processor while downloading a file from the Internet while printing the text of an E-mail message.

译文：微软公司的 Windows 平台和苹果公司的 Mac OS 平台都是单用户、多任务操作系统的例子。例如，Windows 操作系统用户就完全可能做到一边从互联网上下载文件，一边打印电子邮件的信息，一边还在用文字处理器编辑文本。

说明：这是一个长句，可以拆成两句。其中 both examples 作表语，"single-user, multi-tasking operating systems"作表语的定语，由 that 引导定语从句，修饰"single-user, multi-tasking operating systems"。

[5] The operating system must make sure that the requirements of the various users are balanced, and that each of the programs they are using has sufficient and separate resources so that a problem with one user doesn't affect the entire community of users.

译文：多用户操作系统必须保证不同用户的需求的平衡，每一个用户所使用的程序有独立的足够的资源，这样一个用户的问题不会影响到整个用户群体。

说明：本句有两个并列的由 that 引导的宾语从句，so that 引导的是目的状语从句。

5.2　Reading Material 1：Windows Vista

Windows Vista is the latest generation of Microsoft's operating system. It provides several enhancements over its direct predecessor, Windows XP, Service Pack 2. As soon as you start up your computer and log into Windows, you will notice a difference. A new Start menu, new desktop backgrounds, and even the new Sidebar that docks on the right side of your screen will tell you that you are experiencing something new. Then go ahead and click the new Start menu button and navigate the menus to launch a program. No longer does your screen fill with multiple layers of menus that reach to the edge of your screen and back. Instead, each time you drill down in a menu, it overlays the previous menu to make it easier to find the program you want.

Other new features include new programs (such as Windows Calendar, Windows Photo Gallery, Windows Media Center, and Windows Fax and Scan), handy networking tools (the Network and Sharing Center, for example), a redesigned Control Panel, and many more advancements.

As Microsoft's newest operating system, Vista has introduced "Life Immersion" concept for the first time, namely, it integrates many human factors in the system and everything is people-oriented, causing the operating system to close to the user as best as possible, to understand user's feeling, and to make convenience for the user.

Vista's minimum system requirements are significant but still relatively modest. They consist of a "modern" 800MHz processor, 512MB of RAM and a 20GB hard drive. Systems that meet these criteria can run Windows Vista basic and rated as Windows Vista "Capable" by Microsoft.

Microsoft assures you that Windows Vista will bring transparency to your world, so you can more safety and confidently rely on your PC. Get more lively multimedia experience with dynamic audio-video output, music and TV, exclusively on your Windows Vista-based PC.

As far as security is concerned, the older versions of Windows, had a lot of security problems. It is the hackers who invade the system, sitting miles away from the system, and take the data that they need, such as personal information, banking details, that one might have stored in the computer. The default browser Internet Explorer for Windows Vista has been upgraded to 7.0. It contains a lot of security features, like protected mode browsing, antiphishing, outbound and inbound firewall, standard user account functionality, user account control, windows defender, and parental control.

Windows Vista provides new amazing graphics functions. With Windows Vista, Microsoft has surpassed Mac OS in terms of graphics. It is the first time that Microsoft offers a great deal of high-quality graphics including 3D effects. This will help to boost programs meant for 3D games.

In the field of wireless networking function, one will see much more changes in the final release of Windows Vista. In Windows Vista, one can save the wireless connection setting and also one can name it. One can try to reconnect using these settings.

Windows Vista Home Premium is the preferred edition for home desktop and mobile PCs. It is the edition that delivers more ease of use, security, and entertainment to your PC at home and on the go. It includes Windows Media Center, and that makes it easier to enjoy your digital photos, TV and movies, and music.

With Windows Aero, you will experience dynamic reflections, smooth-gliding animations, transparent glass-like menu bars, and the ability to switch between your open windows in a new three-dimensional layout. Instant desktop search capabilities and new ways to organize your information mean that you can instantly find and use the E-mails, documents, photos, music, and other information you want. Windows Defender helps automatically safeguard your PC against pop-ups, slow performance, and security threats caused by spyware and other unwanted software.

Windows Vista Business is the first edition of Windows designed specifically to meet the needs of small businesses. You will spend less time on technology support-related issues—so you can spend more time making your business successful. You can help your business to work more efficiently with the improved, simple-to-use interface, which helps you find the information you need quickly and easily, on a PC or on the Web. With powerful new safety features, you are in control and can protect the key information that is important to your business and that builds the trust of your customers.

Keywords

antiphishing	n.	反钓鱼软件	gallery	n.	画廊，陈列室
boost	v.	帮助，促进，提高	immersion	n.	沉浸，浸没
calendar	n.	日历	invade	v.	侵入，打扰
confidently	adv.	确信地，自信地	outbound	adj.	跳出，外出的
drill	v.	操作，操练	overlay	v.	覆盖
enhancement	n.	增强，增值，提高	transparency	n.	透明度，透明性
exclusively	adv.	除外地，专有地	wireless	adj.	无线的

5.3　Reading Material 2：Dual-core Computing

Dual-core refers to a CPU that includes two complete execution cores. In a dual-core configuration, an integrated circuit contains two computer processors and their caches and cache controllers. Usually, the two identical processors are manufactured so they reside side-by-side on the same die, each with its own path to the system front-side bus.

A dual-core processor has many advantages especially for those looking to boost their system's multitasking computing power. Dual-core processors provide two complete execution cores instead of one, each with an independent interface to the front-side bus. Since each core has its own cache, the operating system has sufficient resources to handle intensive tasks in parallel, which provides a noticeable improvement to multitasking.

Complete optimization for the dual-core processor requires both the operating system and applications running on the computer to support a technology called thread-level parallelism, or TLP. Thread-level parallelism is the part of the operating system or application that runs multiple threads simultaneously, where threads refer to the part of a program that can execute independently of other parts.

Even without a multithread-enabled application, you will still see benefits of dual-core processors if you are running an operating system that support TLP. For example, if you have Microsoft Windows XP, you could have your Internet browser open along with a virus scanner running in the background, while using Windows Media Player to stream your favorite radio station and the dual-core processor will handle the multiple threads of these programs running simultaneously with an increase in performance and efficiency.

Today Windows XP and hundreds of applications already support multithread technology, especially applications that are used for editing and creating music files, videos and graphics because types of programs need to perform operations in parallel. As dual-core technology becomes more common in homes and the workplace, you can expect to see more applications support thread-level parallelism.

Software benefits from dual-core architectures where code can be executed in parallel. Under most common operating systems this requires code to execute in separate threads. Each application running on a system runs in its own thread so multiple applications will benefit from dual-core architectures. Each application may also have multiple threads but must be specifically written to do so. Operating system software also tends to run many threads as a part of its normal operation. Running virtual machines will benefit from adoption of dual-core architectures since each virtual machine runs independently of others and can be executed in parallel.

Keywords

cache	n.	高速缓冲存储器	optimization	n.	最优化
favorite	adj.	中意的，喜欢的	parallelism	n.	并行性
intensive	adj.	加强的，集中的	thread	n.	线程
multitasking	n.	多任务	virtual	adj.	虚拟的
noticeable	adj.	显著的，引人注意的	workplace	n.	工作场所

5.4 计算机专业英语中长句的运用

由于科学的严谨性，专业英语中常常出现许多长句。长句主要是由于修饰语多、并列成分多及语言结构层次多等因素造成的，如名词后面的定语短语或定语从句，以及动词后面或句首的介词短语或状语从句。这些修饰成分可以一个套一个地连用（包孕结构），形成长句结构。显然，英语的一句话可以表达好几层意思，而汉语习惯用一个小句表达一层意思，一般好几层意思要通过几个小句表达。在专业文章中，长句往往是对技术的关键部分的叙述，翻译得不恰当就会造成整个段落甚至通篇文章都不清楚。

对于长句的翻译而言，一遍看不出句子的意思，就多看几遍，在这个过程中步步深入。首先，弄清楚句型、句种、结构和各成分之间的关系，逐渐推进；然后，纵观全局，确切地把握句子所表达的意思。从全局到局部，分清句子的整体结构；再从局部到总体，深入把握句子的细节。

通常分析长句时采用的方法如下。

- ◆ 找出全句的基本语法成分，即主语、谓语和宾语，从整体上把握句子的结构。
- ◆ 找出句子中所有的谓语结构、非谓语动词、介词短语和从句的引导词等。
- ◆ 分析从句和短语的功能，即是否为主语从句、宾语从句、表语从句等；若是状语从句，则分析它是属于时间状语从句、原因状语从句、条件状语从句、目的状语从句、地点状语从句、让步状语从句、方式状语从句、结果状语从句，还是比较状语从句。
- ◆ 分析词、短语和从句之间的相互关系，如定语从句修饰的先行词是哪一个等。
- ◆ 注意分析句子中是否有固定词组或固定搭配。
- ◆ 注意插入语等其他成分。

在英语长句的阅读和翻译过程中，必须清楚句子的逻辑结构、层次关系和所用的语体。常用的翻译方法有以下几种。

1. 顺序法

当英语长句的内容叙述层次与汉语基本一致时，或者英语长句中所描述的一连串动作是按时间顺序安排的，可以按照英语原文的顺序翻译成汉语。

例 1：The close-loop system has a control unit which gets information from a sensing element, compares the real state with that required by the program and, when there is a different between the two, makes the necessary adjustment to the control element so that the desired state is maintained.

译文：闭环系统有一个控制单元，该单元从传感器获得信息，把真实值和程序的预定值进行比较，当两者有区别时，就对控制器进行必要的调整，从而保持预定值。

例 2：Being able to receive information from any one of a large number of separate places, carry out the necessary calculations and give the answer or order to one or more of the same number of places scattered around a plant in a minute or two, or even in a few seconds, computers are ideal for automatic control in process industry.

译文：由于计算机能从工厂大量分散的任何地方获取信息进行必要的运算，并在一两分钟甚至几秒钟内向分散在工厂各处的一处或多处提供响应或发出指令，因此它对加工工业的自动控制是非常理想的。

例 3：Personal computer-based office automation software has become an indispensable part of electron management in many countries. Word processing programs have replaced type-writers; spreadsheet programs have replaced ledger books; database programs have replaced paper-based electoral rolls, inventories and staff lists; personal organizer programs have replaced paper diaries; and so on.

译文：个人计算机办公自动化软件在许多国家已经成为电子管理不可缺少的组成部分。文字处理程序取代了打字机；电子表格取代了账簿；数据库取代了传统的纸选票、库存品和职员列表；个人管理程序取代了纸质日记簿等。

2．逆序法

所谓逆序法，就是从长句的后面或中间译起，把长句的开头放在译文的结尾。这是由于英语和汉语的表达习惯不同：英语习惯于用前置性陈述，先结果后原因；而汉语习惯则相反，一般先原因后结果，层层递进，最后综合。当遇到这些表达次序与汉语表达习惯不同的长句时，就要采用逆序法。

例 1：Instead of paying someone to manually enter reams of data into the computer, you can use a scanner to automatically convert the same information to digital files using OCR (optical character recognition) software.

译文：只要在使用扫描仪的过程中借助于光学字符识别软件就可以将信息转换成数字文件的形式，从而代替人们手工将大量数据输入到计算机中的过程。

例 2：In order to assist users to name files consistently, and, importantly, to allow the original creator and other users to find those files again, it is useful to establish naming conventions.

译文：为了帮助用户统一地命名文件，重要的是使最初的创建者和其他用户能再一次找到那些文件，建立命名公约是很必要的。

3．分句法

有时长句中主语或主句与修饰词的关系并不十分密切，翻译时可以按照汉语多用短句的习惯，把长句的从句或短语化成句子，分开来叙述。而有时英语长句包含多层意思，而汉语习惯于一个小句表达一层意思。为了使行文简洁，翻译时可把长句中的从句或介词短语分开叙述，顺序基本不变，保持前后的连贯。翻译时为了使语意连贯，有时需要适当增加词语。

例 1：The structure design itself includes two different tasks, the design of the structure, in which the sizes and locations of the main members are settled, and the analysis of this structure by mathematical or graphical methods or both, to work out how the loads pass through the structure with the particular members chosen.

译文：结构设计包括两项不同的任务：一是结构设计，确定主要构件的尺寸和位置；二是用数学方法或图解方法或二者兼用进行结构分析，以便在构件选定后计算出各载荷通过结构的情况。

例 2：The loads a structure is subjected to are divided into dead loads, which include the weights of all the parts of the structure, and live loads, which are due to the weights of people, movable equipment, etc..

译文：一个结构受到的载荷可以分为静载和动载两类。静载包括该结构各部分的重量；动载则是由于人和可移动设备等的重量而引起的载荷。

例 3：Television, it is often said, keeps one informed about current events, allow one to follow the latest developments in science and politics, and offers an endless series of programs which are both instructive and entertaining.

译文：人们常说，通过电视可以了解时事，掌握科学和政治的最新动态。从电视里还可以看到层出不穷、既有教育意义又有娱乐性的系列节目。

4．综合法

当一些长句单纯采用上述任何一种方法翻译都不准确时，就需要仔细分析，或者按照时间先后，或者按照逻辑顺序，顺逆结合，主次分明地对全句进行综合处理。

例 1：Noise can be unpleasant to live even several miles from an aerodrome; is you think what it must be like to share the deck of a ship with several squadrons of jet aircraft, you will realize that a modern navy is a good place to study noise.

译文：噪声甚至会使住在远离飞机场几英里的人感到不适。如果你能体会出站在甲板上的几个中队喷气式飞机中间是什么滋味的话,那你就会意识到现代海军是研究噪声的理想场所。

例 2: Modern scientific and technical books, especially textbooks, require revision at short intervals if their authors wish to keep pace with new ideas, observations and discoveries.

译文: 现代科技书籍,特别是教科书,如果作者希望书中内容与新见解、新观察、新发现保持一致,就应该在较短的时间内将内容重新修改。

5.5 Exercises

1. Translate the following phrases.

    Start menu
    real-time operating system
    job management
    command language
    system resource
    primary storage
    secondary storage
    operating system
    security problem
    personal information
    dual-core processor
    multithread technology

2. Fill in the blanks with appropriate words or phrases.

    a. components   b. control machinery   c. programs   d. technology
    e. Windows   f. users   g. operating system   h. languages

    (1) The _____ is also responsible for keeping track of the activities in the computer system.

    (2) Real-time operating systems are used to _____, scientific instruments and industrial systems.

    (3) A multi-user operating system allows many different _____ to take advantage of one computer's resources simultaneously.

    (4) The OS is a large program made up of many smaller _____.

    (5) The OS controls how the CPU communicates with other hardware _____.

    (6) The OS also makes computers easier to operate by people who don't understand programming _____.

    (7) Operating systems are constantly being improved or upgraded as _____ advances.

    (8) _____ is custom-made to run with Intel CPU and Intel-compatible CPUs, such as the Pentium Ⅳ.

3. Match the following terms to the appropriate definition.

    a. certificate authority   b. clock rate   c. crash
    d. CRT   e. cache   f. directory

    (1) It is the fundamental rate in cycles per second, measured in hertz, at which a computer performs its most

basic operations such as adding two numbers or transferring a value from one processor register to another.

(2) A trusted third-party organization or company that issues digital certificates used to create digital signatures and public-private key pairs.

(3) It is the display device used in most computer displays, video monitors, televisions and oscilloscopes. It was used in all television sets until the late 20th century.

(4) It is a common term for a computer fault that brings down a software program or operating system. It is also refer to the failure of a hard disk drive.

(5) A special high-speed storage mechanism. It can be either a reserved section of main memory or an independent high-speed storage device.

(6) A major division on a hard drive used to divide and organize files.

Unit 6   Data Structure

6.1   Text

A data structure is a specialized format for organizing and storing data. General data structure types include the array, the file, the record, the table, the tree, and so on. Any data structure is designed to organize data to suit a specific purpose so that it can be accessed and worked with in appropriate ways.[1] In computer programming, a data structure may be selected or designed to store data for the purpose of working on it with various algorithms.

Data Structures and Algorithms

Data structures organize data in ways that make algorithms more efficient. For example, consider some of the ways we can organize data for searching it. One simplistic approach is to place the data in an array and search the data by traversing element by element until the desired element is found. However, this method is inefficient because in many cases we end up traversing every element. By using another type of data structure, such as a hash table or a binary tree we can search the data considerably faster.[2]

Procedural abstraction, or algorithmic abstraction, is hiding of algorithmic details, which allows the algorithm to be seen or described, at various levels of detail. Building subprograms so that the names of the subprograms describe what the subprograms do and the code inside subprograms shows how the processes are accomplished is an illustration of abstraction in action. Similarly, data abstraction is the hiding of representational details. An obvious example of this is the building of data types by combining together other data types, each of which describes a piece, or attribute, of a more complex object type. An object-oriented approach to data structures brings together both data abstraction and procedural abstraction through the packaging of the representations of classes of objects.

Once problems are abstracted, it becomes apparent that seemingly different problems are essentially similar or even equivalent in a deep sense. For example, the problems of maintaining a list of students taking a lecture course and of organizing a dictionary structure in a compiler have much in common; both require the storage and manipulation of named things and the things have certain attributes or properties. Abstraction allows common solutions to seemingly different problems. By using abstract algorithms and data structures, a solution to a problem has maximum utility and scope for reuse.

Data Type

A data structure is a data type whose values are composed of component elements that are related by some structure. It has a set of operations on its values. In addition, there may be operations that act on its component elements. Thus we see that a structured data type can have operations defined on its component values, as well as on the component elements of those values.

Integer: Integer is amounts to a particular collection of axioms or rules that must be obeyed. The way in which integers are represented is unimportant provided only that all readers understand the notation—binary, octal, decimal, hexadecimal, twos complement or sign and magnitude.

List: The list is a flexible abstract data type. It is particularly useful when the number of elements to be stored is not known before running a program or can change during running. It is well suited to sequential processing of elements because the next element of a list is readily accessible from the current element.[3]

Array and Record: The data types arrays and records are native to many programming languages. By using the pointer data type and dynamic memory allocation, many programming languages also provide the facilities for constructing linked structures. Arrays, records, and linked structures provide the building blocks for implementing what we might call higher-level abstractions.

Stacks and Queues

A stack is a data type whose major attributes are determined by the rules governing the insertion and deletion of its elements. The only element that can be deleted or removed is the one that was inserted most recently. Such a structure is said to have a last-in/first-out (LIFO) behavior, or protocol. When a call is made to a new function, all the variables local to the calling routine need to be saved by the system, otherwise the new function will overwrite the calling routine's variables. Furthermore, the current location in the routine must be saved so that the new function knows where to go after it is done. The variables have generally been assigned by the compiler to machine registers, and there are certain to be conflicts, especially if recursion is involved.

The end of a stack at which entries are inserted and deleted is called the top of the stack. The other end is sometimes called the stack's base. To reflect the fact that access to a stack is restricted to the topmost entry, we use special terminology when referring to the insertion and deletion operations. The process of inserting an object on the stack is called a push operation, and the process of deleting an object is called a pop operation.

To implement a stack structure in a computer's memory, it is customary to reserve a block of contiguous memory cells large enough to accommodate the stack as it grows and shrinks. Determining the size of this block can often be a critical decision. If too little room is reserved, the stack ultimately exceeds the allotted storage space; if too much room is reserved, memory space will be wasted. One end of this block is designated as the stack's base. It is here that the first entry pushed on the stack is stored, with each additional entry being placed next to its predecessor as the stack grows toward the other end of the reserved block.

There are several algorithms that use queues to give efficient running times. For now, we will give some simple examples of queue usage. When jobs are submitted to a printer, they are arranged in order of arrival. Thus, essentially, jobs sent to a line printer, are placed on a queue. Another example concerns computer networks. There are many network setups of personal computers in which the disk is attached to one machine, known as the file server. Users on other machines are given access to files on a first-come first-served basis, so the data structure is a queue.

A common solution is to set aside a block of memory for the queue, start the queue at one end of the block, and let the queue migrate toward the other end of the block. Then, when the tail of the queue reaches the end of the block, we merely start inserting additional entries back at the original end of the block, which by this time is vacant.[4] Likewise, when the last entry in the block finally becomes the head of the queue and is removed, the head pointer is adjusted back to the beginning of the block where other entries are, by this time, waiting. In this manner, the queue chases itself around within the block rather than wandering off through memory. Such a technique results in an implementation that is called a circular queue because the effect is that of forming a loop out of the block of memory cells allotted to the queue. As far as the queue is concerned, the last cell in the block is adjacent to the first cell.

Multilevel Feedback Queue

In computer science, a multilevel feedback queue is a scheduling algorithm. It is intended to meet the following design requirements for multimode systems.

♦ Give preference to short jobs.
♦ Give preference to I/O bound processes.
♦ Quickly establish the nature of a process and schedule the process accordingly.

Multiple FIFO (first-in first-out) queues are used and the operation is as follows:

♦ A new process is positioned at the end of the top-level FIFO queues.
♦ At some stage the process reaches the head of the queue and is assigned the CPU.
♦ If the process is completed it leaves the system.
♦ If the process voluntarily relinquishes control it leaves the queuing network, and when the process becomes ready again it enters the system on the same queue level.
♦ If the process uses all the quantum time, it is pre-empted and positioned at the end of the next lower level queue.
♦ This will continue until the process completes or it reaches the base level queue.

At the base level queue the processes circulate in round robin fashion until they complete and leave the system.

In the multilevel feedback queue, a process is given just one chance to complete at a given queue level before it is forced down to a lower level queue.[5]

Keywords

attribute	n.	属性，特质	migrate	v.		移动
axioms	n.	公理，原理	notation	n.		记号，符号
circular	adj.	循环的，圆环的	octal	adj.		八进制的
compiler	n.	编译器	property	n.		特征，属性
considerably	adv.	相当地，大量地	quantum	n.		定量，定额
customary	adj.	通常的，照惯例的	recursion	n.		递归，递推
flexible	adj.	灵活的	reflect	v.		反映，表现
hexadecimal	adj.	十六进制的	representational	adj.		表现的，再现的
magnitude	n.	量值	scope	n.		范围

| simplistic | *adj.* | 简单化的 | topmost | *adj.* | 最高的，最上的 |
| subprogram | *n.* | 子程序 | traverse | *v.* | 遍历 |

Notes

[1] Any data structure is designed to organize data to suit a specific purpose so that it can be accessed and worked with in appropriate ways.

译文：任何数据结构都是用来组织数据以适应某种特定的目标，从而以适当的方法达到存取和操作的目的。

说明：本句的 to organize data 是动词宾语，to suit a specific purpose 作定语，修饰 data，so that 引导目的状语从句。

[2] By using another type of data structure, such as a hash table or a binary tree we can search the data considerably faster.

译文：使用其他类型的数据结构，如哈希表和二叉树，我们能够相当快速地搜寻数据。

说明：在本句中，By using another type of data structure 作方式状语，such as a hash table or a binary tree 是 data structure 的同位语。

[3] It is well suited to sequential processing of elements because the next element of a list is readily accessible from the current element.

译文：它非常适合元素的顺序处理，因为要访问表中当前元素的下一个元素非常容易。

说明：本句由 because 引导原因状语从句。

[4] Then, when the tail of the queue reaches the end of the block, we merely start inserting additional entries back at the original end of the block, which by this time is vacant.

译文：当队尾到达块的末端，我们开始将新的条目反向于末端的方向插入，此时它是空闲的。

说明：在本句中，when 引导的是时间状语从句，which by this time is vacant 是非限定性定语从句。

[5] In the multilevel feedback queue, a process is given just one chance to complete at a given queue level before it is forced down to a lower level queue.

译文：在多级反馈队列中，一个过程在被强制降到了下一级队列之前只被给予一次在给定队列级别上完成任务的机会。

说明：在本句中，In the multilevel feedback queue 是条件状语，before it is forced down to a lower level queue 是时间状语从句。

6.2　Reading Material 1：.NET Framework

　　.NET is a capable platform on which to develop almost any solution, and it offers substantial support for network programming. In fact, .NET has more intrinsic support for networking than any other platform developed by Microsoft. A network program is any application that uses a computer network to transfer information to and from other applications. Examples range from the ubiquitous Web browser such as Internet Explorer, or the program you use to receive your E-mail. All of these pieces of software share the ability to communicate with other computers, and in so doing, become more useful to the end-user.

　　In the case of a browser, every Web site you visit is actually files stored on a computer somewhere else on the Internet. With your E-mail program, you are communicating with a computer at your Internet service provider or company E-mail exchange, which is holding your E-mail for you. Users generally trust network applications. Therefore, these programs have much greater control over the computers on which they are running than a Web site has over the computers viewing it. This makes it possible for a network application to manage files on the local computer, whereas a Web site, for all practical purposes, cannot do this. More importantly, from a networking perspective, an application has much greater control over how it can communicate with other computers on the Internet.

.NET is not a programming language. It is a development framework that incorporates four official programming languages: C#, VB.NET, Managed C++, and J#.NET. Where there are overlaps in object types in the four languages, the framework defines the framework class library (FCL). All four languages in the framework share the FCL and the common language runtime (CLR), which is an object-oriented platform that provides a runtime environment for .NET applications. The CLR is analogous to the virtual machine in Java, except it is designed for Windows, not cross-platform, use.

.NET languages are object-oriented rather than procedurally based. This provides a natural mechanism to encapsulate interrelated data and methods to modify this data within the same logical construct. An object is a programmatic construct that has properties or can perform actions. A core concept of object orientation is the ability of one class to inherit the properties and methods of another.

An assembly is generally a .DLL file that contains precompiled code for a collection of .NET classes. Unlike standard Win32 DLLs in which developers had to rely on documentation, such as header files, to use any given DLL, .NET assemblies contain metadata, which provides enough information for any .NET application to use the methods contained within the assembly correctly. Metadata is also used to describe other features of the assembly, such as its version number, the originator of the code, and any custom attributes that were added to the classes.

.NET provides a unique solution to the issue of sharing assemblies between multiple applications. Generally, where an assembly is designed for use with only one application, it is contained within the same folder (or bin subfolder) as the application. This is known as a private assembly. A public assembly is copied into a location where all .NET applications on the local system have access too. Furthermore, this public assembly is designed to be versioned, unique, and tamperproof, thanks to a clever security model. This location into which public assemblies are copied is called the global assembly cache.

Applications written in .NET are referred to as managed, or type-safe, code. This means that the code is compiled to an intermediate language that is strictly controlled, such that it cannot contain any code that could potentially cause a computer to crash. Applications written in native code have the ability to modify arbitrary addresses of computer memory, some of which could cause crashes, or general protection faults.

Components designed before the advent of .NET are written in native code and are therefore unmanaged and deemed unsafe. There is no technical difficulty in combining unsafe code with a .NET application. However, if an underlying component has the potential to bring down a computer, the whole application is also deemed unsafe. Unsafe applications may be subject to restrictions; for instance, when they are executed from a network share, they could be prevented from operating.

Keywords

analogous	*adj.*	类似的，相似的	intrinsic	*adj.*	内在的，内部的
encapsulate	*v.*	封装	official	*adj.*	公认的，正式的
framework	*n.*	框架，机构	orientation	*n.*	定位，定向
incorporate	*v.*	结合，合并	originator	*n.*	创作人，创始人
inherit	*v.*	继承	overlap	*v.*	重复，重叠
intermediate	*adj.*	中间的，中介的	substantial	*adj.*	实质的，可靠的

6.3 Reading Material 2: Multiprogramming and Multiprocessing

Early computers were capable of executing a few thousand instructions per second. Modern mainframes are much faster, executing millions of instructions per second. Such speeds are difficult to imagine, let us just say that today's computers are very fast, indeed. Unfortunately, the speeds of peripheral devices have not kept pace.

A modern mainframe is capable of processing data hundreds even thousands of times faster than its peripheral devices can supply them. What does the computer do during input or output? Nothing. A program can't process data it doesn't yet have, and the success of an output operation can't be assumed until the operation is finished, so the program waits. Since the program controls the computer, the computer waits, too.

Why not put two programs in main memory? Then, when program A is waiting for data, the processor can turn its attention to program B. And why stop at two programs? With three, even more otherwise wasted time is utilized. Generally, the more programs in memory, the greater the utilization of the processor. This technique is known as multiprogramming.

Multiprogramming and time-sharing require a resident operating system to deal with the conflicts that arise when multiple concurrent users share limited resources. Both techniques help to improve the efficiency of system, allowing more programs to be processed over the same time period on the same hardware. However, the operating system modules, essential though they may be, occupy main memory and consume processor time, they represent unproductive overhead.

Consider, for example, the problem of controlling I/O channels were developed to relieve the processor of much of this responsibility. Unfortunately, a channel can't do the whole job itself; certain logical functions such as starting, finishing, and checking the status of the I/O operation were, until recently, performed by the main processor working under the control of the operating system. Why not identify the operating system code that performs these functions, and program a microprocessor to do the same things? We might replace the channel with this new I/O processor. Now, because we have two independent processors, the instructions associated with I/O can be executed in parallel with more productive main processor activities. Two processors share the same main memory, forming a multiprocessing system.

For large scientific and engineering problems, an array processor might relieve the main processor of the time-consuming chore of array manipulation. Language processors might allow the direct execution of programs written in a high-level language, thus bypassing the inefficient compilation step. If many of the functions now associated with operating systems and system software are shifted to independent processors, these control programs will no longer tie up the main processor's time. The result is greater efficiency.

Keywords

associate	v.	使联合，使加入	overhead	adv.	在头顶上，在上面
assume	v.	假想，假装	relieve	v.	减轻，缓和
compilation	n.	编译，汇编	resident	adj.	驻留的，内在的，固有的
concurrent	adj.	同时发生的，并发的	utilization	n.	利用，效用

6.4 被动语态的运用

英语中被动语态的使用范围极为广泛，尤其是在科技英语中，被动语态几乎随处可见。凡是在不必、不知道或不愿说出主动者的情况下均可使用被动语态。

专业英语文体在很多情况下是对某个科学论题的讨论，介绍某个科技产品和科学技术，为了表示一种公允性和客观性，往往在句子结构上采用被动语态描述，即以被描述者为主体，或者以第三者的身份介绍文章要点和内容。于是，被动语态反映了专业英语文体中文体的客观性。除了表述作者自己的看法、观点以外，很少直接采用第一人称表述法，但在阅读理解和翻译时，根据具体情况，又可以将一个被动语态句子翻译成主动形式，以便强调某个重点，更适合汉语的习惯。

1．常用被动语态的几种情况

（1）当我们强调的是动作的承受者或给动作的承受者较大关注时，多用被动语态。这时，由于动作的执行者处于次要地位，句子中由 by 引导的短语可以省略。

例：The virus in the computer has been found out.

译文：计算机中的病毒已经找出来了。

（2）当我们不知道或不想说出动作的执行者时，可使用被动语态。这时句子中不带由 by 引导的短语。

例：Electricity was discovered a very long time ago.

译文：电是很久以前发明的。

（3）当动作的执行者是"物"而不是"人"时，常用被动语态。

例：This machine is controlled by a computer.

译文：这台机器由计算机控制。

（4）当动作的执行者已为大家所熟知，而没有必要说出来时，也常常使用被动语态。

例：This factory was built twenty years ago.

译文：这座工厂是二十年前兴建的。

2．专业英语中主要时态的被动语态形式

1）一般现在时

一般现在时的被动语态构成如下：

主语 + am（is，are）+及物动词的过去分词

例：The switches are used for the opening and closing of electrical circuits.

译文：开关是用来开启和关闭电路的。

2）现在进行时

现在进行时的被动语态构成如下：

主语 + is（are）being + 及物动词的过去分词

例：Electron tubes are found in various old products and are still being used in the circuit of some new products.

译文：在各种老产品里看到的电子管，在一些新产品的电路中也还在使用。

3）现在完成时

现在完成时的被动语态构成如下：

主语 + have（has）been + 及物动词的过去分词

例：The letter has not been posted.

译文：信还没有寄出。

4）一般过去时

一般过去时的被动语态构成如下：

主语 + was（were）+ 及物动词的过去分词

例：That plotter was not bought in Beijing.

译文：那台绘图仪不是在北京买的。

5）过去进行时

过去进行时的被动语态构成如下：

主语 + was（were）being + 及物动词的过去分词

例：The laboratory building was being built then.

译文：实验大楼当时正在建造。

6）过去完成时

过去完成时的被动语态构成如下：

主语 + had been + 及物动词的过去分词

例：When he came back, the problem had already been solved.

译文：他回来时，问题已经解决了。

7）一般将来时

一般将来时的被动语态构成如下：

主语 + will be + 及物动词的过去分词

当主语是第一人称时，可用：

主语 + shall be + 及物动词的过去分词

例1：I shall not be allowed to do it.

译文：不会让我做这件事的。

例2：What tools will be needed for the job?

译文：工作中需要什么工具？

3．含被动语态句子的翻译

在汉语中，也有被动语态，通常通过"把"或"被"等词体现出来，但它的使用范围远远小于英语中被动语态的使用范围，因此英语中的被动语态在很多情况下都翻译成主动结构。

（1）英语原文中的主语在译文中仍作主语。在采用此方法时，我们往往在译文中使用"加以""经过""用……来"等词体现原文中的被动含义。

例1：In other words mineral substances which are found on earth must be extracted by digging, boring holes, artifical explosions, or similar operations which make them available to us.

译文：换言之，矿物就是存在于地球上，但须经过挖掘、钻孔、人工爆破或类似作业才能获得的物质。

例2：Nuclear power's danger to health, safety, and even life itself can be summed up in one word: radiation.

译文：核能对健康、安全，甚至对生命本身构成的危险可以用一个词——辐射来概括。

（2）将英语原文中的主语翻译为宾语，同时增补泛指性的词语（大家、人们等）作主语。

例1：Television, it is often said, keeps one informed about current events, allows one to follow the latest developments in science and politics, and offers an endless series of programmes which are both instructive and entertaining.

译文：人们常说，电视使人了解时事，熟悉政治领域的最新发展变化，并能源源不断地为观众提供各种既有教育意义又有趣的节目。

例2：It could be argued that the radio performs this service as well, but on television everything is much more living, much more real.

译文：可能有人会指出，无线电广播同样也能做到这一点，但还是电视屏幕上的节目要生动、真实得多。

（3）将英语原文中的by、in、for等作状语的介词短语翻译成译文的主语，在此情况下，英语原文中的主语一般被翻译成宾语。

例1：And it is imagined by many that the operations of the common mind can be by no means compared with these processes, and that they have to be acquired by a sort of special training.

译文：许多人认为，普通人的思维活动根本无法与科学家的思维过程相比，而且认为这些思维过程必须经过某种专门的训练才能掌握。

例2：A right kind of fuel is needed for an atomic reactor.

译文：原子反应堆需要一种合适的燃料。

（4）专业英语中的一些被动句也可以翻译成汉语的被动句。常用"被""由……""受到""遭""为……所""使"等表示。

例1：Over the years, tools and technology themselves as a source of fundamental innovation have largely been ignored by historians and philosophers of science.

译文：工具和技术本身作为根本性创新的源泉多年来在很大程度上被科学史学家和科学思想家们忽视了。

例2：The behaviour of a fluid flowing through a pipe is affected by a number of factors, including the viscosity of the fluid and the speed at which it is pumped.

译文：流体在管道中流动的情况，受到诸如流体黏度、泵送速度等各种因素的影响。

6.5 Exercises

1. Translate the following phrases.

   data structure

binary tree

data abstraction

object-oriented approach

procedural abstraction

data type

structured data type

pointer data type

dynamic memory allocation

first come first served

scheduling algorithm

first in first out

2. **Fill in the blanks with appropriate words or phrases.**

   a. collection   b. abstracted   c. algorithms   d. the file server

   e. a new function   f. common   g. data structures   h. format

(1) A data structure is a specialized _____ for organizing and storing data.

(2) Data structures organize data in ways that make _____ more efficient.

(3) An object–oriented approach to _____ brings together both data abstraction and procedural abstraction.

(4) Once problems are _____, it becomes apparent that seemingly different problems are essentially similar or even equivalent in a deep sense.

(5) Abstraction allows _____ solutions to seemingly different problems.

(6) Integer is amounts to a particular _____ of axioms or rules that must be obeyed.

(7) When a call is made to _____, all the variables local to the calling routine need to be saved by the system, otherwise the new function will overwrite the calling routine's variables.

(8) Computers in which the disk is attached to one machine, known as _____.

3. **Match the following terms to the appropriate definition.**

   a. direct connection   b. digital compression   c. demodulation

   d. digital certificate   e. distributed computing environment   f. dedicated line

(1) An attachment to an electronic message used for security purposes. The most common use is to verify that a user sending a message is who he or she claims to be, and to provide the receiver with the means to encode a reply.

(2) A suite of technology services developed by the open group for creating distributed applications that run on different platforms.

(3) With this method, your PC is typically issued a static IP address to be used to transfer data back and forth via the Internet. It keeps you continually connected to the Internet.

(4) An engineering technique for converting an analog television signal into a digital format.

(5) The extraction of the modulation or information from a radio-frequency current.

(6) A permanently connected telephone line between two computer systems. Dedicated lines make up the bulk of the Internet.

Unit 7　　Database Principle

7.1　Text

A database consists of a file or a set of files. The information in these files may be broken down into records, each of which consists of one field or more fields. Fields are the basic units of data storage, and each field typically contains information pertaining to one aspect or attribute of the entity described by the database. Using keywords and various sorting commands, users can rapidly search, rearrange, group, and select the fields in many records to retrieve or create reports on particular aggregates of data. Database records and files must be organized to allow retrieval of the information. Many users of a large database must be able to manipulate the information within it quickly at any given time.

Data Abstraction

It should be obvious that between the computer, dealing with bits, and the ultimate user dealing with abstractions such as flights or assignment of personnel to aircraft, there will be many levels of abstraction.

The lowest level, i.e., the physical level has the data stored on hardware devices. User programs cannot access them directly. They have to go through the logical level to access the data. The external level defines the different views of the database as required by the external or user programs. One user program may not require all the data in the database. Hence the user/application programs view only the required information from the database. That means different program will have different views of the database depending on their requirement of data. Such views are external to the database and are specified at the external level. Also it is not necessary that different views should contain altogether different data. There can be common information in different views.

The conceptual level describes the entire database. It is used by database administrators, who must decide what information is to be kept in the database.

Data Models

A data model is a collection of conceptual tools for describing data, data relationships, data semantics and data constraints. The data models are divided into three classes, viz., object-based logical models, record-based logical models and physical data models.

Object-based logical models are used for describing data at the conceptual and view levels. They are very close to human logic. Many different models are available to describe object-based logical models. The most important among them are semantic data model and entity-relationship model. Semantic data model provides a facility for expressing meaning about the data in the database. The entity-relationship model (E-R model) is based on a perception of a real world which consists of a collection of objects called entities and relationships among these objects.[1] An entity is an object, which can be uniquely distinguished from other objects. For instance, the designation, physical dimensions and weight per unit length uniquely describe a particular rolled steel section. The set of all entities of the same type and relationships of the same type are termed as entity set and relationship set respectively.[2]

Entities and relationships are to be distinguished and a database model should specify how this can be carried out. This achieved using the concept of primary key. An entity-relationship model may define certain constraints to which the contents of a database must confirm. One important constraint is the number of entities to which another entity can be associated via a relationship. For relationships involving two entity sets, there can be relationships like one-to-one, one-to-many, many-to one, and many-to-many.

Record-based logical models define the overall logical structure of the database as well as higher level description of its implementation. Three different record-based logical models are widely used. They are hierarchical model, network model, and relational model.

Physical data models are used to describe data at the lowest level. There are very few physical data models in use. Some of the widely known ones are unifying model and frame memory.

Some Terms of Relational Database

Tables: These are the objects that contain the data types and actual raw data.

Fields or columns: These are part of the table that holds the data. Columns must be assigned a data type and unique name.

Data types: There are various data types to choose from, such as character, numeric, date, or other.

Stored procedure: This is like a macro in that SQL code can be written and stored under a name. By running the name, you actually run the code. One use would be to take the SQL code that runs a weekly report and save it as a stored procedure; from then on, you would only have to run the stored procedure in order to generate the report.

Primary key: While not an object, primary keys are essential to relational databases. Primary keys enforce uniqueness among rows.

Foreign key: Again, not an object, a foreign key is a file or column that references another table's primary key. SQL Server uses primary and foreign keys to relate the data back together from separate tables when a query is performed.[3]

Constraints: A constraint is a server-based, system-implemented, data-integrity enforcement mechanism.

Views: A view is basically a query stored in the database that can reference one or many tables. It can be created and saved so that it can be easily used in the future. A view usually either excludes certain columns from a table or links two or more tables together.

Data Security, Integrity and Independence

Data security is the protection of the database to prevent the data disclosure, modification or destruction because of the unlawful use.[4] Database security and computer systems security are closely linked and mutually supportive.

Data integrity is the data's accuracy and reliability. It checks if the database is inconsistent with the semantics of data and information to prevent errors caused by invalid input and output operations or erroneous information. Data integrity is divided into four categories: entity integrity, domain integrity, referential integrity and user-defined integrity.

♦ Entity Integrity

It means each row in a table is the only entity. It can be embodied by using the Primary Key, Unique or Identity constraints.

♦ Domain Integrity

It means each column in a table must meet certain specific data type or bind. Check, Foreign Key constraints and Default, Not Null definitions in a table belong to the area of domain integrity.

♦ Referential Integrity

It is built on two tables that the keywords of one table should be consistent with the other table's foreign keywords. In SQL Server, referential integrity does the following:

(1) Prohibits the insertion of new rows into one table if there is no relation with the data of another table.

(2) Prohibits the deletion of records from one table if there is a relation with the records of another table.

♦ User-defined Integrity

User-defined integrity is a particular relational database constraint; it reflects that a specific application of the data must meet the requirements of semantics.

It can be said that the development of data processing is the history of continuous data independence. In the manual management phase, data and procedures were completely intertwined, with no independence at all.[5] Data independence includes the physical independence and the logical independence.

Physical independence means the user application program and the database data are independent of each other. The user need not know how data are stored in the disk and how they are accessed. Logical independence refers to the independence of the user application program and database logical structure, namely, when the database logical structure changes, the user program can remain unchanged.

Keywords

aggregate	v.	集合，聚集	inconsistent	adj.	不一致的，不调和的
aspect	n.	方面	integrity	n.	完整性，完全
conceptual	adj.	概念的	intertwine	v.	缠绕在一起，使缠结
designation	n.	名称，称呼	invalid	adj.	无用的，无效的
destruction	n.	破坏，消灭	modification	n.	更改，修改
disclosure	n.	泄露，暴露	mutually	adv.	相互地，彼此
distinguish	v.	区别，辨别	perception	n.	感知，感觉
enforcement	n.	强制，强迫，推行	referential	adj.	参照的，参考的
entity	n.	实体，统一体	ultimate	adj.	最终的，最后的
erroneous	adj.	错误的，不正确的			

Notes

[1] The entity-relationship model (E-R model) is based on a perception of a real world which consists of a collection of objects called entities and relationships among these objects.

译文：实体关系模型（E-R 模型）是基于这样的认识：现实世界是由一组称为实体的对象和这些对象之间的关系组成的。

说明：本句由 which 引导定语从句，修饰 real world。

[2] The set of all entities of the same type and relationships of the same type are termed as entity set and relationship set respectively.

译文：相同类型的所有实体的集合和相同类型的各种关系分别称为实体集合和关系集合。

说明：本句的主语是 The set of all entities 和 relationships，这两个主语分别有定语修饰，宾语是 entity set and relationship set。

[3] SQL Server uses primary and foreign keys to relate the data back together from separate tables when a query is performed.

译文：在查询时，SQL 服务器用主键和外键来关联不同表中的数据。

说明：when a query is performed 是时间状语从句，to relate the data back together from separate tables 作目的状语。

[4] Data security is the protection of the database to prevent the data disclosure, modification or destruction because of the unlawful use.

译文：数据的安全性是指保护数据库以防止不合法的使用所造成的数据泄露、更改或破坏。

说明：本句的 to prevent the data disclosure, modification or destruction 是目的状语，because of the unlawful use 是原因状语。

[5] In the manual management phase, data and procedures were completely intertwined, with no independence at all.

译文：在手工管理阶段，数据和程序完全交织在一起，没有独立性可言。

说明：In the manual management phase 是时间状语，而 with no independence at all 也是状语。

7.2　Reading Material 1：Data Warehouse and Data Mining

1. Data Warehouse

A data warehouse is a database that consolidates data extracted from various production and operational systems into one large database that can be used for management reporting and analysis. The data from the organization's core transaction processing systems are reorganized and combined with other information, including historical data so that they can be used for management decision making and analysis.

In most cases, the data in the data warehouse can be used for reporting. They cannot be updated so that the performance of the company's underlying operational system is not affected. The focus on problem solving describes some of the benefits companies have obtained by using data warehouses.

Data warehouses often contain capabilities to remodel the data. A relational database allows views of data into two dimensions. A multidimensional view of data lets users look at data in more than two dimensions. For example, sales by region by quarter. To provide this type of information, organizations can either use a specialized multidimensional database or tool that takes multidimensional views of data in relational databases. Multidimensional analysis enables users to view the same data in different ways using multiple dimensions.

Among the issues to be addressed in building a warehouse are the following:

- When and how to gather data. In a source-driven architecture for fathering data, the data sources transmit new information, either continually, as transaction processing takes place, or periodically, such as each night. Data warehouses typically have slightly out-of-date data. That, however, is usually not a problem for decision-support systems.
- What schema to use. Data sources that have been constructed independently are likely to have different schemas. In fact, they may even use different data models. Part of the task of a warehouse is to perform schema integration, and to convert data to the integrated schema before they are stored. As a result, the data stored in the warehouse are not just a copy of the data at the sources. Instead, they can be thought of as a stored view (or materialized view) of the data at the sources.
- How to propagate updates. Updates on relations at the data sources must be propagated to the data warehouse.
- What data to summarize. The raw data generated by a transaction-processing system may be too large to store on-line. However, we can answer many queries by maintaining just summary data obtained by aggregation on a relation, rather than maintaining the entire relation.

2. Data Mining

Data mining is about analyzing data and finding hidden patterns using automatic or semiautomatic means. Data mining provides a lot of business value for enterprises.

- Increasing competition
- Customer segmentation
- Churn analysis
- Cross-selling
- Sales forecast
- Fraud detection
- Risk management

During the past decade, large volumes of data have been accumulated and stored in databases. Much of this data comes from business software, such as financial applications, enterprise resource planning (ERP), customer relationship management (CRM), and Web logs. The result of this data collection is that organizations have become data-rich and knowledge-poor. The collections of data have become so vast and are increasing so rapidly in size that the practical use of these stores of data has become limited. The main purpose of data mining is to extract patterns from the data at hand, increase its intrinsic value and transfer the data to knowledge.

Data mining applies algorithms, such as decision trees, clustering, association, time series, and so on, to a dataset and analyzes its contents. This analysis produces patterns, which can be explored for valuable information (Fig. 7-1). Depending on the underlying algorithm, these patterns can be in the form of trees, rules, clusters, or simply a set of mathematical formulas. The information found in the patterns can be used for reporting, as a guide to marketing strategies, and, most importantly, for prediction.

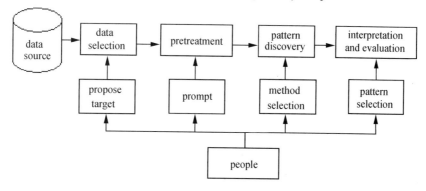

Fig. 7-1   data mining

Keywords

aggregation	n.	聚合，集合，凝聚，聚合体	intrinsic	adj.	内在的，真正的，实在的
			materialize	v.	具体化，实体化
association	n.	合伙，联合，联盟，协会	multidimensional	adj.	多维的
churn	v.	搅拌，翻腾	out-of-date	adj.	过期的
cluster	n.	丛，簇，聚集，组件	periodically	adv.	周期地，定期地
consolidate	v.	统一，整顿，加固，强化	propagate	v.	传播，宣传，使蔓延
data mining	n.	数据挖掘	reorganize	v.	改组，改造，改编
extract	v.	抽出，提取，引用	underlying	adj.	基础的，潜在的
fraud	n.	欺骗，欺诈	warehouse	n.	仓库

7.3   Reading Material 2：Library

In computer science, a library is a collection of subprogram used to develop software. Libraries contain "helper" code and data, which provide services to independent programs. This allows code and data to be shared and changed in a modular fashion. Some executables are both standalone programs and libraries, but most libraries are not executables. Libraries may be classified by how they are shared, how they are linked, and when they are linked.

Dynamic linking means that the data in a library is not copied into a new executable or library at compile time, but remains in a separate file on disk. Only a minimal amount of work is done at compile time by the linker. It only records what libraries the executable needs and the index names or numbers. The majority of the work of linking is done at the time the application is loaded or during the execution of the process. Libraries that are dynamically linked at run time are also called plugins. Software that uses libraries for core functionality can be described as having plugin architecture.

Dynamic linkers/loaders vary widely in functionality. Some depend on explicit paths to the libraries being stored in the executable. Any change to the library naming or layout of the files system will cause these systems to fail. More commonly, only the name of the library is stored in the executable, with the operating system supplying a system to find the library on-disk based on some algorithm.

One of the biggest disadvantages of dynamic linking is that the executables depend on the separately stored libraries in order to function properly. If the library is deleted, moved or renamed, or if an incompatible version of the DLL is copied to a place that is earlier in the search, the executable could malfunction or even fail to load.

Dynamic loading is a subset of dynamic linking where a dynamically linked library loads and unloads at run-time on request. The request to load such a dynamically linked library may be made implicitly at compile-time, or explicitly by the application at run-time. Implicit requests are made by adding library references, which may include file paths or simply file names, to an object file at compile-time by a linker. Explicit requests are made by applications using a run-time linker API. Most operating systems that support dynamically linked libraries also support dynamically loading such libraries via a run-time linker API.

Another solution to the library issue is to use completely separate executables and call them using a remote procedure call (RPC). This approach maximizes operating system re-use: the code needed to support the library is the same code being used to provide application support and security for every other program. Additionally, such systems do not require the library to exist on the same machine, but can forward the requests over the network.

Keywords

architecture	n.	构造，结构	remote	adj.	远程的，遥远的
explicitly	adv.	明显地，显而易见地	standalone	adj.	独立的，独立式，单机模式
malfunction	n.	故障，失灵，机能失常	subset	n.	子集
modular	adj.	模块的			

7.4 专业英语中常用的符号和数学表达式

在专业英语中，通常会用到一些符号和数学表达式，下面简要介绍加、减、乘、除、分数、小数及关系运算符的表达。

1．加法（addition）

$3 + 4 = 7$　　Three plus four equals seven. 或者 Three plus four is seven.

2．减法（subtraction）

$7 - 3 = 4$　　Seven minus three leaves four. 或者 Seven minus three equals four. 或者 Seven minus three is four.

3．乘法（multiplication）

$3 \times 4 = 12$　　Three multiplied by four is twelve.

$2 \times 3 \times 4 = 24$　　Two times three times four is twenty-four.

4．除法（division）

$15 \div 3 = 5$　　Fifteen divided by three is five.

例：14 除以 4 等于 3，余数为 2。

译文：When 14 is divided by 4, the quotient is 3 and there is a remainder of 2.

5．幂运算

5^2 five squared

5^3 five cubed

5^4 five to the power four

6．关系运算符

在等式表达中，常用到等号（＝），而在一些不等关系表达中，则要用到表示不等关系的各种符号（≈、≠、≤、≥、<、>）。下面分别举例说明。

a = b   a equals b 或 a is equal to b 或 a is b

a ≈ b   a is approximately equal to b

a ≠ b   a is not equal to b 或 a is not b

a ≤ b   a is less than or equal to b

a ≥ b   a is more than or equal to b

a < b   a is less than b

a > b   a is greater than b

7．分数

分数词由基数词和序数词搭配而成，分子用基数词，分母用序数词。当分子是 1 时，可以用 one 或 a，分母使用序数词；当分子大于 1 时，分子使用基数词，而分母要用序数词的复数形式。分数的另外一种表示方法是将/说成 over。对于假分数，也就是分数由一个整数和一个真分数组成，则说成"整数部分 and 真小数部分"。

1/2   a half

1/5   one fifth 或 a fifth

1/123   one over one hundred and twenty-three

3/4   three fourths 或 three quarters

45/187   forty-five over one hundred and eighty-seventh

6 又 2/3   six and two thirds

例 1：Nitrogen forms about four-fifths of the atmosphere.

译文：氮约占大气的五分之四。

例 2：At least two-thirds of the class have had colds.

译文：这个班至少有三分之二的学生患重感冒。

8．小数

对于小数的表示，直接将组成小数的各个数字读出。其中小数点可以说成 point。0 的表示可以有多种，可以是 zero、naught 或 o（字母 o 的音）。

0.5   zero point five

0.08   zero point zero eight

302.67   three o two point six seven

例 1：The new engine has a capacity of 4.3 litres and a power out-put of 153 kilowatts at 4400 revolutions per minute.

译文：这台发动机的容积为 4.3 升，转速为每分钟 4400 转时，输出功率是 153 千瓦。

例 2：We known that the weight of a cubic foot of air at 0℃ and 76cm, pressure is 0.081 pound, or 12 cubic feet of air weigh a pound.

译文：我们知道，1 立方英尺的空气在 0 摄氏度和 760 毫米汞柱压力下，重量是 0.081 磅，也就是说 12 立方英尺空气的重量是 1 磅。

9. 百分数

百分数用 percent 表示，既可与 by 连用，也可单独使用。在句子中主要作状语。

例 1：Sixty percent profit was reported.

译文：据报道有 60%的利润。

例 2：Its total output value increased by 10 percent over the previous year.

译文：它的总产值比去年增长了 10%。

例 3：The output of computers in China last year was 25 percent more than in 2000.

译文：去年中国计算机的产量比 2000 年上升了 25%。

10. 倍数

倍数的表达在计算机专业英语中时常出现，一般用 times 表示。而倍数也是"汉译英"及"英译汉"中的难点，稍有不慎，就会产生理解错误。故在翻译倍数时，应仔细推敲。下面通过几个例子来说明倍数的各种翻译方式。

在简单的倍数表示法中，主要使用…x times as…as…x times the size (length, width, height, depth, amount) of…和…x times+比较级+than…几种形式。

例 1：The speed of this new printer is **four times faster** than that old one.

译文：这台新打印机的速度比那台旧打印机**快四倍**。

对于增加倍数的表示，在 increase 等表示"增加"或"提高"词的后面也可用 times。若是 increase by x times，则翻译为"增加了 x 倍"或"增加到原来的 x+1 倍"。另外，在 increase by 后面除可用 x times 外，也可使用其他数词。increase 后面也可跟介词 to，表示"增加到……"。

例 2：Next year the output of computers will **increase by four times**.

译文：明年计算机的产量将**增加到五倍**。

对于减少倍数的表示，在 decrease 或 reduce 等表示"减少"或"降低"的词后面也常用 times。若是 decrease by x times，则常译为"减少到 1/x"或"减少了 1 减 1/x"。decrease by 也可使用其他数词，decrease to 表示"减少到……"。注意，by 后面表示的是净增减数，by 有时可以省略，而 to 后面为增加或减少后达到的数字。

例 3：The new equipment they are designing will **reduce** error probability **by three times** and its speed will **increase by two times.**

译文：他们正在设计的那种新设备将使误差率**降低到三分之一**，而速度将**增加两倍**。

11. 章节、页数的表示法和读法

第一章　　the First Chapter 或 Chapter One

第二节　　the Second Section 或 Section Two

第三课　　the Third Lesson 或 Lesson Three

第 463 页　　page four six three

第 2564 页　　page two five six four 或 page twenty-five sixty-four

7.5 Exercises

1. Translate the following phrases.

database model
physical level
object-based logical model
user program
database administrator
data model
primary key
record-based logical model
foreign key
entity-relationship model
hierarchical model
network model

2. Fill in the blanks with appropriate words or phrases.

a. data storage   b. data types   c. conceptual   d. entity
e. data   f. files   g. database   h. entire

(1) A database consists of a file or a set of _____.
(2) Fields are the basic units of _____.
(3) The external level defines the different views of the _____ as required by the external or user programs.
(4) One user program may not require all the _____ in the database.
(5) The conceptual level describes the _____ database.
(6) Object-based logical models are used for describing data at the _____ and view levels.
(7) For relationships involving two _____ sets, there can be relationships like one-to-one, one-to-many, many-to one, and many-to-many.
(8) There are various _____ to choose from, such as character, numeric, date, or other.

3. Match the following terms to the appropriate definition.

a. download   b. domain name   c. document
d. extranet   e. E-mail   f. E-commerce

(1) To transfer a file from another computer to your computer. It is usually realized by the Internet.
(2) It can be viewed as part of a company's intranet that is extended to users outside the company. The main purpose is to share information with individuals or groups outside a company, such as suppliers, customers and partners.
(3) The unique name that identifies an Internet site.

Database Principle

(4) A article, book, or other work, typically containing text or other media, which includes information content.

(5) The purchasing of goods and services over the Internet. It is the application of technology toward the automation of business transactions and work flow.

(6) A communication system that allows you to send text, files and graphical messages over the Internet.

Unit 8　Software Engineering

8.1　Text

Software engineering first emerged as a popular term in the title of a 1968 NATO conference held in Garmisch, Germany. The digital computer was less than a quarter of a century old, and already we were facing a "software crisis". First we had invented computer programming, and then we taught people to write programs. The next task was the development of large systems that were reliable, delivered on schedule, and within budget. As with every technological advancement, our aspirations were at the boundary of what we could do successfully. As it turned out, we were not very good at building large systems on time.

Early approaches to software engineering insisted on performing analysis, design, implementation, and testing in a strictly sequential manner. As a result, software engineers insisted that the entire analysis of the system be completed before beginning the design and, likewise, that the design be completed before beginning implementation. The result was a development process now referred to as the waterfall model, an analogy to the fact that the development process was allowed to flow in only one direction.

The first phase, requirements definition, refers to the period during which the requirements of the system desired, that is, it's functional characteristics and operational details, are specified.[1] If the system is to be a generic product sold in a competitive market, this analysis would involve a broad-based investigation to identify the needs of potential customers. If, however, the system is to be designed for a specific user, then the process would be a more narrow investigation.[2] As the needs of the potential user are identified, they are compiled to form a set of requirements that the new system must satisfy. These requirements are stated in terms of the application rather than in the technical terminology.

One requirement might be that access to data must be restricted to authorized personnel. Another might be that the data must reflect the current state of the inventory as of the end of the last business day or that the arrangement of the data as displayed on the computer screen must adhere to the format of the paper forms currently in use.[3] After the system requirements are identified, they are converted into more system technical specifications. An error in requirements, for example, a misstated function, leads to a faulty design and an implementation that does not do what is required. If this is allowed to proceed undetected, say, until the testing phase, the cost of repairing this error (including redesign and re-implementation) can be substantial.

The second phase is design. Design concentrates on how the system will accomplish the goals. It is here that the structure of the software system is established. The input to this phase is a (debugged and validated) requirements document: the output is a design expressed in some appropriate form (for example, pseudo-code). Validation of a design is important. Each requirement in the requirements document must have a corresponding design fragment to meet it. Formal verification, while possible to a limited extent, can be exceedingly difficult. More informal revolve the entire design team, management, and even the client.

It is a well-established principle that the best structure for a large software system is a modular one. Indeed, it is by means of this modular decomposition that the implementation of large systems becomes a possibility. Without such a breakdown, the technical details required in the implementation of a large system would exceed a human's comprehensive powers. With a modular design, however, only the details pertaining to the module under consideration need be mastered. This same modular design is also conductive to future maintenance because it allows changes to be made on a modular basis.

The third phase, implementation, is the actual coding of the design developed in the second phase. The design must be translated into a machine-readable form. The coding step performs this task. If design is performed in a detailed manner, coding can be accomplished mechanistically. The lure of this phase is strong, and many a foolhardy programmer has been drawn to it before adequately laying the groundwork in the first two phases. As a result, requirements are incompletely understood and the design is flawed. The implementation proceeds blindly, and many problems arise as a result.

The forth phase is software testing. Testing is closely associated with implementation, because each module of the system is normally tested as it is implemented. In the development of a large system, testing involves several stages. First, each program module is tested as a single program, usually isolated from the other programs in the system. Such testing is defined as module testing or unit testing. Unit testing is done in a controlled environment whenever possible so that the test team can feed a predetermined set of data to the module being tested and observe what output data are produced.[4] In addition, the test team checks the internal data structures, the logic, and the boundary conditions for the input and output data.

When collections of modules have been unit-tested, the next step is to insure that the interfaces among the modules are defined and handled properly. Integration testing is the process of verifying that the components of a system work together as described in the program design and system design specifications. Once we are sure that information is passed among modules according to the design prescriptions, we test the system to assure that it has the desired functionality. A function test evaluates the system to determine if the functions described by the requirements specification are actually performed by the integrated system.

Unfortunately, the testing and debugging of a system is extremely difficult to perform successfully. Experience has shown that large software systems can contain numerous errors. Even after significant testing. Many of these errors may go undetected for the life of the system, but others may cause major malfunctions. The elimination of such errors is one of the goals of software engineering.

The term "software maintenance" is used to describe the software engineering activities that occur following delivery of a software product to the customer. Maintenance activities involve making enhancements to software product, adapting products to new environments, and correcting problems. Software product enhancement may involve providing new functional capabilities, improving user displays and modes of interaction, upgrading external documents and internal documentation, or upgrading the performance characteristics of a system. Adaptation of software to a new environment may involve moving the software to a different machine, or for instance, modifying the software to accommodate a new telecommunication protocol or additional disk drives. Problem correction involves modification and revalidation of software to correct errors. Some errors require immediate attention, some can be corrected on a scheduled, periodic basis, and others are known but never corrected.

Software maintenance is a microcosm of the software development cycle. Enhancement and adaptation of software reinitiate development in the analysis phase, while correction of a software problem may reinitiate the development cycle in the analysis phase, the design phase, or the implementation phase. Thus, all of the tools and techniques used to develop software are potentially useful for software maintenance.

The activity of software quality assurance is closely related to verification and validation activities carried out at each stage of the software life cycle. Indeed, in many organizations there is no distinction made between these activities. However, quality assurance and other verification and validation activities are actually quite separate, with quality assurance being a management function and verification and validation being part of the process of software development.

The development of software engineering project standards is an extremely difficult process. A standard is some abstract representation of a product which defines the minimal level of performance, robustness, organization, etc., which the developed product must attain. The problem with national software standards is that they tend to be very general in nature. This is inevitable as, unlike hardware, we are not yet capable of quantifying most software characteristics. Effective quality assurance within an organization thus requires the development of more specific organizational standards. Of course, the problem which arises in developing software standards for quality assurance and which makes the assessment of the level of excellence of a software product difficult to assess is the elusive nature of software quality.[5]

Software quality assurance is now an emerging sub-discipline of software engineering. Quality assurance is not the same as system testing. It is the development or testing team's responsibility to validate the system, with the quality assurance team reporting on both the validation and the adequacy of the validation effort. This naturally involves quality assurance being closely associated with the final integration testing of the system.

Keywords

aspiration	n.	渴望，愿望	decomposition	n.	分解，解体
assess	v.	估定，评定	elusive	adj.	无从捉摸的
assurance	n.	保证，担保	faulty	adj.	不完善的，有错误的
blindly	adv.	盲目地	foolhardy	adj.	蛮干的
boundary	n.	边界，界限	inevitable	adj.	必然的，照例的
breakdown	n.	崩溃，损坏	investigation	n.	调查，研究
budget	n.	预算，经费	isolate	v.	隔离，绝缘
conference	n.	会议	mechanistically	adv.	机械地

microcosm	n.	缩影，微观世界	quarter	n.	四分之一
misstate	v.	虚称，谎报	validated	v.	证实，确认
pseudo	adj.	假，伪，冒充的			

Notes

[1] The first phase, requirements definition, refers to the period during which the requirements of the system desired, that is, it's functional characteristics and operational details, are specified.

译文：第一阶段，即需求定义阶段，是指系统期望的需求分析，即描述功能特征和操作细节的阶段。

说明：requirements definition 作主语 The first phase 的同位语，"that is, it's functional…" 是定语。

[2] If, however, the system is to be designed for a specific user, then the process would be a more narrow investigation.

译文：但是，如果系统是为特殊用户设计的，那么这个过程就是一个更专业的调查。

说明：本句由 If 引导条件状语从句。

[3] Another might be that the data must reflect the current state of the inventory as of the end of the last business day or that the arrangement of the data as displayed on the computer screen must adhere to the format of the paper forms currently in use.

译文：另一种可能是当最后一个工作日结束时，数据必须反映目前的清单状态，或者可能是在计算机屏幕上的数据组织必须按照目前使用的纸质形式的格式来显示。

说明：本句是有两个并列的表语从句 that the data must…和 that the arrangement…，而 as of the end of the last business day 是时间状语。

[4] Unit testing is done in a controlled environment whenever possible so that the test team can feed a predetermined set of data to the module being tested and observe what output data are produced.

译文：在任何可能的时候，单元测试是在控制环境下进行的，这是为了让测试小组能提供被测试模块先前确定的数据并观察产生的输出数据。

说明：本句由 so that 引导目的状语从句。

[5] Of course, the problem which arises in developing software standards for quality assurance and which makes the assessment of the level of excellence of a software product difficult to assess is the elusive nature of software quality.

译文：当然，为保证质量而开发软件标准所带来的问题及软件产品的优良程度难以评估是软件质量难以把握的本质。

说明：which arises in developing…和 which makes the assessment…都是定语，修饰主语 the problem。

8.2　Reading Material 1：Enterprise Resource Planning

　　Enterprise resource planning (ERP) software attempts to integrate all departments and functions across a company into a single computer system that can serve all those different departments' particular needs.

　　ERP systems can become the place where the customer order lives from the time a customer service representative receives it until the loading dock ships the merchandise and finance sends an invoice. By having this information in one software system, rather than scattered among many different systems that can't communicate with one another, companies can keep track of orders more easily, and coordinate manufacturing, inventory and shipping among many different locations at the same time.

ERP helps the manufacturing process flow more smoothly, and it improves visibility of the order fulfillment process inside the company. That can lead to reduced inventories of the stuff used to make products, and it can help users planning deliveries to customers, reducing the finished good inventory at the warehouses and shipping docks.

ERP takes a customer order and provides a software road map for automating the different steps along the path to fulfilling it. When a customer service representative enters a customer order into an ERP system, he has all the information necessary to complete the order. For example, the customer's credit rating and order history from the finance module, the company's inventory levels from the warehouse module and the shipping dock's trucking schedule from the logistics module, for example.

People in these different departments all see the same information and can update it. When one department finishes with the order it is automatically routed via the ERP system to the next department. To find out where the order is at any point, you need only log in to the ERP system and track it down. With luck, the order process moves like a bolt of lightning through the organizations, and customers get their orders faster and with fewer errors than before. ERP can apply that same magic to the other major business processes, such as employee benefits or financial reporting.

ERP was an attractive solution for many large companies because it offered so many potential uses. For example, the same system could be used to demand for a product, order the necessary raw materials, establish production schedules, track inventory, allocate costs, and project key financial measures. ERP acts as a planning backbone for a company's core business processes. In addition to directing many of them, the system also ties together these varied processes using data from across the company. For instance, a typical ERP system manages functions and activities as different as the bills of materials, order entry, purchasing, human resources, and inventory control, to name just a few of the 60 modules available. As needed, ERP is also able to share the data from these processes with other corporate software systems. Another important benefit of ERP systems was that they allowed companies to replace of complex computer applications with a single, integrated system.

Despite these potential benefits, however, traditional ERP systems also had a number of drawbacks. For instance, the early systems tended to be large, complicated, and expensive. Implementation required an enormous time commitment from a company's information technology department or outside professionals. In addition, because ERP systems affected most major departments in a company, they tended to create changes in many business processes. Putting ERP in place thus required new procedures, employee training, and both management and technical support. As a result, many companies found to ERP a slow and painful process. Once the implementation phase was complete, some businesses had trouble quantifying the benefits they gained from ERP.

Finally, as technology began shifting toward speedy Internet connections, Web-based business-to-business relationships, and electronic commerce, some companies worried that their mainframe-based ERP systems were too slow. As it turned out, though, many companies found that their ERP systems provided a solid technological foundation for future growth by standardizing business procedures, facilitating information sharing across the company, and creating an organization to change. ERP systems may feel like to companies that have expensively and painfully installed them. Nonetheless, they constitute a valuable foundation for a wide range of new value-enhancing applications.

Keywords

commitment	*n.*	许诺，义务	logistics	*n.*	后勤
customer	*n.*	客户，顾客	merchandise	*n.*	商品，货物
finance	*n.*	金融，资金	painfully	*adv.*	痛苦地，费力地
fulfillment	*n.*	完成，履行	smoothly	*adv.*	平滑地，顺利地
inventory	*n.*	清单，报表	visibility	*n.*	显著，明白
invoice	*n.*	发票，发货清单			

8.3 Reading Material 2：Software Testing

Software testing can be defined as: testing is an activity that helps in finding out bugs, defects or errors in a software system under development, in order to provide a bug free and reliable system to the customer.

White box testing strategy deals with the internal logic and structure of the code. White box testing is also called as glass, structural or open box testing. The tests written based on the white box testing strategy incorporate coverage of the code written, branches, paths, statements and internal logic of the code etc.

In order to implement white box testing, the tester has to deal with the code and hence is needed to possess knowledge of coding and logic i.e. internal working of the code. White box test also needs the tester to look into the code and find out which unit/statement/chunk of the code is malfunctioning.

Types of testing under white box testing strategy:

- Unit testing
- Static and dynamic analysis
- Statement coverage
- Branch coverage
- Security testing
- Mutation testing

Black box testing is not a type of testing; it instead is a testing strategy, which does not need any knowledge of internal design or code etc. As the name "black box" suggests, no knowledge of internal logic or code structure is required. The types of testing under this strategy are totally focused on the testing for requirements and functionality of the work product application.

The base of the black box testing strategy lies in the selection of appropriate data as per functionality and testing it against the functional specifications in order to check for normal and abnormal behavior of the system. Nowadays, it is becoming common to route the testing work to a third party as the developer of the system knows too much of the internal logic and coding of the system, which makes it unfit to test the application by the developer.

In order to implement black box testing strategy, the tester is needed to be thorough with the requirement specifications of the system and as a user, should known, how the system should behave in response to the particular action.

Various testing types that fall under the black box testing strategy are: functional testing, stress testing, recovery testing, volume testing, system testing, load testing etc.

Keywords

abnormal	*adj.*	反常的，不规则的	strategy	*n.*	战略，策略
chunk	*n.*	大块，相当大的部分	stress	*n.*	重点，重要
coverage	*n.*	范围，规模	thorough	*adj.*	彻底的，全面的
mutation	*n.*	变化，变异			

8.4 专业英语的阅读

所谓阅读，实际上就是语言知识、语言技能和智力的综合运用。在阅读过程中，这3个方面的作用总是浑然一体、相辅相成的。词汇和语法结构是阅读所必备的语言知识，但仅此是难以进行有效阅读的，学生还需具备运用这些语言知识的能力。即根据上下文来确定准确词义和猜测生词词义的能力；辨认主题和细节的能力；正确理解连贯的句与句之间、段与段之间的逻辑关系的能力。这里所指的智力是学生的认知能力，包括记忆、判断和推理的能力。因为在阅读专业英语文章中常常出现一些要求领悟文章的言外之意和作者的态度、倾向等情况。

阅读理解能力的提高是由多方面因素决定的，学生应从以下三方面进行训练。

1. 打好语言基本功

扎实的语言基础是提高阅读能力的先决条件。首先，词汇是语言的建筑材料。提高专业英语资料的阅读能力必须扩大词汇量，尤其是掌握一定量的计算机专业词汇。如果词汇量掌握得不够，阅读时就会感到生词多，不但影响阅读的速度，而且影响理解的程度，因此不能进行有效的阅读。其次，语法是语言中的结构关系，用一定的规则把词或短语组织到句子中，表示一定的思想。熟练掌握英语语法和惯用法也是阅读理解的基础。在阅读理解中必须运用语法知识来辨认出正确的语法关系。如果语法基础知识掌握得不牢固，在阅读中遇到结构复杂的难句、长句，就会不知所措。

2. 在阅读实践中提高阅读能力

阅读能力的提高离不开阅读实践。在打好语言基本功的基础上，还要进行大量的阅读实践。词汇量和阅读能力的提高是一种辩证关系：要想读得懂，读得快，就必须扩大词汇量；反之，要想扩大词汇量，就必须大量阅读。同样，语法和阅读之间的关系也是如此：有了牢固的语法知识就能够促进阅读的顺利进行，提高阅读的速度和准确率；反之，通过大量的阅读实践又能够巩固已掌握的语法知识。只有在大量的阅读中，才能培养语感，掌握正确的阅读方法，提高阅读理解能力。同时在大量的阅读中，还能巩固计算机专业知识，以及了解到计算机专业的发展趋势，对于跟踪计算机技术的发展很有好处。

3．掌握正确的阅读方法

阅读时，注意每次视线的停顿应以一个意群为单位，而不应以一个单词为单位。要是一个单词一个单词地读，当读完一个句子或一个段落时，前面读的是什么早就忘记了。这样读不仅速度慢，还影响理解。因此，正确的阅读方法可以提高阅读速度，同时也能增强阅读理解能力。

常用的有效的阅读方法有 3 种，即略读（Skimming）、查读（Scanning）和精读（Reading for full understanding）。

略读是指以尽可能快的速度进行阅读。了解文章的主旨和大意，对文章的结构和内容获得总的概念和印象。一般来说，400 字左右的短文要求在 6~8 分钟内完成。进行略读时精力必须特别集中，还要注意文中各细节分布的情况。略读过程中，学生不必去读细节，遇到个别生词及难懂的语法结构也应略而不读。不要逐词逐句读，力求一目数行而能知大体含义。略读时主要注意以下几点。

- ◆ 注意短文的开头句和结尾句，力求抓住文章的主旨大意。
- ◆ 注意文章的体裁和写作特点，了解文章结构。
- ◆ 注意了解文章的主题句及结论句。
- ◆ 注意支持主题句或中心思想的信息句，其他细节可以不读。

查读的目的主要是要有目的地去找出文章中某些特定的信息，也就是说，在对文章有所了解的基础上，在文章中查找与某一问题、某一观点或某一单词有关的信息。查读时，要以很快的速度扫视文章，确定所查询的信息范围，注意所查信息的特点，如有关日期、专业词汇、某个事件、某个数字、某种观点等，寻找与此相关的关键词或关键段落。注意与所查信息无关的内容可以略过。

精读是指仔细地阅读，力求对文章有深层次的理解，以获得具体的信息。包括理解衬托主题句的细节，根据作者的意图和中心思想进行推论，根据上下文猜测词义等。对难句、长句，要借助语法知识，对其进行分析，达到准确的理解。

总之，要想提高阅读理解能力，必须掌握以下 6 项基本的阅读技能。

- ◆ 掌握所读材料的主旨大意。
- ◆ 了解阐述主旨的事实和细节。
- ◆ 根据上下文判断某些词汇和短语的意义。
- ◆ 既理解个别句子的意义，也理解上下文之间的逻辑关系。
- ◆ 根据所读材料进行一定的判断、推理和引申。
- ◆ 领会作者的观点、意图和态度。

没有习惯于阅读科技文章的学生们常常会觉得阅读之后还总是停留在文字表面意义的理解上，而弄不清作者究竟要谈什么，不能形成一个总的、清晰的概念。科技文章的内容是浓缩的，在任何一页上都会发现大量的信息。可以把一篇文章分成几个小部分，甚至一次就读一个自然段。在阅读过程中不要忘了经常复习所读过的段落。

在阅读计算机科技文章时，还要注意在阅读中及时获取细节信息。在所有的文章中，作者都使用细节或事实来表达和支持他们的观点。阅读要想有效果，就要能够辨认并记住文章中重要的细节。一个细节就是一个段落中的一条信息或一个事实。它们或者给段落的主题提供证据，或者为其提供例子。有些细节或事实是完整的句子，而有一些只是简单的短语。在很多情况下，还必须能区分哪些是重要细节，哪些是次要细节。记住所有的细节是不可能的，但是在阅读过程中要尽量发现重要的细节并记住它们。

8.5 Exercises

1. **Translate the following phrases.**

   software engineering
   potential user
   technical terminology
   software crisis
   waterfall model
   development process
   testing phase
   pseudo-code
   boundary condition
   internal data structure
   integration testing
   software maintenance

2. **Fill in the blanks with appropriate words or phrases.**

   a. implementation   b. difficult   c. design   d. each module
   e. goals   f. sequential manner   g. important   h. consideration

   (1) Formal verification, while possible to a limited extent, can be exceedingly _____.
   (2) Early approaches to software engineering insisted on performing analysis, design, implementation, and testing in a strictly _____.
   (3) Testing is closely associated with implementation, because _____ of the system is normally tested as it is implemented.
   (4) The second phase is design. Design concentrates on how the system will accomplish the _____.
   (5) Validation of a design is _____.
   (6) With a modular design, however, only the details pertaining to the module under _____ need be mastered.
   (7) Each requirement in the requirements document must have a corresponding _____ fragment to meet it.
   (8) The third phase, _____, is the actual coding of the design developed in the second phase.

3. **Match the following terms to the appropriate definition.**

   a. face recognition system   b. fiber-optic cable   c. file sharing network
   d. firmware   e. genetic algorithm   f. gateway

   (1) Dozens or hundreds of thin strands of glass or plastic that use light to transmit signals.
   (2) It is based on a biological metaphor. It views learning as a competition among a population of evolving candidate program solutions. A "fitness" function evaluates each solution to decide whether it will contribute to the next generation of solutions.

(3) Type of peer-to-peer network on which users access each other's hard disks and exchange files directly over the Internet. Refer to "P2P".

(4) Software (programs or data) that has been written onto read-only memory. Firmware is a combination of software and hardware.

(5) A computing machine which is both connected to one or more networks and is capable of passing network information from one network to another.

(6) Biometric device that captures a live face image and compares it with a stored image to determine if the person is a legitimate user.

Unit 9　Multimedia

9.1　Text

音频

Multimedia is any combination of text, graphic art, sound, animation, and video delivered to you by computer or other electronic means. It is richly presented sensation. When you weave together the sensual elements of multimedia-dazzling pictures and animations, engaging sounds, compelling video clips, and raw textual information you can electrify the thought and action centers of people's minds. Multimedia excites eyes, ears, fingertips, and, most importantly, the head.

Elements of Multimedia

Multimedia is a set of more than one media element used to produce a concrete and more structured way of communication. Multimedia is simultaneous use of data from different sources. These sources in multimedia are known as media elements. With growing and very fast changing information technology, multimedia has become a crucial part of computer world.

- Text

The Windows operating environment gives the user an almost infinite range of expressing text. By displaying text in more than one format, the message a multimedia application is trying to portray can be made more understandable.

- Audio Sound

The integration of audio sound into a multimedia application can provide the user with information not possible through any other method of communication. Some types of information can't be conveyed effectively without using sound. It is nearly impossible, for example, to provide an accurate textual description of the bear of a heart or the sound of the ocean.

- Static Graphics Images

When you imagine graphics images you probably think of "still" images-that is, images such as those in a photograph or drawing. There is no movement in these types of picture. Static graphics images are an important part of multimedia because humans are visually oriented. Windows is also a visual environment. This makes displaying graphics images easier than it would be in a DOS-based environment.

- Animation

Animation refers to moving graphics images. Just as a static graphics image is a powerful form of communication, such is the case with animation. Animation is especially useful for illustrating concepts that involve movement. Such concepts as playing a guitar or hitting a golf ball are difficult to illustrate using a single photograph, or even s series of photographs, and even more difficult to explain using text. Animation makes it easier to portray these aspects of your multimedia application.

Full-motion video, such as the images portrayed in a television, can add even more to a multimedia application. Although full-motion video may sound like an ideal way to add a powerful message to a multimedia application, it is nowhere near the quality you would expert after watching television.[1]

Application of Multimedia

Business applications for multimedia include presentations, training, marketing, advertising, product demos, databases, catalogues, and networked communications.[2] Voice mail and video conferencing will soon be provided on many local and wide area networks.

◆ Computer-based Training

Multimedia is enjoying widespread use in training programs. Many corporations are turning to multimedia applications to train their employees. A major telephone company has put together a multimedia application that simulates major emergencies and trains the employees on what to do in these situations. By using a multimedia application, the company found it has saved expenses and trained employees more effectively than anything else it had tried.[3]

◆ Education

Educational uses of multimedia are rapidly increasing. Its power to present information and its integration of resources, allow for the creation of rich learning environments. In the past, a teacher or student might have to consult many sources and use several media to access the needed information. By integrating media and utilizing hypermedia we are able to create user controlled, information on-demand learning environments.

Multimedia will provoke radical changes in the teaching process during the coming decades, particularly as smart students discover they can go beyond the limits of traditional teaching methods. Indeed, in some instances, teachers may become more like guides and mentors along a learning path, not the primary providers of information and understanding—the students, not teachers, become the core of the teaching and learning process. This is a sensitive and highly politicized subject among educators, so educational software is often positioned as "enriching" the learning process, not as a potential substitute for traditional teacher-based methods.

◆ Entertainment

There is absolutely nothing wrong with a litter fun. In many cases, the graphics technology used in today's best games will show up in tomorrow's business applications. If that is a way to expand technology to the next level, it definitely deserves attention from multimedia developers. Besides, writing entertainment applications can be a lot of fun.

This is a new type of software category that mixes education with entertainment. The idea is to make learning fun while providing some type of entertainment.

◆ Others

From gardening to cooking to home design, remodeling, and repair to genealogy software, multimedia has entered the home. Eventually, most multimedia projects will reach the home via television sets or monitors with built-in interactive user inputs. The multimedia viewed on these sets will likely arrive on a pay-for-use basis along the data highway. In hotels, train stations, shopping malls, museums, and grocery stores, multimedia will become available at stand-alone terminals or kiosks to provide information and help. Such installations reduce demand on traditional information booths and personnel, and they can work around the clock, even in the middle of the night, when live help is off duty.

As companies and businesses catch on to the power of multimedia, and the cost of installing multimedia capability decreases, more applications will be developed both by themselves and by third parties to allow businesses to run more smoothly and efficiently.[4]

Multimedia Networking

Many applications, such as video mail, video conferencing, and collaborative work systems, require networked multimedia. In these applications, the multimedia objects are stored at a server and played back at the client's sites. Such applications might require broadcasting multimedia data to various remote locations or accessing large depositories of multimedia sources. Multimedia networks require a very high transfer rate or bandwidth, even when the data is compressed. Traditional networks are used to provide error-free transmission. However, most multimedia applications can tolerate errors in transmission due to corruption or packet loss without retransmission or correction. In some cases, to meet real-time delivery requirements or to achieve synchronization, some packets are even discarded. As a result, we can apply lightweight transmission protocols to multimedia networks. These protocols cannot accept retransmission, since that might introduce unacceptable delays.

Multimedia networks must provide the low latency required for interactive operation. Since multimedia data must be synchronized when it arrives at the destination site, networks should provide synchronized transmission with low jitter.[5]

In multimedia networks, most communications are multipoint as opposed to traditional point-to-point communication. For example, conferences involving more than two participants need to distribute information in different media to each participant. Conference networks use multicasting and bridging distribution methods. Multicasting replicates a single input signal and delivers it to multiple destinations. Bridging combines multiple input signals into one or more output signals, which then deliver to the participants.

In the past, multimedia was difficult to apply on a wide scale because the available technology was not sufficient to support it. With today's fast computers and Internet connections, multimedia implementation is much more feasible and its use is increasing exponentially.

Keywords

animation	n.	动画	destination	n.	目的地，目标
consult	v.	商量，协商	emergency	n.	紧急情况，非常时期
convey	v.	传递，输送	entertainment	n.	娱乐
corruption	n.	误传，讹误	exponentially	adv.	指数地，幂地
crucial	adj.	极紧要的，决定性的	genealogy	n.	家谱，系统
depository	n.	存放处，寄存处	grocery	n.	杂货店

guitar	n.	吉他	radical	adj.	根本的，重要的
kiosk	n.	书报摊，广告亭	sensation	n.	感觉，知觉
lightweight	adj.	不重要的	sensual	adj.	感觉上的
participant	n.	参与者，参加者	substitute	n.	替代品，代用品
portray	v.	描绘，描述	textual	adj.	文本的
provoke	v.	引起，激发	weave	v.	构成，设计

Notes

[1] Although full-motion video may sound like an ideal way to add a powerful message to a multimedia application, it is nowhere near the quality you would expert after watching television.

译文：虽然全运动影像听起来像是一个往多媒体程序中加入强有力信息的理想方法，但它无法达到人们看电视一样的效果。

说明：本句由 Although 引导让步状语从句。

[2] Business applications for multimedia include presentations, training, marketing, advertising, product demos, databases, catalogues, and networked communications.

译文：多媒体的商业应用包括展示、培训、市场、广告、产品演示、数据库、产品目录及网络通信。

说明：本句的宾语包括多项，即"presentations, training, marketing, advertising, product demos, databases, catalogues, and networked communications."都作宾语。

[3] By using a multimedia application, the company found it has saved expenses and trained employees more effectively than anything else it had tried.

译文：通过使用多媒体，该公司发现既节省了开支又比其他培训员工的方法有效。

说明：本句的 By using a multimedia application 是状语，it has saved…是宾语从句。

[4] As companies and businesses catch on to the power of multimedia, and the cost of installing multimedia capability decreases, more applications will be developed both by themselves and by third parties to allow businesses to run more smoothly and efficiently.

译文：当公司和商业机构了解了多媒体的功能，并且多媒体设施的成本降低时，更多的能使企业更平稳、更有效地运作的多媒体应用程序将由企业自己或第三方开发出来。

说明：本句较长，As companies and businesses catch on to the power of multimedia, and the cost of installing multimedia capability decreases 作条件状语，to allow businesses to run more smoothly and efficiently 作目的状语。

[5] Since multimedia data must be synchronized when it arrives at the destination site, networks should provide synchronized transmission with low jitter.

译文：因为当多媒体数据到达目的站点时必须同步，所以网络必须用低抖晃提供同步。

说明：本句由 Since 引导原因状语从句。

9.2　Reading Material 1：Computer Aided Design

In the broadest sense, computer aided design (CAD) refers to any application of a computer to the solution of design problems, the engineer may communicate with the computer in many forms, either via the visual display screen, keyboard, graph plotter or many more man-machine interfaces. They can ask a question and receive an answer from the computer in a matter of seconds. More specifically, CAD is a technique in which the engineer and a computer work together as a team, utilizing the best characteristics of each.

CAD allowed the designer to interact graphically with the computer, and enables the engineer to test a design idea and to rapidly see its effect, the design idea can then be modified and reassessed. The process being repeated until a good design is achieved. Following each iteration, the design solution hopefully improves. Therefore, the more cycles that can be carried out within the financial, material and time constraints, the better the result should be.

CAD is so effective because the computer communicate with the designer not with numbers but pictures, which are much easier for the human mind to absorb. The methodology also permits the simulated testing and evaluation of the components being designed. The test results are superimposed, pictorially or numerically, on the television screen. If, for instance, the model is being heated, portions of it will be tinted in different colors each representing a specific temperature range. This powerful pictorial representation helps the designer visualize exactly what is taking place.

CAD should involve the development of a central design description on which all applications in design and manufacture should feed. This implies that computer-based techniques for the analysis and simulation of the design, and for the generation of manufacturing instructions, should be closely integrated with the techniques for modeling the form and structure of the design. In addition, a central design description forms an excellent basis for the concurrent development of all aspects of a design in simultaneous engineering activities. In principle, CAD could be applied throughout the design process, but in practice its impact on the early stages, where very imprecise representation such as sketches are used extensively, has been limited. It must also be stressed that at present CAD does not help the designer in the more creative parts of design, such as the generation of possible design solutions, or in those aspects that involve complex reasoning about the design.

The driving force behind the provision of computer assistance for conventional modeling techniques has been the desire to improve the productivity of the designer by the automation of the more repetitive and tedious aspects of design, and also to improve the precision of the design models. New techniques have been developed in an attempt to overcome perceived limitations in conventional practice—particularly in dealing with complexity—for example in the complexity of form of some design such as automobile bodies, or the intricacy of structure of products such as integrated circuits. CAD should therefore enable the designer to tackle a task more quickly and accurately, or in a way that could not be achieved by other means.

The general CAD system was developed by considering a wide range of possible uses of such a system. The following were considered in detail:
- Mechanical engineering design
- Building design
- Structural engineering design
- Electronic circuit design
- Animation and graphic design

It was thought that for most practical applications the general drawing system would be incorporated in a much larger specific applications system. For this reason the drawing system was as simple as possible consistent with reasonable running efficiency, so that it could be incorporated into an application system with the minimum of effort. For both the production of drawing items by analysis and the analysis of drawings, it is essential that there is a simple efficient link between data produced by the drawing system and analysis programs. It is also essential that graphic data can be annotated in a way which is recognized by analysis programs but which does not affect the drawing system.

Keywords

imprecise	*adj.*	不精确的，不精密的	provision	*n.*	设备，准备，预备
intricacy	*n.*	纷乱，复杂	simulate	*v.*	模拟
iteration	*n.*	迭代法，反复	sketch	*n.*	草图，略图
methodology	*n.*	方法学，方法论	superimpose	*v.*	附加，添加
numerically	*adv.*	数值地，用数字	tedious	*adj.*	单调沉闷的
pictorially	*adv.*	绘图地，用图片表示地	tint	*v. n.*	着色，色彩

9.3 Reading Material 2: Creating an Effective Web Presence

A business creates a presence in the physical world by building stores, factories, warehouses, and office buildings. An organization's presence or reputation is the public image it conveys.

As companies grow, they tend to worry much about the image they project. Managers of growing companies are often completely occupied with the day-to-day tasks of managing growth. Although the growing company's salespersons work to maintain a good reputation with customers and its purchasing agents do the same with suppliers, very few growing firms make a concerted, centrally organized effort to build the reputation of the company as a whole.

When companies reach a significant size, they again turn their attentions and resources to building and maintaining a presence as a means of maintaining their market share and customer base. Most large companies have a well-staffed public relations department to communicate with customers and the media. However, very few rapidly growing firms will put the same emphasis on public relations because they are focused primarily on growth.

On the Web, presence can be much more important for businesses of all sizes, instead of just for large businesses. Often, the only contact that customers and suppliers have with a firm on the Web is through its presence on the Internet. Creating an effective Web presence is equally critical for the smallest companies, the most rapidly growing companies, and the largest companies when those companies are operating on the Web. Thus, every business operating on the Web must identify its Web presence goals.

When a business creates a physical space in which to conduct its activities, its managers focus on very specific objectives. On the Web, however, businesses and other organizations can intentionally build a space that creates a distinctive presence. A Web site can perform many image-creation and image-enhancement tasks very effectively; it can serve as a sales brochure, a product showroom, a financial report, an employment ad, or a customer contact point. Each entity that establishes a Web presence must decide which tasks the Web site must accomplish and which tasks are the most important to include. Different companies—even those in the same industry—might establish very different Web presence goals.

An effective Web site creates an attractive presence that meets the objectives of the business or other organization. These objectives include:

- Attractive visitors to the Web site.
- Encouraging visitors to follow the site's links to obtain information.
- Malting the site interesting so that visitors stay and explore it.
- Creating an impression consistent with the organization's desired image.
- Reinforcing positive images about the organization that the visitor might already have.

Keywords

attractive	*adj.* 有吸引力的，引人注目的		presence	*n.*	存在，出席
brochure	*n.*	小册子	reputation	*n.*	名望，信誉
distinctive	*adj.*	有特色的，独特的	salesperson	*n.*	推销员，营业员
intentionally	*adv.*	有意地，故意地			

9.4 专业术语的翻译及专业英语翻译的基本方法

1. 专业术语的翻译方法

随着社会的进步和科技的发展，新的发明创造不断涌现，随之也就出现了描述这些事物的新术语。常用翻译词汇的方法有以下几种。

1）音译

音译就是根据英语单词的发音译成读音与原词大致相同的汉字。一般地，表示计量单位的词和一些新发明的材料或产品，它们的汉语名称在刚开始时基本就是音译的。或者当由于某些原因不便采用意译法时，也可采用音译法或部分音译法。

例如：Radar 是取 radio detection and ranging 等词的部分字母拼成的，如译成"无线电探测和测距设备"，显得十分啰嗦，故采用音译法，译成"雷达"。

又如：

baud	波特（发报速率单位）
bit	比特（二进制的单位）
hertz	赫[兹]（频率单位）
penicillin	盘尼西林（青霉素）
vaseline	凡士林（石油冻）

2）意译

意译就是对原词所表达的具体事物和概念进行仔细推敲，以准确译出该词的科学概念。这种译法最为普遍，科技术语在可能情况下应尽量采用意译法。采用这种方法便于读者顾名思义，不加说明就能直接理解新术语的确切含义。

例如：

loudspeaker	扬声器

semiconductor	半导体
videophone	可视电话
E-mail = Electronic mail	电子邮件
modem = modulator + demodulator	调制解调器

3）形译

形译是指用英语常用字母的形象来为形状相似的物体定名，翻译时也可以通过具体形象来表达原义，也称为"象译"。科技文献常涉及型号、牌号、商标名称及代表某种概念的字母。这些一般不必译出，直接抄下即可。另外，对于人名及公司名等名称类的词汇，翻译时可直接使用原文。

例如：

I-shaped	工字形
T square	丁字尺
X ray	X 射线
Y-connection	Y 形连接

4）准技术词汇

由于准技术词汇的扩展意义（即技术意义）在科技文献中出现的频率要比其一般意义大得多。因此对这种词的翻译应充分考虑其技术语境，不能望词生义，翻译为一般意义。有些语境提供的暗示并不充分，这要靠译者从更广阔的篇章语境中去寻找合理的解释，消除歧义。

例如，table 和 form 这两个词在普通英语中都翻译为"表，表格"，但在 Access 数据库中，二者的界限是严格区分的，table 是指数据库中的子表格，而 form 则指用于简化数据输入和查询的窗体。如果像有些译文那样把两个词都译为"表"，难免会引起译文不准确，造成阅读困难。

5）专业英语中新词的翻译

科技的发展还创造了大量的新词汇。这些词往往通过复合构词（compounding）或缩略表达全新的概念。它们由于在词典中缺乏现成的词项，一个词往往会有两个甚至两个以上的译名，造成很大混乱。全国科学技术名词审定委员会为了规范译名，定期发表推荐译名，因此我们还必须跟踪计算机行业的发展，掌握新出现的词汇。以下词汇表列出了计算机专业的几个新词。

例如：

Internet——由多个网络互联而成的计算机网，网络互联需要遵循 TCP/IP 协议，可译为"互联网"。

virtual reality——可译为"虚拟现实"，目前没有一个公认的虚拟现实定义，但其主要特征是以人为核心，使人身临其境，并能进行相互交流、实时操作，有如在真实世界中的感觉。

MPEG——一种压缩比率较大的活动图像和声音的压缩标准。

MFC——一个为基于 Windows 应用程序开发和创建 ActiveX 控件而建立的框架。

.NET——Microsoft XML Web 服务平台。XML Web 服务允许应用程序通过 Internet 进行通信和共享数据，而不管所采用的是哪种操作系统、设备或编程语言。

6）词义的选择与引申

（1）词义的选择。在翻译过程中，若英汉双方都是相互对应的单义词时则汉译不成问题，如 ferroalloy（铁合金）。然而，由于英语词汇来源复杂，一词多义和一词多性的现象十分普遍。例如，power 在数学中译为"乘方"，在光学中译为"率"，在力学中译为"能力"，在电学中译为"电力"。

例 1：The electronic microscope possesses very high resolving power compared with the optical microscope.

译文：与光学显微镜相比，电子显微镜具有极高的分辨率。

例 2：Energy is the power to do work.

译文：能量是指做功的能力。

（2）词义的引申。英汉两种语言在表达方式方法上差异较大，英语一词多义现象使得在汉语中很难找到绝对相同的词。如果仅按词典意义原样照搬，不仅使译文生硬晦涩，而且可能会词不达意，造成误解。因此，有必要结合语言环境透过外延看内涵，把词义做一定程度的扩展、引申。

例 1：Two and three make five.

译文：二加三等于五。（make 本意为"制造"，这里扩展为"等于"）

例 2：The report is happily phrased.

译文：报告措辞很恰当。（happily 不应译为"幸运地"）

7）词语的增减与变序

（1）词语的增加。由于英语和汉语各自独立演变发展，因而在表达方法和语法结构上有很大的差别。在英译汉时，不可能要求二者在词的数量上绝对相等。通常应该依据句子的意义和结构适当增加、减少或重复一些词，以使译文符合汉语习惯。

例 1：The more energy we want to send, the higher we have to make the voltage.

译文：想要输电越多，电压也就得越高。（省略 we）

例 2：This condenser is of higher capacity than is actually needed.

译文：这只电容器的容量比实际所需要的容量大。（补译省略部分的 capacity）

（2）词序的变动。英语和汉语的句子顺序通常都是按主语+谓语+宾语排列的，但修饰语的区别却较大。英语中各种短语或定语从句作修饰语时，一般都是后置的，而汉语的修饰语几乎都是前置的，因而在翻译时应改变动词的顺序。同时，还应注意英语几个前置修饰语（通常为形容词、名词和代词）中最靠近被修饰词的为最主要的修饰语，翻译时应首先译出。此外，英语中的提问和强调也大都用倒装词序，翻译时应注意还原。

例如：The transformer is a device of very great practical important which makes use of the principle of mutual induction.

译文：变压器是一种利用互感原理的在实践中很重要的装置。（从句）

8）词性及成分的转换

（1）词性的转换。因为两种语言的词源不同，语法相异，因而会经常遇到两种语言中缺乏词性相同而词义完全一样的词汇。这时，常常需要改变原文中某些词的词性以适应汉语的表达习惯，如实词之间、虚词之间，以及实词和虚词之间都可以互换。

例 1：Extreme care must be taken to the selection of algorithm in program design.

译文：在程序设计中必须注意算法的选择。

上例中将作主语的名词 care 转为动词并作为谓语来翻译，这样更符合汉语表达的习惯。

例 2：The information here proves invaluable in reaching a conclusion.

译文：这里的资料对于获得结论价值极高。（here 转为形容词）

（2）成分的转换。在英译汉时，不能仅依赖语法分析去处理译文，还必须充分考虑汉语的习惯及专业科技文献的逻辑性和严密性。这就要求在翻译过程中，视具体情况将句子的某一成分（主语、谓语、宾语、定语、表语、状语或补语）译成另一成分，或者将短语与短语、主句与从句、短语与从句进行转换。

例：Electronic computers must be programmed before they can work.

译文：必须先为电子计算机编好程序，它才能工作。（从句译为主句）

2．专业英语的翻译方法

科技文献本身体裁多样，而且随着时代的推进，越来越多的科技文献开始显示出其内容及文体风格上的创新和活跃。如果没有一定的翻译技巧，是翻译不好具有不同特点的科技文章的。一般来说，科技文献主要是述说事理、描写现象、推导公式及论证规律的，因此其特点是结构严谨、逻辑严密、行文规范、用词准确、技术术语正确、修辞手段较少。

科技文献翻译需要熟悉和掌握如下的知识。

- 掌握一定的词汇量。
- 具备科技知识，熟悉所翻译专业的知识背景。
- 了解中西方文化背景的异同，以及科技英语中词汇的特殊含义。
- 日常语言和文本的表达方式及科技英语的翻译技巧。

在翻译的过程中，理解是第一位的，表达是第二位的。

1）理解原文

透彻理解原文是确切表达的前提。理解原文必须从整体出发，不能孤立地看待一词一句。每种语言几乎都存在着一词多义的现象。因此，同样一个词或词组，在不同的上下文搭配中，在不同的句法结构中就可能有不同的意义。

（1）领略全文大意，分析语法关系。任何一篇文章都是一个有机整体，在翻译的过程中一定要在通读全文的基础上，领略原文的大意。透过各种语言现象对句子进行微观分析，即弄清句中各词、各成分之间的种种关系，如通常所说的主从关系、主谓关系、动宾关系等。

（2）理解原作事理，注意逻辑判断。有些句子在语法上可以有几种不同的解释，在语义上也可以有几种不同的理解，要判断这些句子的真正含义，还必须通过逻辑判断，对句子结构、语言环境、上下文和事理关系进行综合分析才能有效。理解原作的事理和逻辑关系时，必须仔细推敲，反复斟酌原文的含义，有时还要估计实际情况，根据自己的生活体验和客观道理来验证自己的理解是否合乎情理。

2）汉语表达

译文表达的好坏取决于译者对原文的理解程度、译者的逻辑思维能力和运用汉语的熟练程度。为了使译文准确通顺，在表达阶段一般要注意以下几个方面。

（1）表达的规范性。科技语体讲究论证的逻辑性，要求语言规范。对于科技文章来说，更要求运用规范的汉语来表达。

（2）表达的逻辑性。任何一位作者在著书立说时，都要运用概念、判断和推理这一思维形式，而人的思维要反映客观规律，必须符合逻辑，因此，表达思维的语言也要符合逻辑。

（3）表达的主动权。英汉两种语言在句法结构、表意方式等方面存在着很大的差异。因此，为了得到准确而流畅的译文，译者就不能把自己局限在原文的语言形式中。出于汉语表达的需要，译者可以甚至必须跳出原文的框框，对原文的句子成分、结构形式进行必要的调整，从容自如地按照汉语的特点和习惯组织自己的译文。

3）校对阶段

校对阶段是理解和表达的进一步深化，是使译文符合标准的一个必不可少的阶段，是对原文内容的进一步核实，对译文的进一步推敲。

（1）体会下面句子的翻译。

例 1：The binary system of representing numbers will now be explained by making reference to the familiar decimal system.

译文：下面参照所熟悉的十进制系统来说明表示数目的二进制系统。

例 2：The paper addresses an important problem in data-base management systems.

译文：本文讨论了基本数据管理系统的一个重要问题。

例 3：The author proposes an approach to the creation of an integrated method of investigation and designing objects, based on a local computer system.

译文：作者创立一种综合法，以当地计算机系统为基础来进行对象研讨与设计。

（2）在下面计算机专业英语资料的翻译中，给出了必要的分析。

例 1：The technical possibilities could well exist, therefore, of nation-wide integrated transmission network of high capacity, controlled by computers, interconnected globally by satellite and submarine cable, providing speedy and reliable communications throughout the world.

这句话看起来挺难理解，但若采用分解归类的语法分析方法则不难对付。首先，能够充当句子谓语的只能是"could well exist"，而不可能是其他非限定动词，如"integrated""controlled" "interconnected"。接下来主语自然是"possibilities"，由于"exist"是不及物动词，因此，不存在宾语。进一步分析将发现用作定语的介词短语"of nation-wide…cable"除了修饰"possibilities"以外没有其他名词可以承受，而分词短语"controlled…"和"interconnected…"又进一步修饰介词短语中的"network"。至于"providing…the world"则显然是表示结果的状态。这样一来，句子的骨架就比较清楚了。本句之所以将定语和待修饰的词分开，是因为定语太长而谓语较短，将谓语提前有助于整个句子结构的平衡。

译文：因此，在技术上完全有可能实现全国性的集成传输网，这种网络容量大，可由计算机控制，并能通过卫星和海底电缆与全球相联系，提供全世界范围内高速、可靠的通信。

例 2：A computer is an inanimate device that has no intelligence of its own and must be supplied with instructions so that it knows what to do and how and when to do it. These instructions are called software. The importance of software can't be overestimated.

Software is made up of a group of related programs, each of which is a group of related instructions that perform very specific processing tasks. Software acquired to perform a general business function is often referred to as a software package. Software packages, which are usually created by professional software writers, are accompanied by documentation—that explains how to use the software.

这篇短文讲述的是计算机软件，涉及计算机软件的定义及功能。在对一些词的理解上就要注意准确性及合理性。例如，inanimate device、intelligence、instructions、software、programs、software package 及 documentation，可分别翻译成无生命的装置、智能、指令、软件、程序、软件包及文档。

在理解这篇短文的关键词的基础上，可以给出每句话的表达。到此，可能还存在对原文把握不准确的地方，因此需要对原文内容进行进一步的核实和推敲。

译文：计算机本是一个无生命的装置。它没有自身的智能，并且必须给它指令，它才能知道去做什么、如何做，以及何时去做。这些指令就称为软件。软件的重要性无论如何估计，都不会过分。

软件是由一组相互有关的程序组成的，其中的每一个程序都是一组相关指令，而每组指令都将执行极为特定的处理任务。用来完成一般事务性功能任务的软件通常称为软件包。软件包一般由软件专业人员编写。软件包都带有文档，文档是用来解释如何使用这个软件的。

9.5 Exercises

1. Translate the following phrases.

static graphics image

visual environment

full-motion video

multimedia application

traditional teaching method

educational software

multimedia project

real-time delivery requirement

computer aided design

mechanical engineering design

building design

electronic circuit design

2. Fill in the blanks with appropriate words or phrases.

    a. delivered   b. business   c. infinite   d. graphics images

    e. simultaneous   f. enjoying   g. networked   h. the head

(1) Animation refers to moving _____.

(2) Multimedia is _____ use of data from different sources.

(3) Multimedia is any combination of text, graphic art, sound, animation, and video _____ to you by computer or other electronic means.

(4) Many applications, such as video mail, video conferencing, and collaborative work systems, require _____ multimedia.

(5) Multimedia is _____ widespread use in training programs.

(6) Multimedia excites eyes, ears, fingertips, and, most importantly, _____.

(7) The graphics technology used in today's best games will show up in tomorrow's _____ applications.

(8) The Windows operating environment gives the user an almost _____ range of expressing text.

3. Match the following terms to the appropriate definition.

    a. host   b. home page   c. HTTP

    d. hyperlink   e. hypertext   f. HTML

(1) Any computer on a network that is a repository for services available to other computers on the network.

(2) The first page that you intend people to see at your web site.

(3) This will take you from one Internet site to the next with a simple click of your mouse. Also called a link.

(4) The protocol for moving hypertext files across the Internet. It is the most important protocol used in the WWW.

(5) The coding scheme used to format text for use on the WWW.

(6) Generally, any text that contains links to other documents-words or phrases in the document that can be chosen by a reader and which cause another document to be retrieved and displayed.

Unit 10　Computer Graphics and Images

音频

10.1　Text

  The three main fields of computer imagery are computer graphics, image processing and computer vision. They are beginning to merge in many applications. Image synthesis or computer graphics is normally three-dimensional model in, two-dimensional image out. Image processing is normally three-dimensional reality as two-dimensional image (or two-dimensional reality) in, two-dimensional image out. Computer vision is normally three-dimensional reality as two-dimensional projection in, numbers, classification or action out. Such a classification is to some extent ephemeral and depends on current technology. It is simply the case that at the moment the most common graphics displays are two-dimensional. The most used input device in computer vision is a TV camera which gathers information from a three-dimensional scene as a two dimensional projection.[1]

Graphics

  Image synthesis or computer graphics is the methodology of the creation of images using a computer. In three-dimensional computer graphics the image is generated by a program from a mathematical description or a model. Computer graphics takes a three-dimensional model and calculates a two-dimensional projection for display. Algorithms in computer graphics mostly function in a three-dimensional domain and the creations in this space are then mapped into a two-dimensional display or image plane at a late stage in the overall process. Traditionally computer graphics has created pictures by starting with a very detailed geometric description, subjecting this to a series of transformations that orient a viewer and objects in three-dimensional space, then imitating reality by marketing the objects look solid and real—a process known as rendering.

  Typically, graphics programming packages provide functions to describe a scene in terms of these basic geometric structures, referred to as output primitives, and to group sets of output primitives into more complex structures.[2] Each output primitive is specified with input coordinate data and other information about the way that the object is to be displayed. Points and straight-line segments are the simplest geometric components of pictures. Additional output primitives that can be used to construct a picture include circles and other conic sections, quadric surfaces, spline curves and surfaces, polygon color areas, and character strings.

  Point plotting is accomplished by converting a single coordinate position furnished by an application program into appropriate operations for the output device in use. With a CRT monitor, for example, the electron beam is turned on to illuminate the screen phosphor at the selected location. How the electron beam is positioned depends on the display technology.

Line drawing is accomplished by calculating intermediate positions along the line path between two specified endpoint positions. An output device is then directed to fill in these positions between the endpoints. For analog devices, such as a vector pen plotter or a random-scan display, a straight line can be drawn smoothly from one endpoint to the other. Linearly varying horizontal and vertical deflection voltages are generated that are proportional to the required changes in the x and y directions to produce the smooth line. Digital devices display a straight-line segment by plotting discrete points between the two endpoints. Discrete coordinate positions along the line path are calculated from the equation of the line.

Image Processing

Image processing is the manipulation of an image to produce another image which is in some way different from the input image.[3] The source image can be an image file or the program that may operate directly on an image that is output by a TV-type camera. We also include image analysis as part of image processing. Image processing is characterized nowadays more by its migration into new application areas than the emergence of new algorithms. In particular we have seen the emergence of so-called image-based rendering techniques in virtual reality applications which have more in common with the two-dimensional world of image processing than the three-dimensional algorithms of computer graphics.[4] Another popular application of image processing is its use in graphic design to manipulate photographic images for presentation in advertisements and the like.

For display, the image is stored in a rapid-access buffer memory which refreshes the monitor at 30 frames/s to produce a visibly continuous display. Mini-computers or micro-computers are used to communicate and control all the digitization, storage, processing, and display operations via a computer network (such as the Ethernet). Program inputs to the computer are made through a terminal, and the outputs are available on a terminal, television monitor, or a printer/plotter.

Images acquired by satellites are useful in tracking of earth resources; geographical mapping; prediction of agricultural crops, urban growth, and weather; flood and fire control; and many other environmental applications. Space image applications include recognition and analysis of objects contained in images obtained from deep space-probe missions. Image transmission and storage applications occur in broadcast television, teleconferencing, transmission of facsimile images (printed documents and graphics) for office automation, communication over computer networks, closed-circuit television based security monitoring systems, and in military communications. Radar and sonar images are used for detection and recognition of various types of targets or in guidance and maneuvering of aircraft or missile systems.

Computer Vision

Computer vision is the extraction of information from an image. It is different from image analysis in its goal which most ambitiously attempts to emulate the human visual system. The source image in computer vision is usually a two-dimensional projection of a real scene. The goal of the process may be to recover three-dimensional information from the two-dimensional projections as in, for example, depth from stereo projections. Alternatively the output of the process may be a number, or a label as in the case, for example, of character recognition or it may be an action as in computer vision in robotics. The reason computer vision is such a demanding field is because we try to extract three-dimensional information from a series of two-dimensional projections. Despite early optimism it has proved to be enduringly difficult to find solutions for most of the potential application areas.

24-Bit Color

Computers are generally structured to work in groups of eight bits (called a byte). These eight numbers can be used to count up to 256, and so can describe 256 shades of grey from black to white, which is more than enough to satisfy the eye. A computer which can assign eight bits to describe each pixel will produce perfect black and white photographs on its monitor.[5] A monitor than can show all these shades is called a grayscale display.

Now your eye can detect those 150-200 shades in all three of the colors it can see: red, green, and blue. If you use eight bits to describe color, you only get 256 colors, which aren't enough—you get a mildly painting effect, although the dithering process can simulate more colors at the expense of quality.

To get the full color photographic effect on a computer monitor, you need to be able to generate 256 shades for each color. This takes eight bits of information per color, giving a total of 24 bits. This is the 24-bit color that you keep reading about in computer magazines. If you take all the possible variations of 256 shades of three colors, you end up with a possible 16.7 million color shades.

You only really need 24-bit color if you are going to do color photographic retouching on-screen or similar 'painting' on-screen. For line work and picture placing, an 8-bit color monitor is perfectly adequate, as you can still define colors for print even if you can't show them on the screen.

Keywords

conic	n.	圆锥形	military	adj.	军事的，军队的
dither	v.	发抖，抖动	painting	n.	画，画作
enduringly	adv.	持久地，永久地	phosphor	n.	磷光体，荧光粉
ephemeral	adj.	短暂的	polygon	n.	多边形，多角形
facsimile	n.	传真，复制，摹本	primitive	n.	原始事物，本原，图元
grayscale	n.	灰度，灰色标度	projection	n.	投影，透射
imagery	n.	像，画像	proportional	adj.	比例的，相称的
imitate	v.	模仿，仿效	quadric	adj. n.	二次的，二次曲面
mathematical	adj.	数学上的，数理上的	render	v.	渲染，描写
migration	n.	移动，迁移	synthesis	n.	合成，综合
mildly	adv.	柔和地，温和地	urban	adj.	城市的

Notes

[1] The most used input device in computer vision is a TV camera which gathers information from a three-dimensional scene as a two dimensional projection.
译文：在计算机视觉中，使用最多的输入设备是把从三维景物中采集的信息作为二维投影图的 TV 摄像机。
说明：本句的"a TV camera"是表语，"which gathers…"修饰表语。

[2] Typically, graphics programming packages provide functions to describe a scene in terms of these basic geometric structures, referred to as output primitives, and to group sets of output primitives into more complex structures.

译文：通常，图形编程软件包提供了一些功能来描述场景，这些功能使用了称为输出图元的基本几何结构并将输出图元组合成更复杂的结构。

说明：本句的"to describe a scene…"作宾语补足语，"referred to as output primitives"是插入语。

[3] Image processing is the manipulation of an image to produce another image which is in some way different from the input image.

译文：图像处理是对一幅图像进行处理以产生另一幅图像，该图像不同于输入图像。

说明：本句的"to produce another image"作目的状语，"which is in some…"作定语，修饰"another image"。

[4] In particular we have seen the emergence of so-called image-based rendering techniques in virtual reality applications which have more in common with the two-dimensional world of image processing than the three-dimensional algorithms of computer graphics.

译文：我们已经看到，在虚拟现实的应用中，基于图像的渲染技术的出现，比起三维计算机图形学的算法来，与图像处理的二维世界有着更多的共同之处。

说明：这是一个长句，"of so-called image-based rendering techniques"作定语，修饰"emergence"，"which have more in…"是定语从句，修饰"applications"。

[5] A computer which can assign eight bits to describe each pixel will produce perfect black and white photographs on its monitor.

译文：使用8位描述1像素的计算机就能在其显示器上产生完美的黑白照片。

说明：本句由"which"引导的定语从句修饰主语。

10.2 Reading Material 1：Digital Photography

Digital photography is photography using a camera that uses an electronic sensor to record the image as a piece of electronic data rather than as chemical changes on a photographic film. The sensor is either a light sensitive CCD, or a CMOS semiconductor device. A digital memory device is usually used for storing images, which may then be transferred to a computer later.

The advantages of this method over traditional film include the greatly reduced cost per image, the potential to make taken images instantly available for appraisal, the greater number of images that can be conveniently transported, and the removal of the requirement to develop the film in a photo lab. In addition, digital cameras can be smaller and lighter than film cameras.

Recent digital cameras from leading manufacturers such as Nikon and Canon have promoted the adoption of digital single-lens reflex cameras (SLRs) by photojournalists. Images captured at 2-mega pixels are deemed to be of sufficient quality for small images in newspaper or magazine reproduction. Six mega pixel images, found in modern digital SLRs, when combined with high-end lenses can match or even exceed the detail of film prints taken with 35mm film based SLRs, and the latest 12-mega pixel models can produce astoundingly detailed images better than almost all 35mm images.

With the acceptable image quality and the other advantages of digital photography, an increasing number of professional news photographers use these devices.

It has also been adopted by many amateur snapshot photographers, who take advantage of the convenience of the form when to send images by E-mail and to place on the World Wide Web.

Digital photography was used in astronomy long before its use by the general public and had almost completely displaced photographic plates by the early 1980. Not only are CCD more sensitive to light than plates, but the information can be downloaded onto a computer for data analysis. Other commercial photographers, and many amateurs, have enthusiastically embraced digital photography, as they believe that its flexibility and lower long-term costs outweigh its initial price disadvantages.

Exchangeable image file format (EXIF) is a set of file formats specified for use in digital cameras. This specifies the use of TIFF for the highest quality format and JPEG as a space-saving but lower quality format. Many low-end cameras can deliver only JPEG files. Another format that may be encountered is CCD-RAW, which is no-standardized.

Digital cameras generally include dedicated digital image processing chips to convert the raw data from the image sensor into a color-corrected image in a standard image file format. Images from digital cameras have over film cameras. The digital image processing is typically done by special software programs that can manipulate the images in many ways. Many digital cameras also enable viewing of histograms of images, as an aid for the photographer to better understand the intensity range of each shot.

Photo editor is an image editing application that is specialized for digital cameras. It is used to crop and touch up photos from the camera and organize them into albums and slide shows.

Digital image processing allows the use of much more complex algorithms for image processing, and hence can offer both more sophisticated performance at simple tasks, and the implementation of methods which would be impossible by analog means. The most common kind of digital image processing is digital image editing.

Image editors typically deal with only bitmapped images such as GIFs, JPEGs and BMPs; however, some editors support both bitmaps and illustrations. Common functions are manually cropping and resizing the image and using "filters" to adjust brightness, contrast and color. A myriad of filters are available for special effects. Image filter is a routine that changes the appearance of an image or part of an image by altering the shades and colors of the pixels in some manner. Filters are used to increase brightness and contrast as well as to add a wide variety of textures, tones and special effects to a picture.

Keywords

amateur	adj.	业余的，爱好……的	intensity	n.	（底片）的明暗度
appraisal	n.	评价，鉴定	myriad	adj.	无数，众多的
astoundingly	adv.	使人震惊地	photography	n.	摄影术，照相术
astronomy	n.	天文学	photojournalist	n.	摄影记者
camera	n.	照相机，摄像机，暗房，暗箱	promote	v.	推广，发起
embrace	v.	接受，采用	reflex	adj.	反射的
enthusiastically	adv.	热心地，热情地，热烈地	removal	n.	移动，迁移
histogram	n.	直方图，矩形图	snapshot	v.	快拍

10.3 Reading Material 2: Recent Advances in Computer Vision

Computer vision is the branch of artificial intelligence that focuses on providing computers with the functions typical of human vision. To date, computer vision has produced important applications in fields such as industrial automation, robotics, biomedicine, and satellite observation of Earth. In the field of industrial automation alone, its applications include guidance for robots to correctly pick up and place manufactured parts, nondestructive quality and integrity inspection, and on-line measurements.

Until a few years ago, chronic problems affected computer-vision systems and prevented their widespread adoption. Since its start, computer vision has appeared as a computationally intensive and almost intractable field because its algorithms require a minimum of hundreds of MIPS to be executed in acceptable real time. Even the input-output of high-resolution images at video rate was traditionally a bottleneck for common computing platforms such as personal computers and workstations. To solve these problems, the research community has produced an impressive number of dedicated computer-vision systems. One such famous system was the massively Parallel Processor (MPP), designed at the Goddard Space Flight Center in 1983 and operated there until 1991. the MPP used an array of 16,384 single-bit processors and was capable at peak performance of 250 million floating-point operations/s—an impressive feat at the time.

Dedicated computers such as the MPP have always received a cold reception from industry because they were expensive, cumbersome, and difficult to program. In recent years, however, increased performance at the system level—faster microprocessors, faster and larger memories, and faster and wider buses—has made computer vision affordable on a wide scale. Fast microprocessors and digital-signal processors are now available as off-the-shelf solutions, and some of them can execute calculations at rates of thousands of MIPS.

The availability of affordable hardware and software has opened the way for new, pervasive applications of computer vision. These applications have one factor in common. They tend to be human-centered; that is, either humans are the targets of the vision system or they wander about wearing small cameras, or sometimes both. Vision systems have become the central sensor in applications such as:

- Human-computer interface, the links between computers and their users;
- Augmented perception, tools that increase normal perception capabilities of humans;
- Automatic media interpretation, which provides an understanding of the content of modern digital media, such as videos and movies, without the need for human intervention or annotation;
- Video surveillance and biometrics.

Keywords

annotation	n.	注释者	massively	adv.	大规模地，大量地
biomedicine	n.	生物药剂学	nondestructive	adj.	无害的，无破坏性的
biometrics	n.	生物统计学	peak	n.	峰值，最高点
bottleneck	n.	瓶颈	pervasive	adj.	普及的，扩大的
cumbersome	adj.	麻烦的，讨厌的	robotics	n.	机器人
intervention	n.	调解，介入	surveillance	n.	监视，管制

10.4 科技论文写作

科技论文用于介绍有一定学术水平的研究成果，讲究叙述准确、概念明确、判断恰当、推理严格及具有严肃的科学性。科技论文的读者一般都具有一定的科技知识和专业水平，因此论文不必写得太通俗，不必包括过多烦琐的介绍。但对于不同学科的论文及论文的不同部分，要求也不完全一样。

科技论文一般由以下几部分组成。

1. 标题（title）

一篇论文的标题要简单明了，既能概括全文内容又能引人入胜。

例1：Algebraic Solutions of Linear Partial Differential Equation.

译文：线性偏微分方程的代数解。

例2：An Improved PSO Algorithm and Its Application to Grid Scheduling Problem.

译文：改进的粒子群优化算法及其在网格调度问题中的应用。

例3：A Discrete PSO Algorithm for Partner Selection of Virtual Enterprise.

译文：离散粒子群优化算法用于虚拟企业的伙伴选择。

2. 摘要（abstract）

论文摘要要求写得简明扼要，具有相当的独立性。

摘要一般要概括如下内容。

- 从事该研究工作的起因。
- 完成了哪些工作。
- 突出的成果和发现。
- 成果和发现的意义。
- 还有哪些工作待解决。

例：On the basis of analyzing the particle swarm optimization (PSO) algorithm and support vector machine (SVM), the PSO algorithm with last out mechanism is applied to optimize the parameters of SVM, then the PSO-SVM model about a practical soft-sensor of gasoline endpoint of delayed coking plant is constructed. The method takes advantages of the minimum structure risk of SVM and the quickly globally optimizing ability of PSO for soft sensor modeling. The simulation results demonstrate that the model has effective generalization performance and higher precision.

译文：在分析基本微粒群优化算法（PSO）和支持向量机（SVM）原理的基础上，采用带有末位淘汰机制的微粒群优化算法优化支持向量机的参数，建立了延迟焦化装置粗汽油干点软测量的微粒群支持向量机模型。该方法利用支持向量机结构风险最小化原则和PSO算法快速全局优化的特点，用于软测量建模。仿真实验表明，所建模型的泛化性较好，模型具有较高的精度。

3. 关键词（keywords）

科技论文中常有关键词，这主要是方便读者通过关键词检索出该论文，一般在用计算机查找有关论文时能够用到。关键词的选择要求能很好地体现该篇论文的研究内容，一般要求少而精，以3～5个为宜。

对应上面例子的关键词如下。

Key words: particle swarm optimization algorithm, support vector machine, soft-sensor, optimization

关键词：微粒群优化算法，支持向量机，软测量，优化

4. 引言（introduction）

引言与摘要一样在论文中起着十分重要的作用，摘要是全篇论文的缩影，引言则起着向读者解释论文的主题、目的和总纲的作用。一般情况下引言要包括以下内容。

- 阐明论文的主题和目的。
- 说明写作论文的背景和情况。
- 介绍达到理想解决方案的方法。

对于长篇论文或某些论述不常见内容的论文，引言还要使读者易于阅读该论文，这样引言中会有论文课题的历史、背景及相关著作回顾；引起研究的现实情况；相关资料的来源；该研究课题的现实目的和意义；实验采用的方法；论文展开研究的计划及可行性分析等。

引言只是一个开头，许多内容只要稍微提到就行，不要面面俱到。如果感觉引言太长，有关历史、背景、定义、资料等内容可以独立成一节。

5. 正文（text）

正文是科技论文的主体。正文的内容要合乎逻辑，就是要做到观点和材料的统一，以明确的观点来统领材料，以恰当的材料来支持观点，形成一个统一的整体。观点必须准确、鲜明，材料必须经过提炼，能很好地反映事物本质和内在联系。在材料的筛选、提炼过程中，要去伪存真、去粗取精、保留理性成分。

有些论文较为专业，需要从正文中了解种种专业细节，进行深入钻研，并要求能重复论文中的实验及相关结论，因此正文中实验材料、设备、过程、结果、讨论等必须详细写明。

正文内容要合乎思维规律，顺理成章。人们在认识事物时，总是通过"实践—认识—再实践—再认识"这种规律不断提升、加深认识的，所以，在正文中也要分感性认识和理性认识两个阶段。第一阶段是通过感觉获得大量感性材料的阶段；第二阶段是综合材料加以整理和提炼，进行判断、推理的阶段。在这些过程完成之后，又回到实践中去，检验和发展理论。

关于实验部分的描述，需要考虑以下三方面。

- 实验材料的说明。
- 实验过程的说明。
- 实验结果的分析。

6. 总结（conclusions）

总结是论文内容的概述。其写作要求大体上和摘要相似，一般放在正文的后面。总结的内容一般包括论文的中心思想、研究问题的提出、研究方法、突出的成果及意义等。其作用是帮助读者回顾全文，加深理解。

例如：Based on the research of existing scheduling technique, a discrete PSO algorithm was presented to solve grid scheduling problem within reasonable time. In the grid environment, the scheduling problem is to schedule a stream of tasks to a set of nodes. During the execution, there are some communications between nodes. The function of DPSO is to find the best tasks scheduling strategy and to obtain the optimal makespan.

We tested the DPSO algorithm against the MaxMin method. The results obtained from the simulation study illustrated that DPSO algorithm outperformed the MaxMin heuristic. So, it is a very promising runtime optimizer for grid scheduling problem.

7. 参考文献（references）

科技论文列举参考文献是传统惯例，反映作者严肃的科学态度和广泛的研究依据。凡是引用其他作者的观点或研究成果，都应该用数码标明，并在参考文献部分说明出处。

每篇参考文献都应包括作者姓名、文献名称、正式刊出时间、相关章节或页码等信息。

例如：

[1] J. Robinson, Y. Rahmat. Particle Swarm Optimization in Electromagnetics. IEEE Trans. on Antennas and Propagation. Vol. 52, No. 2. Feb. 2004, 397-407.

[2] T. Arslan, N. Haridas, E. Yang, et al. ESPACENET: s Framework of Evolvable and Reconfigurable Sensor Networks for Aerospace-based Monitoring and Diagnostics. In Proc. 1st NASA/ESA Conf. Adapt. Hardware and Systems, 323-329, Istanbul, Turkey, 2006.

[3] S. M. Guru, S. K. Halgamuge, S. Fernando. Particle Swarm Optimizers for Cluster Formation in Wireless Sensor Networks. In Proc. 2005 Int. Conf. Intel. Sensors, sensor Networks and Info. Processing Comf., 319-324, Melbourne, Australia, 2005.

10.5 Exercises

1. **Translate the following phrases.**

   three-dimensional model
   image processing
   virtual reality application
   closed-circuit television
   24-bit color
   digital photography
   digital camera
   standard image file format
   photo editor
   bitmapped image
   image filter
   artificial intelligence

2. **Fill in the blanks with appropriate words or phrases.**

   a. geometric   b. electron beam   c. information   d. methodology
   e. computer graphics f. endpoint positions   g. Positions   h. projection

   (1) Algorithms in _____ mostly function in a three-dimensional domain.
   (2) How the _____ is positioned depends on the display technology.
   (3) Discrete coordinate _____ along the line path are calculated from the equation of the line.
   (4) Computer graphics takes a three-dimensional model and calculates a two-dimensional _____ for display.
   (5) Line drawing is accomplished by calculating intermediate positions along the line path between two specified _____.
   (6) Points and straight-line segments are the simplest _____ components of pictures.
   (7) Image synthesis or computer graphics is the _____ of the creation of images using a computer.

(8) Each output primitive is specified with input coordinate data and other _____ about the way that the object is to be displayed.

3. **Match the following terms to the appropriate definition.**

   a. instruction   b. initialized   c. ISP
   d. interlace style   e. ink jet printer   f. ISDN

(1) In this style, electron-beam sweeps elements in odd rows first time and does elements in even rows second time. A frame to be renewed needs sweeping two times.

(2) In programming, it means to assign a starting value to a variable. It can also refer to the process of starting up a program or system.

(3) A company that provides access to the Internet. For a monthly fee, the company gives you a software package, username, password and access phone number.

(4) It uses a nozzle and sprays ink onto the paper to form the appropriate characters. In order to get the correct character, the ink is directed with a valve and one or more electronic deflectors that control the vertical and horizontal position of the jet stream of ink.

(5) It allows you to send digital information at speeds of 128Kb over the normal telephone network.

(6) It is a form of communicated information that is both command and explanation for how an action, behavior, method, or task is to be begun, completed, conducted, or executed.

Unit 11 Animation

11.1 Text

Computer animation generally refers to any time sequence of visual changes in a scene. In addition to changing object position with translations or rotations, a computer-generated animation could display time variations in object size, color, transparency, or surface texture.[1]

Many applications of computer animation require realistic displays. An accurate representation of the shape of a thunderstorm or other natural phenomena described with a numerical model is important for evaluating the reliability of the model.

Entertainment and advertising applications are sometimes more interested in visual effects. Thus, scenes may be displayed with exaggerated shapes and unrealistic motions and transformations. There are many entertainment and advertising applications that do require accurate representations for computer-generated scenes. And in some scientific and engineering studies, realism is not a goal. For example, physical quantities are often displayed with pseudo-colors or abstract shapes that change over time to help the researcher understand the nature of the physical process.[2]

The storyboard is an outline of the action. It defines the motion sequence as a set of basic events that are to take place. Depending on the type of animation to be produced, the storyboard could consist of a set of rough sketches or it could be a list of the basic ideas for the motion.[3]

An object definition is given for each participant in the action. Object can be defined in terms of basic shapes, such as polygons or splines. In addition, the associated movements for each object are specified along with the shape.

A key frame is a detailed drawing of the scene at a certain time in the animation sequence. Within each key frame, each object is positioned according to the time for that frame. Some key frames are chosen at extreme positions in the action; others are spaced so that the time interval between key frames is not too great. More key frames are specified for intricate motions than for simple, slowly varying motions.

In-betweens are the intermediate frames between the key frames. The number of in-betweens needed is determined by the media to be used to display the animation. Film requires 24 frames per second, and graphics terminals are refreshed at the rate of 30 to 60 frames per second. Typically, time intervals for the motion are set up so that there are from three to five in-betweens for each pair of key frames.[4] Depending on the speed specified for the motion, some key frames can be duplicated. For a 1-minute film sequence with no duplication, we would need 1440 frames. With five in-betweens for each pair of key frames, we would need 288 key frames. If the motion is not too complicated, we could space the key frames a little farther apart.

There are several other tasks that may be required, depending on the application. They include motion verification, editing, and production and synchronization of a soundtrack. Many of the functions needed to produce general animations are now computer-generated.

Some steps in the development of an animation sequence are well suited to computer solution, these include object manipulations and rendering, camera motions, and the generation of in-betweens. Animation packages, such as Wave-front, for example, provide special functions for designing the animation and processing individual objects.[5]

One function available in animation packages is provided to store and manage the object database. Object shapes and associated parameters are stored and updated in the database. Other object functions include those for motion generation and those for object rending. Motions can be generated according to specified constraints using two-dimensional or three-dimensional transformations. Standard functions can then be applied to identify visible surfaces and apply the rendering algorithms.

Another typical function simulates camera movements. Standard motions are zooming, panning, and tilting. Finally, given specification for the key frames, the in-betweens can be automatically generated.

Flash

Micromedia's Flash has become one of the premier Web animation tools and formats in the short time that it is been on the scene. Some of Flash's success comes from its ambidextrous nature: It's both an authoring tool and a file format. Not only is it much easier to learn than, say, DHTML, but it comes chock-full of important animation features, such as keyframe interpolation, motion paths, animated masking, shape morphing, and onion skinning. Quite the versatile program, you can use it not only to create Flash movies, but also to import an animation in a QuickTime file or in a number of different file formats. And Flash just keeps getting better with every release. On the Web even the very simple animation has a long download time. But Flash has changed all that with its streaming technology and vector graphics.

Flash movies are graphics and animation for Web sites. They consist primarily of vector graphics, but they can also contain imported bitmap graphics and sounds. Flash movies can incorporate interactivity to permit input viewers, and you can create nonlinear movies that can interact with other Web applications. Web designers use Flash to create navigation controls, animated logos, long-form animations with synchronized sound, and even complete, sensory-rich Web sites.

Standard Flash work environment includes menu bar, tool bar, stage, time-axis window, and tool palette. Besides these several major parts, with Windows menu opened, some small windows like material window can be called on.

Work area refers to a Flash work platform; it is a rather big area, actually covering all stages as mentioned below and work objects for drawing pictures or editing movie clips. It can be seen as combination of backstage and stage. Stage is a platform for demonstrating all elements of Flash movie clip, displaying the content of the currently selected frame. Different from work area, only the content of stage is visible after the movie clip is played back, while content in the work area beyond the stage is invisible, just like players and work staffs in the backstage which are invisible to the audience. Just like drama having several scenes, the stage can have several scenes.

3ds Max

3ds Max is a full-featured 3D graphics application developed by Autodesk Media and Entertainment. It runs on the Win32 and Win64 platforms. 3ds Max is one of the most widely-used 3D animation programs by content creation professionals. It has strong modeling capabilities, a flexible plug-in architecture and a long heritage on the Microsoft Windows platform. It is mostly used by video game developers, TV commercial studios and architectural visualization studios. It is also used for movie effects and movie previsualisation.

In addition to its modeling and animation tools, the latest version of 3ds Max also features advanced shaders (such as ambient occlusion and rendering), dynamic simulation, particle systems, radiocity, normal map creation and rendering, global illumination, an intuitive and fully-customizable user interface, its own scripting language and much more.

Polygon modeling is more common with game design than any other modeling technique as the very specific control over individual polygons allows for extreme optimization. Also, it is relatively faster to calculate in real-time. Usually, the modeler begins with one of the 3ds Max primitives, and using such tools as bevel, extrude, and polygon cut, adds detail to and refines the model. Versions 4 and up feature the Editable Polygon object, which simplifies most mesh editing operations, and provides subdivision smoothing at customizable levels.

Keywords

ambidextrous	adj.	两面性的，非常熟练的	outline	n.	轮廓，草稿
drama	n.	剧本，戏剧	pan	v.	摇镜头，拍摄全景
duplication	n.	重复，复制	permit	n. v.	许可，准许
exaggerated	adj.	夸张的，过大的	phenomenon	n.	现象
heritage	n.	继承，传统	reliability	n.	可靠性，安全性
interpolation	n.	插入，插值法	rotation	n.	旋转，轮转
intricate	adj.	复杂的，错综的	storyboard	n.	剧本
mesh	n.	网眼，网状物	thunderstorm	n.	雷雨
morph	v.	变换图像，变来变去	variation	n.	变化，变量
nonlinear	adj.	非线性的	zoom	v.	拉镜头

Notes

[1] In addition to changing object position with translations or rotations, a computer-generated animation could display time variations in object size, color, transparency, or surface texture.

译文：除了通过平移、旋转来改变对象的位置外，计算机生成的动画还可以随时间的进展而改变对象的大小、颜色、透明度和表面纹理等。

说明："In addition to changing object position with translations or rotations" 是状语，"in object size, color, transparency, or surface texture" 也是状语。

[2] For example, physical quantities are often displayed with pseudo-colors or abstract shapes that change over time to help the researcher understand the nature of the physical process.

译文：例如，物理量经常使用随时间而变化的伪彩色或抽象形体来显示，以帮助研究人员理解物理过程的本质。

说明：本句的"that change over time"作定语，修饰"pseudo-colors or abstract shapes"，"to help the researcher…"是目的状语。

[3] Depending on the type of animation to be produced, the storyboard could consist of a set of rough sketches or it could be a list of the basic ideas for the motion.

译文：依赖于要生成的动画类型，剧本可能包含一组粗略的草图或运动的一系列基本思路。

说明：本句中的"Depending on the type of animation to be produced"作状语，后面跟两个并列的句子。

[4] Typically, time intervals for the motion are set up so that there are from three to five in-betweens for each pair of key frames.

译文：一般情况下，运动的时间间隔设定为每一对关键帧之间有 3~5 个插值帧。

说明：本句中的"so that"引导目的状语从句。

[5] Animation packages, such as Wave-front, for example, provide special functions for designing the animation and processing individual objects.

译文：动画软件包，如 Wave-front，提供了设计动画和处理单个对象的专门功能。

说明：本句的"such as Wave-front"是主语的同位语，"for example"是插入语，"for designing the animation and processing individual objects"是宾语补足语。

11.2　Reading Material 1：Video Compression

Video compression is a collection of techniques used to shrink video files. Embodied by products called codecs (compression/decompression), these methods fall into two general categories: interframe and intraframe compression.

Interframe compression uses a system of key and delta frame, while delta, or "difference", frames record only the interframe changes. During decompression, the CPU builds frames from the key frames and accumulated deltas.

Intraframe compression is performed entirely within individual frames. During intraframe compression, codecs use a variety of techniques to convert pixels to more compact mathematical formulas. The simplest technique is called run length encoding (RLE), in which rows of adjacent identical pixels are grouped together.

Intraframe technologies range from simple RLE to documented standards such as JPEG to exotic mathematical disciplines such as wavelets and fractal transform. Not all codecs use both interframe and intraframe techniques—some use only intraframe. Those that use both apply intraframe compression on key frames and the information remaining in delta frames after removing interframe redundancies.

A standard video source, such as a camcorder, VCR, or laserdisc player, transmits the analog video signal to the video-capture board, while analog audio is sent to a sound board inside the PC.

The capture board utilizes analog-to-digital converters (ADCs) to transform the analog video signal into binary code. The video footage can be captured as a raw sequence of video frames, which is sent to and held in system RAM where software compression is performed.

Meanwhile, the audio signal undergoes analog-to-digital conversion by the second board's converters.

This information is also sent to the PC's main system RAM.

After the video and sound tracks have been captured, the captured signal can be stored directly to the hard disk or software-based compression can be applied. Generally, the digital video and audio signals are stored as a synchronized, or interleaved. AVI file on the hard disk.

The MPEG (Moving Picture Experts Group) standards are the main algorithms used to compress videos and have been international standards since 1993. It is the name of the family of standards used for coding audio-visual information (e.g., movies, video, and music) in a digital compressed format. The major advantage of MPEG compared to other video and audio coding formats is that MPEG files are much smaller. The major MPEG standards include the following.

MPEG-1: Its goal is to produce video recorder-quality output using a bit rate of 1.2 Mbps. It can be transmitted over twisted pair transmission lines for modest distances, and also used for storing movies on CD-ROM in CD-I and CD-Video format.

MPEG-2: It was originally designed for compressing broadcast quality video into 4 to 6 Mbps, so it could fit in a NTSC or PAL broadcast channel. It is a standard for "the generic coding of moving pictures and associated audio information". MPEG-2 is a superset of MPEG-1, with additional features, frame formats and encoding options.

MPEG-3: It supports higher resolutions, including HDTV.

MPEG-4: It is an ISO/IEC standard developed by MPEG. It is for medium-resolution video-conferencing with low frame rates and at low bandwidths. MPEG-4 files are smaller than JPEG or QuickTime files, so they are designed to transmit videos and images over a narrower bandwidth and can mix video with text, graphics and 2D and 3D animation layers.

MPEG-7: It is the multimedia standard for the fixed and mobile web, enabling integration of multiple paradigms. MPEG-7 is formally called multimedia content description interface. It provides a tool set for completely describing multimedia content and is designed to be generic and not targeted to a specific application.

MPEG-21: Unlike other MPEG standards that describe compression coding methods, MPEG-21 describes a standard that defines the description of content and also processes for accessing, searching, storing and protecting the copyright of content.

Keywords

adjacent	*adj.*	毗邻的，邻近的	fractal	*n.*	分形
camcorder	*n.*	摄像机	interframe	*n.*	帧间
capture	*v.*	捕获，获取	intraframe	*n.*	帧内
channel	*n.*	信道，频道	laserdisc	*n.*	光盘
converter	*n.*	转换器，变换器	narrow	*adj.*	窄的，狭小的
discipline	*n.*	学科	wavelet	*n.*	小波，基元波
footage	*n.*	（影片等）片段			

11.3 Reading Material 2：Lossy Audio Compression

Audio compression is a form of data compression designed to reduce the size of audio files. Audio compression algorithms are implemented in computer software as audio codecs. Generic data compression algorithms perform poorly with audio data, seldom reducing file sizes much below 87% of the original, and are not designed for use in real time. Consequently, specific audio "lossless" and "lossy" algorithms have been created. Lossy algorithms provide far greater compression ratios and are used in mainstream consumer audio devices.

Lossy audio compression is used in an extremely wide range of applications. In addition to the direct applications, digitally compressed audio streams are used in most video DVDs; digital television; streaming media on the Internet; satellite and cable radio; and increasingly in terrestrial radio broadcasts. Lossy compression typically achieves far greater compression than lossless compression, but discarding less-critical data.

The innovation of lossy audio compression was to use psychoacoustics to recognize that not all data in an audio stream can be perceived by the human auditory system. Most lossy compression algorithms reduce perceptual redundancy by first identifying sounds which are considered perceptually irrelevant, that is, sounds that are very hard to hear. Typical examples include high frequencies, or sounds that occur at the same time as other louder sounds. Those sounds are coded with decreased accuracy or not coded at all.

In psychoacoustics based lossy compression, the real key is to "hide" the noise generated by the bit savings in areas of the audio stream that can not be perceived. This is done by, for instance, using very small numbers of bits to code the high frequencies of most signals, not because the signal has little high frequency information, but rather because the human ear can only perceive very loud signals in this region, so that softer sound "hidden" there simple aren't heard.

Because data is removed during lossy compression and can not be recovered by decompression, some people may not prefer lossy compression for archival storage. Hence, as noted, even those who use lossy compression may wish to keep a losslessly compressed archive for other applications.

Due to the nature of lossy algorithms, audio quality suffers when a file is decompressed and recompressed. This makes lossy compression unsuitable for storing the intermediate results in professional audio engineering applications, such as sound editing and multi-track recording. However, they are very popular with end users (particularly MP3), as a megabyte can store about a minute's worth of music at adequate quality.

Keywords

archival	adj.	档案的	mainstream	n.	主流
auditory	adj.	听觉的，耳朵的	original	adj.	原始的，原本的
consequently	adv.	结果	psychoacoustics	n.	心理声学
discard	v.	放弃，抛弃	terrestrial	adj.	地面的，地球上的
irrelevant	adj.	不恰当的，不相干的			

11.4 句子成分简介

英语的句子成分一般包括主语、谓语、宾语、定语、状语、补语及同位语、插入语等。

1. 主语（subject）

主语是句子的主体，是句子所要说明的人或事物。主语通常是一些代表事物性或实体性的词语。除了名词外，代词、数词、动词不定式、动名词和从句均可作主语。

例：**She** works in a big company.

译文：她在一家大公司工作。

2. 谓语（predicate）

谓语说明主语"做什么""是什么"或"怎么样"。

例：They **decided** not to sell their computers.

译文：他们决定不卖掉他们的计算机。

3. 宾语（object）

宾语表示动作的对象，是主语的动作的承担者，有宾语的动词称为及物动词，宾语一般在及物动词之后，作宾语的词有名词、代词宾格、数词、动词不定式、动名词、复合结构、从句等。

例：They have already finished **the work**.

译文：他们已经完成了那项工作。

4. 定语（attribute）

定语是用来修饰名词或代词的。除形容词外，数词、名词所有格、动词不定式、介词短语、分词短语、动名词、副词和从句等，都可以作定语。

例：There is some exciting news **on the newspaper today**.

译文：今天报上有令人兴奋的消息。

5. 状语（adverbial）

状语是用来修饰动词、形容词、副词的，表示时间、地点、原因、方式、程度等。作状语的词有副词或相当于副词的其他词、短语和从句。

例：There is a printer **in the room**.

译文：房间里有一台打印机。

6. 补语（complement）

英语中有些及物动词虽然有了宾语，但句子的意思仍不完整，还需要在宾语之后增加一个成分以补足其意义，这种成分称为宾语补语。能作宾语补语的有名词、形容词、介词短语、副词、动词不定式和分词等。

例：They believe this printer to be **the best**.

译文：他们相信这台打印机是最好的。

7. 同位语（appositive）

同位语用来对一个词或词组的内容加以补充和说明。它通常位于其说明的词或词组之后。

例1：Our teacher, **Mr. Smith**, has developed a kind of new software.

译文：我们老师，史密斯先生，已经开发了一种新软件。

例2：We **Chinese** are working hard to make our country rich and strong.

译文：我们中国人民正努力工作使我们的国家富强起来。

8. 插入语

插入语通常是对一句话作一些附加的解释。常用来作这类附加成分的结构有 I think、I hope、I suppose、I guess、you know、don't you think、it seems、you see、it is said、it is suggested 等。它们一般放在句子末尾，也可放在句子中间。

例1：Her design, **I think**, is the best of all.

译文：我认为她的设计是最好的。

例2：This is the best printer in this company, **I suppose**.

译文：我想这是这个公司最好的打印机。

11.5　Exercises

1. **Translate the following phrases.**

    computer animation
    animation sequence
    rendering algorithm
    key frame
    vector graphics
    synchronized sound
    navigation control
    video compression
    video-capture board
    lossy audio compression
    audio compression algorithm
    streaming media

2. **Fill in the blanks with appropriate words or phrases.**

    a. realistic displays　　b. studies　　　c. participant　　　　d. advertising applications
    e. animation　　　　　f. key frame　　g. intermediate frames　h. animation tools

    (1) An object definition is given for each _____ in the action.
    (2) Micromedia's Flash has become one of the premier Web _____ and formats in the short time that it is been on the scene.
    (3) Entertainment and _____ are sometimes more interested in visual effects.
    (4) Many applications of computer animation require _____.
    (5) In-betweens are the _____ between the key frames.
    (6) On the Web even the very simple _____ has a long download time.
    (7) In some scientific and engineering _____, realism is not a goal.
    (8) A _____ is a detailed drawing of the scene at a certain time in the animation sequence.

3. **Match the following terms to the appropriate definition.**

    a. ISP　　　　　　　b. IP　　　　　　　c. Intranet
    d. interconnect　　　e. infrared　　　　f. Java applet

    (1) It can run in a web browser using a Java virtual machine (JVM), or in Sun's AppletViewer, a stand alone tool to test applets.
    (2) It is electromagnetic radiation of a wavelength longer that visible light, but shorter than microwave radiation. The name means "below red", red being the color of visible light of longest wavelength.
    (3) An instruction that provides access to the Internet in some form, usually for money.

(4) A private network inside a company or organization that uses the same kinds of software that you would find on the public Internet, but that is only for internal use.

(5) The computer network protocol that all machines on the Internet must know so that they can communicate with one another.

(6) Two or more cable systems distributing a programming or commercial signal simultaneously.

Unit 12　Big Data

12.1　Text

音频

　　Big data usually refers to data in the petabyte and exabyte range—in other words, billions to trillions of records, all from different sources. Big data are produced in much larger quantities and much more rapidly than traditional data. These data may be unstructured or semi-structured and thus not suitable for relational database products that organize data in the form of columns and rows. The popular term "big data" refers to this avalanche of digital data flowing into firms around the world largely from Web sites and Internet click stream data. The volumes of data are so large that traditional DBMS cannot capture, store, and analyze the data in a reasonable time. Even though "tweets" are limited to 140 characters each. Twitter generates more than 8 terabytes of data daily. According to the IDC technology research firm, data is more than doubling every two years, so the amount of data available to organizations is skyrocketing[1]. Making sense out of it quickly in order to gain a market advantage is critical.

　　Up until about five years ago, most data collected by organizations consisted of structured transaction data that could easily fit into rows and columns of relational database management systems[2]. Since then, there has been an explosion of data from Web traffic, e-mail messages, and social media content (tweets, status messages), even music playlists, as well as machine-generated data from sensors.

　　According to a recent study by IBM, 90 percent of all the information ever created has been produced in the last two years. That is big data, much of it so large that it is beyond the capacity of conventional database management systems to process efficiently. It includes data produced by social media, e-mail, video, audio, text, and web pages. The data, in turn, comes from smart phones, digital cameras, global positioning systems (GPS), industrial sensors, social networks, and even public surveillance and traffic monitoring systems, among other sources. Every time you use a cell phone, do a Google or Baidu search, use your credit rating, or buy something from Amazon, you are expanding your digital footprint and creating more data.

　　An Internet companies such as Google, Baidu, and Amazon have sought ways to improve their network planning, advertising placement, and customer care, they have created innovative new applications of big data analysis. Today, the biggest generators of data are Twitter, Facebook, LinkedIn, and other emerging social networking services.

　　Businesses are interested in big data because they contain more patterns and interesting anomalies than smaller data sets, with the potential to provide new insight into customer behavior, weather patterns, financial market activity, or other phenomena[3]. However, to derive business value from these data, organizations need new technologies and tools capable of managing and analyzing nontraditional data along with their traditional enterprise data.

Big data analysis requires a cloud-based solution—a networked approach capable of keeping up with the accelerating volume and speed of data transmission. The enormous growth of data has coincided with the development of the "cloud", which is really just a new alias for the global Internet. As data proliferates, especially with the growth of mobile data applications using Apple and Android smart phones and operating systems, the requirements for data transmission and storage also grow. No single site server configuration offers sufficient capacity for the enormous amounts of data derived from sites such as Microsoft.com or FedEx.com, which serve global communities of customers, with transactions numbering in the millions. Users want fast response times and immediate results, especially in the largest and most complex web-based business environments, operating at Internet speed.

The largest companies making the most sophisticated industrial equipment are now able to go from merely detecting equipment failures to predicting them. This allows the equipment to be replaced before a serious problem develops.

Large Internet commerce companies such as Amazon and Taobao use big data analytics to predict buyer activity as well as to understand warehousing requirements and geographic positioning.

Government medical agencies and medical scientists use big data for early discovery and tracking of potential epidemics. A sudden increase in emergency room visits, or even increased sales of certain over-the-counter drugs, can be early warnings of communicative disease, allowing doctors and emergency response officials to activate control and containment procedures.

Many big data applications were created to help Web companies struggling with unexpected volumes of data. Only the biggest companies had the development capacity and budgets for big data. But the continuing decline in the costs of data storage, bandwidth, and computing means that big data is quickly becoming a useful and affordable tool for medium-sized and even some small companies. Big data analytics are already becoming the basis for many new, highly specialized business models.

Just as major corporations such as General Electric, Walmart and Google use data analytics today, small companies will use the same processes on a smaller scale to improve their competitive positions and use scarce resources more effectively.

Cloud services allow individuals and businesses to use software and hardware that are managed by third parties at remote locations. Examples of cloud services include online file storage, social networking sites, webmail, and online business applications. The cloud computing model allows access to information and computer resources from anywhere that a network connection is available. Cloud computing provides a shared pool of resources, including data storage space, networks, computer processing power, and specialized corporate and user applications.

The characteristics of cloud computing include on-demand self service, broad network access, resource pooling, rapid elasticity and measured service[4]. On-demand self service means that customers (usually organizations) can request and manage their own computing resources. Broad network access allows services to be offered over the Internet or private networks. Pooled resources mean that customers draw from a pool of computing resources, usually in remote data centers. Services can be scaled larger or smaller; and the use of a service is measured and customers are billed accordingly.

The biggest advantages of cloud computing include the ability to access data from anywhere the user has access to an active Internet connection and, since data is stored online instead of on the device being used, the data is safe if the device is lost, stolen, or damaged[5]. In addition, Web-based applications are often less expensive than installed software. Disadvantages of cloud computing include a possible reduction in performance of applications if they run more slowly via the cloud than they would run if installed locally, and the potentially high expense related to data transfer for companies and individuals using high-bandwidth applications. There are also security concerns about how safe the stored online data is from unauthorized access and data loss. Despite the potential risks, many believe that cloud computing is the wave of the future.

Cloud computing allows companies to outsource, or contract to third-party providers, elements of their information technology infrastructure. They pay only for the computing power, storage, bandwidth, and access to applications that they actually use. As a result, companies need not make large investments in equipment, or the staff to support it.

Keywords

affordable	adj.	付得起的	nontraditional	adj.	非传统的
anomalies	n.	异常，反常	outsource	v.	（将业务、工程等）外包
avalanche	n.	雪崩，崩塌	proliferate	v.	扩散，激增
coincided	v.	相符，极为类似	scarce	adj. adv.	缺乏的，仅仅，几乎不
elasticity	n.	灵活性，伸缩性	semi-structured	adj.	半结构化的
epidemics	n.	流行病，蔓延	skyrocketing	v.	突升，猛涨
explosion	n.	扩张，爆发	sophisticated	adj.	复杂的，精致的
footprint	n.	脚印，占用的空间	surveillance	n.	监督，监视
infrastructure	n.	基础设施，基础建设	Twitter	n.	推特

Notes

[1] According to the IDC technology research firm, data is more than doubling every two years, so the amount of data available to organizations is skyrocketing.

译文：根据 IDC 技术研究公司的调研，数据每两年都会增长一倍以上，因此对组织而言可用的数据量不断激增。

说明：本句的"According to the…"作条件状语，"so the amount of…"是结果状语。

[2] Up until about five years ago, most data collected by organizations consisted of structured transaction data that could easily fit into rows and columns of relational database management systems.

译文：在大约五年以前，各种企业收集的大多数数据都是结构化的事务处理数据。这些数据可以简单地填充到关系数据库管理系统的行和列当中。

说明：本句的"Up until about five years ago"是时间状语从句，"that could easily…"是定语从句。

[3] Businesses are interested in big data because they contain more patterns and interesting anomalies than smaller data sets, with the potential to provide new insight into customer behavior, weather

patterns, financial market activity, or other phenomena.

译文：企业对"大数据"感兴趣，是因为它们与更小的数据集相比包含了更多的模式与有意思的特殊情况，"大数据"具有能够提供新的针对顾客行为、天气模式、金融市场活动或其他现象的洞察力的潜力。

说明：本句的主句很简单，由"because"引导原因状语从句，"with the potential to…"是补语。

[4] The characteristics of cloud computing include on-demand self service, broad network access, resource pooling, rapid elasticity and measured service.

译文：云计算的特征包括按需自助服务、宽带网络、资源缓冲池、快速的伸缩性及可测量的服务。

说明：本句的"on-demand self service, broad network access, resource pooling, rapid elasticity and measured service" 4个部分是并列，共同作宾语。

[5] The biggest advantages of cloud computing include the ability to access data from anywhere the user has access to an active Internet connection and, since data is stored online instead of on the device being used, the data is safe if the device is lost, stolen, or damaged.

译文：云计算的最大优势主要体现在，由于数据是在线存储而不是本地存储，用户可以在任何互联网覆盖的地区获取数据，并且当设备丢失、被盗或损坏时，数据依然安全。

说明：本句的"to access data…"作定语，修饰"ability"，"since data…"引导原因状语从句，"if the device is lost, stolen, or damaged"作条件状语。

12.2 Reading Material 1: Using Instant Messaging for Business

Instant Messaging is a form of communication over the Internet that offers an instantaneous transmission of text-based messages from sender to receiver. In push mode between two or more people using personal computers or other devices, along with shared clients, instant messaging basically offers real-time direct written language-based online chat. The user's text is conveyed over a network, such as the Internet. It may address point-to-point communications as well as multicast communications from one sender to many receivers. More advanced instant messaging allows enhanced modes of communication, such as live voice or video calling, video chat and inclusion of hyperlinks to media.

Instant messaging has become a popular way to communicate with clients and coworkers, and for good reason. Communicating over the Internet in real time can streamline communications and save you time and money. Plus, since most Instant Messaging software is free, you don't need to invest more money in technology.

But before you rush into using instant messaging, think about whether it will really improve your business communications or just serve as another distraction. First, consider reasons to use an instant messenger.

- Cost savings for long distance telephone calls. Using Instant Messaging to communicate with employees and clients in other parts of the world can reduce your long-distance phone bills.
- Conduct real-time interaction. When you work on a project with coworkers who have busy schedules or work in different offices, Instant Messaging facilitates quick and simple communication.
- Reduce spam. Instant messengers get a lot fewer unwanted messages than e-mail.
- Host chats and conferences. Public chat rooms are often chaotic, and it is difficult to conduct a

focused conversation especially for business purposes. Although instant messengers are most commonly used for two-way conversation, most programs offer a conference or chat setting where your workgroup can meet.

Now that you have considered the advantages of instant messaging, take a look at the disadvantages. Like most technologies, Instant Messaging does have some drawbacks.

- Proprietary software. While it is now possible to use software that lets you connect to multiple Instant Messaging services, most services only let you sent messages to other users or add them to your contact list using the same instant messenger you do.
- Word limits. Most messaging programs have a character limit for messages. If you need to communicate a large amount of information, an e-mail or phone call is a better choice.
- Security. Use caution with the information you share over chat or instant messengers. Take a look at your profile to see what information you have made public, and find out if the software creates public directories of its users.
- Misuse. Employees may be tempted to communicate with their friends during work hours because Instant Messaging is easy and funny.

Keywords

chaotic	adj.	混乱的	inclusion	n.	包含，含义，掺杂
chat	v.	聊天	instantaneous	adj.	及时的，瞬间的
convey	v.	传达，表达	instant messaging		即时通信
coworker	n.	同事	mandate	v.	授权
directory	n.	目录	misuse	v.	滥用
distraction	n.	心烦，分心	proprietary	adj.	专有的，独占的，专利的
drawback	n.	弊端	spam	n.	电子垃圾
facilitate	v.	促进	streamline	v.	精简
hyperlink	n.	超链接			

12.3 Reading Material 2：Applications of Big Data and IoT in the Health Domain

With the advancement of IoT (Internet of Things) technologies, personalized healthcare is no longer a luxury idea for the general public.

Personalized health capability is limited to the available data from the patient, which is usually dynamic and incomplete. Therefore, it is presenting a critical issue for knowledge mining, analysis and trending. IoT is the result of the efforts to provide connectivity and intelligence to convert small devices and common things into Smart Objects. These Smart Objects present high capabilities to integrate and transfer enriched data from environmental sensors, parking, activities, behaviours and clinical devices from mobile health and ambient assisted living (AAL) environments.

The huge and enriched data is what defines the big data. Big data brings a lot of opportunities for a wide range of application areas. Particularly, we focus on the personalized healthcare, which enables patients to monitor their own environments, i.e. the deployment of custom remote monitoring (remote assistance) and mobile health solutions. This offers the capability to send alerts, predict possible anomalies in real time, and transfer the collected data to an information system, in order to allow the subsequent study of patients and even, thanks to the large amount of data collected may help future researchers improve treatments for diseases or finding new cures. These systems and solutions require more intelligent physiological sensors.

A variety of technologies are making it feasible to identify sense, locate, and connect all the people, machines, devices and things surrounding us. These new capabilities for linking Internet with everyday sensors and devices, forms of communication among people and things, and exploitation of data capture, define an extension of the usual Intranet of Things to a more Internet of Things.

IoT technologies are changing the way we live. For example, it is well known that the quality of sleep has a great impact on daily life and on the performance one can obtain at work, in sports, etc. Daytime somnolence can be the cause of serious incidents at work or while driving. Further, many researchers have proven how sleep deprivation or a bad quality of sleep for prolonged periods could be related to the rise of hypertension, cardiovascular pathologies, obesity, diabetes and to a decrease in the efficiency of the immune system.

For all those reasons, an indication on the sleep quality may really constitute a good parameter for prevention also in healthy subjects. However, the usual tool for a complete sleep evaluation is the polysomnography that requires a well-equipped laboratory in a sleep clinic and specialized medical personnel. Thus only people with well-defined symptoms are assigned to such an evaluation, which is not indicated for prevention because of its costs, the need of at least one night in the clinic, and the limited number of sleep centers.

Keywords

ambient assisted living		环境辅助生活	
anomaly	n.	不规则，异常	
cardiovascular	adj.	心血管性的	
clinical	adj.	临床的，诊断的	
deprivation	n.	丧失，剥夺，废止	
diabetes	n.	糖尿病	
enriched	adj.	丰富的	
exploitation	n.	利用，宣传，开拓	
feasible	adj.	行得通的，适宜的，可用的	
healthcare	n.	卫生保健	
hypertension	n.	高血压，过度紧张	
immune	adj.	免疫的，有免疫力的	
obesity	n.	肥胖症，肥胖	
pathology	n.	病理学，病理，病状	
polysomnography	n.	多导睡眠图	
prolong	v.	延长，延期	
somnolence	n.	困倦，嗜睡状态，嗜睡症	
symptom	n.	症状，征兆	

12.4 时态简介

1. 一般现在时

一般现在时用于如下情况。

- 经常性或习惯性的动作，常与表示频度的时间状语连用。时间状语有 every…、sometimes、at…、on Sunday 等。
- 表示客观真理、客观存在或科学事实。
- 表示格言或警句，此用法如果出现在宾语从句中，即使主句是过去时，从句也要用一般现在时。
- 现在时刻的状态、能力、性格和个性等。

例 1：I go to school every day.

译文：我每天去上学。

例 2：Pride goes before a fall.

译文：骄者必败。

例 3：The adult mosquito usually lives for about thirty days, although the life span varies widely with temperature, humidity, and other factors of the environment.

译文：成年蚊子的寿命约 30 天，不过其寿命的长短还随温度、湿度和其他环境因素而变化很大。

2．一般过去时

一般过去时用于如下情况。

- 在确定的过去时间里所发生的动作或存在的状态。时间状语有 yesterday、last week、an hour ago、the other day、in 1999、before 等。
- 表示在过去一段时间里，经常性或习惯性的动作。
- wish、wonder、think、hope 等用过去时，作试探性的询问、请求、建议等。
- 用过去时表示现在，表示委婉语气。动词有 want、hope、wonder、think、intend，情态动词有 could、would。

例 1：How many subjects did you study last term?

译文：上学期你学了多少门课程？

例 2：I used to enjoy gardening, but I don't like it any more.

译文：以前我很喜欢园艺，现在一点儿也不喜欢。

3．一般将来时

一般将来时的结构为 will（shall）+动词原形，可用于如下情况。

- 一般将来时的时间状语有 shall 和 will。shall 用于第一人称，常被 will 所代替，will 在陈述句中用于各人称，在征求意见时常用于第二人称。
- be going to + 不定式，表示将来。
- be + 不定式表示将来，按计划或正式安排将发生的事。
- be about to + 不定式，意为马上做某事。但要注意，be about to 不能与 tomorrow、next week 等表示明确将来时的时间状语连用。

例 1：I am going to travel around the world.

译文：我计划周游世界。

例 2：Someday people will go to the moon.

译文：总有一天，人们会到月球上去的。

4．现在进行时

现在进行时的结构为 be + 现在分词，可用于如下情况。

- 表示现在（指说话人说话时）正在发生的事情。
- 习惯进行：表示长期的或重复性的动作，说话时动作未必正在进行。
- 表示渐变的动词有 get、grow、become、turn、run、go、begin 等。
- 与 always、constantly、forever 等词连用，表示反复发生的动作或持续存在的状态，往往带有说话人的主观色彩。

例1：I am speaking with him on the phone.

译文：我正在和他通电话。

例2：What are you studying?

译文：你在学什么？

5．过去进行时

过去进行时的结构为 were（was）+现在分词，可用于如下情况。

- 过去进行时的主要用法是描述一件事发生的背景。常用的时间状语有 this morning、the whole morning、all day yesterday、from nine to ten last evening、when 及 while 等。
- 一个长动作发生的时候，另一个短动作发生。

例1：What were you doing at nine last night?

译文：昨晚9点时你在做什么？

例2：Some students were playing football, while others were running round the track.

译文：一些学生在踢足球，同时别的学生正在跑道上跑步。

6．将来进行时

将来进行时的结构为 will/be +现在分词，可用于如下情况。

- 表示将来某时进行的状态或动作，常用的时间状语有 soon、tomorrow、on Sunday、by this time、this evening、in two days 及 tomorrow evening 等。
- 预测将来会发生的事情。

例：He will be playing football soon.

译文：他一会儿将去踢足球。

7．现在完成时

现在完成时的构成为 have（has）+ 过去分词，可用于如下情况。

- 用来表示之前已发生或完成的动作或状态，其结果的确和现在有联系。常用的时间状语有 for、since、so far、ever、never、just、yet、till/until、up to now、in past years 及 always 等。
- 动作或状态发生在过去但它的影响现在还存在，强调过去的事情对现在的影响。
- 也可表示持续到现在的动作或状态，动词一般是延续性的，如 live、teach、learn、work、study 及 know 等。

例1：I have visited your school before.

译文：我以前曾经去过你们学校。

例2：I have finished my homework now.

译文：现在我已经做完作业了。

8．过去完成时

过去完成时的构成形式是 had + 过去分词，可用于如下情况。

- 表示过去的过去，常用时间状语有 before、by、until、when、after、once 和 as soon as 等。
- 在 told、said、knew、heard、thought 等动词后的宾语从句。

- 状语从句：在过去不同时间发生的两个动作中，发生在先，用过去完成时；发生在后，用一般过去时。
- 表示意向的动词，如 hope、wish、expect、think、intend、mean 及 suppose 等，用过去完成时表示原本……，未能……。

例 1：He said that he had never been to Paris.

译文：他说他从未去过巴黎。

例 2：I did not know what he had done it for?

译文：我以前不知道他做这个究竟为了什么？

9. 将来完成时

将来完成时的结构是 will have + 过去分词，可用于如下情况。
- 表示某事继续到将来某一时为止一直有的状态。
- 表示将来某一时或另一个将来的动作之前，已经完成的动作或获得的经验。

例：Mary has been to Sumatra and Iran as well as all of Europe. By the time she is twenty, she will have been almost everywhere.

译文：玛丽到过苏门答腊、伊朗及整个欧洲。到她 20 岁的时候她将几乎走遍世界各地。

12.5 Exercises

1. **Translate the following phrases.**

big data

social media

industrial sensors

traffic monitoring systems

traditional enterprise data

data transmission

Apple and Android smart phones

buyer activity

competitive positions

cloud services

online file storage

online business applications

2. **Fill in the blanks with appropriate words or phrases.**

   a. data   b. analytics   c. generators   d. smaller
   e. large   f. predicting   g. potential   h. larger

(1) Big data _____ are already becoming the basis for many new, highly specialized business models.

(2) Today, the biggest _____ of data are Twitter, Facebook, LinkedIn, and other emerging social networking services.

(3) The largest companies making the most sophisticated industrial equipment are now able to go from

merely detecting equipment failures to _____ them.

(4) Government medical agencies and medical scientists use big data for early discovery and tracking of _____ epidemics.

(5) Many big data applications were created to help Web companies struggling with unexpected volumes of _____.

(6) Services can be scaled larger or _____; and the use of a service is measured and customers are billed accordingly.

(7) Big data are produced in much _____ quantities and much more rapidly than traditional data.

(8) That is big data, much of it so _____ that it is beyond the capacity of conventional database management systems to process efficiently.

3. Match the following terms to the appropriate definition.

   a. encryption   b. EDI   c. Ethernet
   d. end user   e. FTP   f. firewall

(1) The most popular LAN technology in use today. It is a 10 Mbps, CSMA/CD baseband network that runs over thin coax, thick coax, twisted pair or fiber optic cable.

(2) A person, organization, or telecommunications system that accesses the network in order to communicate via the services provided by the network.

(3) This program will let you transfer files from your computer onto another computer server.

(4) The translation of data into a secret code. It is the most effective way to achieve data security.

(5) It can be used for standardized and repeated transactions that do not fall within the usual definition of trade exchanges. It can be used for pre-sales and after-sales transactions.

(6) A system designed to prevent unauthorized access to or from a private network.

Unit 13 Cloud Computing

13.1 Text

Cloud Computing Basics

The term cloud computing refers to computing in which tasks are performed by a "cloud" of servers, typically via the Internet. This type of network has been used for several years to create the supercomputer-level power needed for research and other power-hungry applications, but it was more typically referred to as grid computing in this context. Today, cloud computing typically refers to accessing Web-based applications and data using a personal computer, mobile phone, or any other Internet-enabled device. The concept of cloud computing is that apps and data are available any time, from anywhere, and on any device. For example, you use cloud computing capabilities when you store or access documents, phones, videos, and other media online; use programs and apps online (i.e., email, productivity, games, etc.); and share ideas, opinions, and content with others online (i.e., social networking sites).

The name cloud computing was inspired by the cloud symbol that is often used to represent the Internet in flow charts and diagrams. Clouds can be classified as public, private or hybrid. Businesses often use applications available in the public cloud; they also frequently create a private cloud just for data and applications belonging to their company.

Cloud computing is a marketing term for technologies that provide computation, software, data access, and storage services that do not require end-user knowledge of the physical location and configuration of the system that delivers the services[1]. A parallel to this concept can be drawn with the electricity grid, wherein end-users consume power without needing to understand the component devices or infrastructure required to provide the service.

Cloud computing providers deliver applications via the Internet, which are accessed from Web browsers, desktop and mobile apps, while the business software and data are stored on servers at a remote location. In some cases, legacy applications (line of business applications that until now have been prevalent in thin client Windows computing) are delivered via a screen-sharing technology, while the computing resources are consolidated at a remote data center location; in other cases, entire business applications have been coded using Web-based technologies such as AJAX.

Cloud computing relies on sharing of resources to achieve coherence and economies of scale, similar to a utility (like the electricity grid) over a network. At the foundation of cloud computing is the broader concept of converged infrastructure and shared services.

Cloud computing also focuses on maximizing the effectiveness of the shared resources. Cloud resources are usually not only shared by multiple users but are also dynamically reallocated per demand. This can work for allocating resources to users. For example, a cloud computer facility that servers European users during European business hours with a specific application (e.g. email) may reallocate the same resources to serve North American users during North America's business hours with a different application (e.g., a Web server). This approach should maximize the use of computing power, thus reducing environmental damage as well since less power, air conditioning, rack space, etc. are required for a variety of functions. With cloud computing, multiple users can access a single server to retrieve and update their data without purchasing licenses for different applications.

Then How Does Cloud Computing Work?

Let's say you are an executive at a large corporation. Your particular responsibilities include making sure that all of your employees have the right hardware and software they need to do their jobs[2]. Buying computers for everyone is not enough—you also have to purchase software or software licenses to give employees the tools they require. Whenever you have a new hire, you have to buy more software or make sure your current software license allows another user. It is so stressful that you find it difficult to go to sleep on your huge pile of money every night[3].

Soon, there may be an alternative for executives like you. Instead of installing a suite of software for each computer, you'd only have to load one application. That application would allow workers to log into a Web-based service which hosts all the programs the user would need for his or her job. Remote machines owned by another company would run everything from e-mail to word processing to complex data analysis programs and so on. It's called cloud computing, and it could change the entire computer industry.

In a cloud computing system, there is a significant workload shift. Local computers no longer have to do all the heavy listing when it comes to running applications[4]. The network of computers that make up the cloud handles them instead. Hardware and software demands on the user's side decrease. The only thing the user's computer needs to be able to run is the interface software of cloud computing system, which can be as simple as a Web browser, and the cloud's network takes care of the rest.

There's good chance you have already used some form of cloud computing. If you have an e-mail account with a Web-based e-mail service like Hotmail, Yahoo! Mail or Gmail, then you have had some experience with cloud computing. Instead of running an e-mail program on your computer, you log into a Web e-mail account remotely. The software and storage for your account doesn't exist on your computer—it's on the service's computer cloud.

Services of Cloud Computing

Cloud computing providers offer their services according to three fundamental models: infrastructure as a service (IaaS), platform as a service (PaaS) and software as a service (SaaS).

Infrastructure as a service like Amazon Web Services provides virtual server instances with unique IP addresses and blocks of storage on demand. Customers use the provider's application program interface (API) to start, stop, access and configure their virtual servers and storage. In the enterprise, cloud computing allows a company to pay for only as much capacity as is needed, and bring more online as soon as required. Because this pay-for-what-you-use model resembles the way electricity, fuel and water are consumed; it is sometimes referred to as utility computing.[5]

Platform as a service in the cloud is defined as a set of software and product development tools hosted on the provider's infrastructure. Developers create applications on the provider's platform over the Internet. PaaS providers may use APIs, website portals or gateway software installed on the customer's computer. GoogleApps are examples of PaaS. Some providers will not allow software created by their customers to be moved off the provider's platform.

In the Software as a Service cloud model, the vendor supplies the hardware infrastructure, the software product and interacts with the user through a front-end portal. SaaS is a very broad market. Services can be anything from Web-based email to inventory control and database processing. Because the service provider hosts both the application and the data, the end user is free to use the service from anywhere.

Keywords

coherence	n.	统一，一致性，凝聚	mobile apps		移动应用程序
consolidate	v.	使坚固，巩固，合并，统一	pile	n. v.	软毛，绒毛，堆积，大量，大批
consume	v.	消费掉，用完，消磨	portal	n.	门户网站，入口，正门
dynamically	adv.	动态地	prevalent	adj.	流行的，盛行的，优势的，普遍的
electricity grid		电网			
hire	n. v.	租金，酬金，雇用	remotely	adv.	遥远地，边远地，边缘地
hybrid	adj. n.	混合的，混杂作用	resemble	v.	像，类似，比拟
inspire	v.	鼓舞，激发，注入	stressful	adj.	紧张的，压力重的
legacy	n.	遗赠，遗产	unique	adj.	唯一的，独特的
legacy application		遗留应用	wherein	adv. conj.	其中，在哪一点上
license	n.	许可证			

Notes

[1] Cloud computing is a marketing term for technologies that provide computation, software, data access, and storage services that do not require end-user knowledge of the physical location and configuration of the system that delivers the services.

译文：云计算是一种技术的营销用语，它提供计算、软件、数据访问、存储的服务而不需要为终端用户提供服务系统的物理位置和配置信息。

说明："that provide computation, software, data access, and storage services"作定语从句，修饰先行词"technologies"，"that do not require end-user knowledge of the physical location and configuration of the system"也是一个定语从句，修饰先行词"services"，最后"that delivers the services"也作定语，修饰"system"。

[2] Your particular responsibilities include making sure that all of your employees have the right hardware and software they need to do their jobs.

译文：你的特定的职责包括确保所有员工有合适的、可供他们做好自己工作所需的硬件和软件。

说明："they need to do their jobs"是一个限定性定语从句，从句的主语"they"之前省略了关系代词"which"。

[3] It is so stressful that you find it difficult to go to sleep on your huge pile of money every night.

译文：如果你要面对这样一大笔开销，你会有非常大的压力，每天晚上难以入睡。

说明：本句由"so…that…"引导一个结果状语从句。

[4] Local computers no longer have to do all the heavy listing when it comes to running applications.

译文：本地计算机在运行应用程序的时候不再需要去完成所有繁重的任务。

说明：本句由"when"引导一个时间状语从句。

[5] Because this pay-for-what-you-use model resembles the way electricity, fuel and water are consumed; it is sometimes referred to as utility computing.

译文：因为这种"支付你所用的"的模式很像消费电、燃料和水的方式；它有时称为公共服务设施计算。

说明：本句中的"pay-for-what-you-use"可以译为"支付你所用的"，"Because this……"是原因状语从句。

13.2　Reading Material 1：Cloud Computing and Cloud Storage

1. Cloud Computing

Cloud computing is computing in which large groups of remote servers are networked to allow centralized data storage and online access to computer services. Cloud computing is the delivery of computing services over the Internet.

The cloud computing service models are software as a service (SaaS), platform as a service (PaaS) and infrastructure as a service (IaaS). In a software as a service model, a pre-made application, along with any required software, operating system, hardware, and network are provided. In PaaS, an operating system, hardware, and network are provided, and the customer installs or develops its own software and applications. The IaaS model provides just the hardware and network; the customer installs or develops its own operating system, software and applications.

A cloud service has three distinct characteristics that differentiate it from traditional hosting. It is sold on demand, typically by the minute or the hour; it is elastic—a user can have as much or as little of a service as they want at any given time; and the service is fully managed by the provider (the consumer needs nothing but a personal computer and Internet access). Significant innovations in virtualization and distributed computing, as well as improved access to high-speed Internet and a weak economy, have led to a growth in cloud computing. Cloud vendors are experiencing growth rates of 50% per annum.

The goal of cloud computing is to provide easy, scalable access to computing resources and IT services.

Cloud services are typically made available via a private cloud, community cloud, public cloud or hybrid cloud.

Generally speaking, services provided by a public cloud are offered over the Internet and are owned and operated by a cloud provider. Some examples include services aimed at the general public, such as online photo storage services, e-mail services, or social networking sites. However, services for enterprises can also be offered in a public cloud.

In a private cloud, the cloud infrastructure is operated solely for a specific organization, and is managed by the organization or a third party.

In a community cloud, the service is shared by several organizations and made available only to those groups. The infrastructure may be owned and operated by the organizations or by a cloud service provider.

A hybrid cloud is a combination of different methods of resource pooling (for example, combining public and community clouds).

2. Cloud Storage

Cloud storage is a model of data storage where the digital data is stored in logical pools, the physical storage spans multiple servers (and often locations), and the physical environment is typically owned and managed by a hosting company. These cloud storage providers are responsible for keeping the data available and accessible, and the physical environment protected and running. People and organizations buy or lease storage capacity from the providers to store user, organization, or application data (Fig. 13-1).

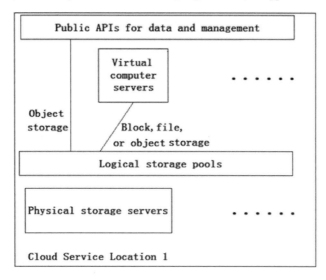

Fig. 13-1 Cloud storage architecture

Cloud storage services may be accessed through a co-located cloud computer service, a Web service application programming interface (API) or by applications that utilize the API, such as cloud desktop storage, a cloud storage gateway or Web-based content management systems.

Cloud storage is based on highly virtualized infrastructure and is like broader cloud computing in terms of accessible interfaces, near-instant elasticity and scalability, multi-tenancy, and metered resources. Cloud storage services can be utilized from an off-premises service or deployed on-premises.

Keywords

annum	n.	年	premise	n.	前提，根据，（企业的）房屋建筑及附属场地
community cloud		社区云			
differentiate	v.	区分，区别，划分	scalability	n.	可量测性，可扩展性
elasticity	n.	弹力，弹性，灵活性	solely	adv.	独自，单独
innovation	n.	革新，创新	span	n.	很小的间隔，片刻
lease	v.	出租，租借，租约	vendor	n.	自动售货机
multi-tenancy	n.	多租户	virtualization	n.	虚拟化

13.3 Reading Material 2: More about the Cloud Computing

In the most basic terms, the cloud refers to the Internet. Lots of engineers might yell at us for being that basic, but when someone says, "I stored it in the cloud," they mean they stored it on an Internet service. For example, e-mail or Webmail services are in "the cloud".

Now, of course, it isn't really that simple. The cloud or cloud computing refers to an application that is hosted on or run on Internet services. All the companies that have these services—Google, Facebook, Yahoo, Flickr, Apple—have servers or server farms. That is where your media is actually stored. Apple even built a huge server farm in North Carolina for its iCloud service. But, hey, if you want to think of it as a floating bubble or hard drive in the sky, we won't tell anyone.

There are a lot of advantages to using Cloud-based services or Cloud computing. The first is that your media lives in a place not on your own device, so you can get it on multiple devices. You can check your e-mail on any computer or access your photos on Flickr or iCloud on any computer. And with cloud storage services like Google Drive, Dropbox, or SugarSync, you can store your files.

That also means that those photos, files or songs are being backed up on those various services. So if your hard drive crashed, you'd have a backup of all your files instantaneously. You'd just log in on another computer to one of those services.

Another benefit? Less computing power is required. The games you play on Facebook are using some of your computer's power, but they are also running on Facebook's or Zynga's (the company that makes most of those games) servers. Similarly, because you might listen to more music through Cloud services like Pandora or Spotify, you might not need as much hard-drive space.

But there are, of course, some things to be mindful of when using these Cloud services. The first is bandwidth and Internet costs. Sending e-mail or looking at photos through these services don't use a lot of bandwidth, but playing games and streaming movies can eat up your monthly data or bandwidth allocation. Make sure you don't go over your monthly allocation, if you have one.

The second is that you will need that Internet connection to get to your data. So if you are on a plane with no WiFi or in an area with no connectivity you won't have access to those files or photos in the cloud.

Third: be mindful or privacy settings and make sure you are using strong passwords with these services. Your data now lives someplace other than the hard drive in your computer—it lives in that fluffy blue cloud—so you just want to be safe in case it decides to turn gray and dark.

Keywords

backed up		备份	mindful	adj.	注意，注意……的，
bubble	n.	泡沫，幻想，妄想			留心……的
float	v.	漂浮	privacy setting		隐私设置
fluffy	adj.	错乱的，糊涂的	yell	v.	叫嚷
log in		登录			

13.4 分词和动名词

1. 分词

分词是非谓语动词的一种。分词有现在分词和过去分词两种。规则动词的现在分词由动词原形加-ing 构成，过去分词由动词原形加-ed 构成；不规则动词的分词形式，其构成是不规则的。分词没有人称和数的变化，具有形容词和副词的作用，同时还保留着动词的特征，只是在句中不能独立作谓语。

现在分词所表示的动作具有主动的意义，而及物动词的过去分词表示的动作具有被动的意义。现在分词与过去分词在时间关系上，前者表示动作正在进行，后者表示的动作往往已经完成。现在分词表示的动作与谓语动词表示的动作相比，具有同时性，而过去分词则具有先时性。分词在各种时态、语态下的表现形式如下表。

时态	与主语动词同时	doing
	先于主语动词	having done
语态	表示主动	现在分词 doing
	表示被动	过去分词 done
	表示动作已经发生	不及物动词的过去分词

分词在句子中具有形容词词性和副词词性，可以充当句子的定语、表语、状语和补足语。下面分别举例说明现在分词和过去分词在句子中的作用。

（1）现在分词的用法。

♦ 作定语。

例 1：They insisted upon their device being tested under operating conditions.

译文：他们坚持他们的装置要在运转条件下检测。

例 2：An atom contains small particles carrying two kinds of electricity.

译文：原子含有带两种电荷的粒子。

♦ 作表语。

例：The result of the experiment was encouraging.

译文：实验结果令人鼓舞。

♦ 作补足语。

例：You'd better start the computer.

译文：你最好打开计算机。

♦ 作状语。

例：While making an experiment on an electric circuit, they learned of an important electricity law.

译文：他们在做电路实验时，学到了一条重要的电学定律。

（2）过去分词的用法。

♦ 作定语。

例 1：The heat energy produced is equal to the electrical energy utilized.

译文：产生的热量与所用的电能相等。

例 2：The charged capacitor behaves as a secondary battery.

译文：充了电的电容就像一个蓄电池一样。

♦ 作表语。

例：Some substances remain practically unchanged when heated.

译文：有几种物质受热时几乎没有变化。

♦ 作补足语。

例：I don't know if we can get the computer repaired in time.

译文：我不知道我们能否按时修好计算机。

♦ 作状语。

例：Given the voltage and current, we can determine the resistance.

译文：已知电压和电流，我们就可以求出电阻。

2．动名词

动名词是一种非谓语动词，由动词原形加词尾-ing 构成，形式上和现在分词相同。由于动名词和现在分词的形成历史、意义和作用都不一样，通常把它们看作是两种不同的非谓语动词。它没有人称和数的变化。动名词具有动词词性和名词词性，因而又可以把它称为"动词化的名词"和"名词化的动词"。在句中充当主语、表语、定语和宾语等，动名词也可以有自己的宾语和状语，构成动名词短语。动名词在各种时态、语态下的形式如下表所示。

时态	语态	
	主动	被动
一般式	doing	being done
完成式	having done	having been done

例 1：Excuse me for coming late.

译文：对不起，我来晚了。

句中，coming 是动名词，late 作 coming 的状语。

例 2：This dual quality of being sensitive to, and curious about, small accidental occurrences, and of possessing a frame of reference capable of suggesting their true significance, is probably what he meant when he said "Change benefits only the prepared mind".

译文：应该具有双重的素质，即对偶然发生的事情很好奇、敏感，而且拥有自己一套能揭示这些偶然事件真正重要性的学术参考体系，也许就是当他说"机会只垂青有备之士"时的意思吧。

下面分别举例说明动名词在句子中的作用。

（1）作主语。动名词作主语表示一件事或一个行为，其谓语动词用第三人称单数。

例 1：Learning computer science is very important now.

译文：现在学习计算机很重要。

例 2：Changing resistance is a method for controlling the flow of the current.

译文：改变电阻是控制电流流动的一种方法。

动名词作主语时，也可用 it 作形式主语，放在句首，而将真正的主语——动名词短语放在谓语之后。

例：It's no good using this kind of material.

译文：采用这类材料是毫无用处的。

（2）作宾语。动名词可以在一些及物动词和介词后作介词宾语。要求动名词作宾语的常用及物动词有 finish、enjoy、avoid、stop、need、start、mean 等。

例 1：This printer needs repairing.

译文：这台打印机需要修理一下。

例 2：I remember having repaired this machine.

译文：我记得曾经修过这部机器。

英语中，suggest、finish、avoid、stop、admit、keep、require、postpone、practice、fancy、deny 等动词都用动名词作宾语，不能用不定式作宾语。但是在 love、like、hate、begin、start、continue、remember、forget、regret 等词后面可以用动名词作宾语，也可以用动词不定式作宾语。

例：Do you like watching/to watch TV?

译文：你喜欢看电视吗？

动名词作宾语时，如本身带有补足语，则常用 it 作形式宾语。而将真正的宾语——动名词放在补足语的后面。

例：I found it useless arguing with her.

译文：我发现与她辩论没有用。

下例是动名词作介词的宾语。

例：Thank you for giving me so much help.

译文：谢谢你给我那么多帮助。

（3）作表语。动名词作表语为名词性表语。表示主语的内容，而不说明主语的性质。主语常为具有一定内涵的名词，这与不定式作表语相似。动名词作表语与进行时的区别在于主语能否执行该词的行为。能执行，即为进行时；否则，即为动名词作表语（系表结构）。

例 1：The function of a capacitor is storing electricity.

译文：电容器的功能是存储电能。

例 2：Seeing is believing.

译文：眼见为实。

句中，动名词 Seeing 作主语，believing 作表语。

（4）作定语。动名词作定语为名词性定语，说明名词的用途，与所修饰名词之间没有逻辑主谓关系，这是与现在分词作定语相区别的关键。动名词作定语只能使用单词，不可用动名词短语；只能放在所修饰名词前面，不可后置。

例 1：Rubber is found to be a good insulating material.

译文：橡胶是一种良好的绝缘材料。

例 2：English is one of the official languages at international meeting.

译文：英语是国际会议上使用的工作语言之一。

（5）作宾语补足语。动名词在句中的作用相当于名词，故可作宾语补足语。动名词只能在少数动词后作宾语补足语，补充说明宾语的性质、行为或状态，与宾语具有逻辑主谓关系。

例：We call this process testing.

译文：我们称这个过程为检测。

句中，动名词 testing 作宾语 this process 的补足语。

13.5 Exercises

1. **Translate the following phrases.**

   cloud computing

   grid computing

   electricity grid

   private cloud

   screen-sharing technology

   cloud resources

   gateway software

   utility computing

   remote machines

   data analysis programs

   front-end portal

   service provider

2. **Fill in the blanks with appropriate words or phrases.**

   a. cloud symbol   b. hybrid   c. system   d. form
   e. maximizing   f. Internet   g. Apps   h. licenses

   (1) The term cloud computing refers to computing in which tasks are performed by a "cloud" of servers, typically via the _____.

   (2) The name cloud computing was inspired by the _____ that is often used to represent the Internet in flow charts and diagrams.

   (3) Clouds can be classified as public, private or _____.

   (4) Cloud computing providers deliver applications via the Internet, which are accessed from Web browsers, desktop and mobile_____.

   (5) Cloud computing also focuses on _____ the effectiveness of the shared resources.

   (6) With cloud computing, multiple users can access a single server to retrieve and update their data without purchasing _____ for different applications.

   (7) In a cloud computing _____, there is a significant workload shift.

   (8) There's good chance you have already used some _____ of cloud computing.

3. **Match the following terms to the appropriate definition.**

   a. keyboard   b. laser   c. layer
   d. LAN   e. line speed   f. link

   (1) A device that generates coherent electromagnetic radiation in, or near, the visible part of the spectrum.

   (2) In networks, it refers to software protocol levels comprising the architecture, with each layer performing functions for the layers above it.

(3) It is a device that converts keystrokes into special codes that can be electronically manipulated by the computer. It may contain 70 to 105 keys.

(4) Expressed in bps, the maximum rate at which data can reliably be transmitted over a line using given hardware.

(5) A computer network limited to the immediate area, usually the same building or floor of a building.

(6) The text or graphic used in an HTML document to jump from one document to another. Typically underlined.

Unit 14 Computer Network Basics

音频

14.1 Text

If the user were to connect his computer to other computers, he could share data on other computers, including high-quality printers.[1] A group of computers and other devices connected together is called a network, and the technical concept of connected computers sharing resources is called networking. Computers that are part of a network can share data, messages, graphics, printers, fax machines, modems, CD-ROMs, hard disks, and other data storage equipment.[2]

Computer network can be used for numerous services, both for companies and for individuals. For companies, networks of personal computers using shared servers often provide flexibility and a good price/performance ratio. For individuals, networks offer access to a variety of information and entertainment resources.

Roughly speaking, networks can be divided up into LANs, MANs, WANs, and Internet works, each with their own characteristics, technologies, speeds, and niches. LANs cover a building, MANs cover a city, and WANs cover a country or continent. LANs and MANs are un-switched (i.e., do not have routers); WANs are switched.

Network software consists of protocols, or rules by which processes can communicate. Protocols can be either connectionless or connection-oriented. Most networks support protocol hierarchies, with each layer providing services to the lower layers. Protocol stacks are typically based either on the OSI model or the TCP/IP model. Both of these have network, transport, and application layers, but they differ on the other layers.

- ◆ Local Area Network (LAN)

A local area networks, or LAN, is a communication network that is privately owned and that covers a limited geographic area such as an office, a building, or a group of building.[3] The LAN consists of a communication channel that connects either a series of computer terminals together with a minicomputer or, more commonly, a group of personal computers to one another. Very sophisticated LANs can connect a variety of office devices such as word processing equipment, computer terminals, video equipment and personal computers.

Two common applications of local area networks are hardware resource sharing and information resource sharing. Hardware resource sharing allows each personal computer in the network to access and use devices that would be too expensive to provide for each user. Information resource sharing allows anyone using a personal computer on the local area network to access data stored on any other computer in the network. In actual practice, hardware resource sharing and information resource sharing are often combined.

- Wide Area Network (WAN)

A WAN is a group of LANs in different locations. They will be interconnected by VPN over the Internet or leased communication methods. A WAN is geographic in scope (as opposed to local) and uses telephone lines, microwaves, satellites, or a combination of communication channels. WANs are used to connect local area networks together, so that users and computers in one location can communicate with users and computers in other locations. Many WANs are built for one particular organization and are private. Others, built by Internet service providers (ISP), provide connections from an organization's LAN to the Internet.

WANs are most often built using leased lines. At each end of the leased line, a router connects to the LAN on one side and a hub within the WAN on the other. Leased lines can be very expensive. Instead of using leased lines, WANs can also be built using less costly circuit switching or packet switching methods. Network protocols including TCP (Transmission Control Protocol) deliver transport and addressing functions. Protocols including packet over SONET (synchronous optical network), MPLS (multi-protocol label switching), ATM (asynchronous transfer mode) and Frame Relay are often used by service providers to deliver the links that are used in WANs. X.25 was an important early WAN protocol, and is often considered to be the "grandfather" of Frame Relay as many of the underlying protocols and functions of X.25 are still in use today (with upgrades) by Frame Relay.

- Metropolitan Area Network (MAN)

The definition of a metropolitan area network cannot be precise. It falls between a LAN and a WAN and has some of the features of each. A LAN operates over a restricted geographical are and almost always operates within one organization. A number of LAN technologies have evolved offering very high performance at low cost over distances of under 10km. The WAN on the other hand uses technologies that do not limit the geographical reach and expects to connect nodes under different management domains.[4] A MAN aims to serve a geographic area beyond the scope of LAN technologies, yet is restricted by some well-defined community of interest, often a city. A MAN will often provide interconnection between sites of the same organization as well as interconnecting organizations. This means that MANs have to provide performance close to that obtained on LANs yet cope with the interaction of multiple management domains.

There are three important features which discriminate MAN from LAN or WAN:

(1) The network size falls intermediate between LAN and WAN. A MAN typically covers an area of between 5 and 50km diameter. Many MANs cover an area the size of a city.

(2) A MAN is not generally owned by a single organization. The MAN, its communication links and equipment are generally owned by either a consortium of users or by a single network provider who sells the service to the users. The level of service provided to each user must therefore be negotiated with the MAN operator.

(3) A MAN often acts as a high-speed network to allow sharing of regional resources (similar to a

large LAN). It is also frequently used to provide a shared connection to other networks using a link to WAN.

In the past few years, a number of communications companies have begun business by installing optical fiber cable in metropolitan areas and then selling it to other companies that use it to establish MANs by interconnecting their locations within the city.[5] Rights-of way are obtained in many ways, such as on telephone poles, in subway tunnels, and under city streets. These companies primarily sell high-speed digital bandwidths and typically offer no value-added or other services.

The primary market for this service is the customer who needs a lot of high-speed communication service within a relatively small, often metropolitan, geographic area. These customers are often large companies with multiple offices or other locations within a city. The MAN providers typically offer power prices than other communications carries and diverse routing that provides backup in emergencies. They also claim to offer quicker installation and better service than the other carriers.

MAN technology has been standardized by a committee of the Institute of Electrical and Electronics Engineers, Inc. (IEEE), which developed the MAN standard known as IEEE 802.6. The work was produced by the committees that defined LANs and was a result of their unresolved concerns about how LANs could be connected across distances of a few kilometers or greater.

Keywords

carrier	n.	负载，载体	minicomputer	n.	微型计算机
consortium	n.	组合，共同体，财团	negotiate	v.	议定，商定
diameter	n.	直径	niche	n.	范围，位置
discriminate	v.	区别，鉴别，识别	numerous	adj.	许多的，大群的，大批的
diverse	adj.	不同的，多种多样的，形形色色的	ratio	n.	比率，比值
			regional	adj.	区域性的，局部的
flexibility	n.	机动性，灵活性	router	n.	路由器
lease	v.	租借，出租	sophisticated	adj.	复杂的，高级的
microwave	n.	微波	roughly	adv.	粗略地，概略地

Notes

[1] If the user were to connect his computer to other computers, he could share data on other computers, including high-quality printers.

译文：如果用户能把他们的计算机与其他计算机互联在一起，他们就可以共享计算机上的数据，包括高性能的打印机。

说明：本句由"if"引导条件状语从句，"including high-quality printers"作补语。

[2] Computers that are part of a network can share data, messages, graphics, printers, fax machines, modems, CD-ROMs, hard disks, and other data storage equipment.

译文：联网的计算机可以共享数据、消息、图形、打印机、传真机、调制解调器、光盘、硬盘及其他数据存储设备。

说明："that are part of a network"作定语，修饰主语，"data, messages…"作宾语。

[3] A local area networks, or LAN, is a communication network that is privately owned and that covers a

limited geographic area such as an office, a building, or a group of building.

译文：局域网是专有的通信网络，它可以覆盖一个有限的地域，如一个办公室、一幢建筑或一群建筑等。

说明："or LAN"是主语的同位语，后面两个并列的由"that"引导的定语从句修饰表语。

[4] The WAN on the other hand uses technologies that do not limit the geographical reach and expects to connect nodes under different management domains.

译文：另一方面，广域网使用不受地理范围限制的技术，并且希望连接位于不同管理区域中的各节点。

说明：本句是并列句，主语都是"The WAN"，谓语是"uses"和"expects"，由"that"引导的定语从句修饰宾语"technologies"，"under different management domains"修饰宾语"nodes"。

[5] In the past few years, a number of communications companies have begun business by installing optical fiber cable in metropolitan areas and then selling it to other companies that use it to establish MANs by interconnecting their locations within the city.

译文：在过去几年中，一些通信公司已经开始城域网业务了，他们先在城市范围内安装光缆，然后将其售给使用它的其他公司，这些公司将他们在市内的各个地点互联起来以建设城域网。

说明："In the past few years"是时间状语，"by installing … and then selling…"作方式状语。"that"引导的定语从句修饰"companies"。

14.2　Reading Material 1：Communication Media

A communication medium carries one or more communications channels and provides a link between transmitting and receiving devices. The media most frequently used in today's communications systems include twisted-pair cable, coaxial cable, fiber-optic cable, and so on.

♦ Twisted-pair Cable

Throughout the world, telephone systems and local area networks use miles and miles of cables that consist of pairs of copper wire twisted together. These twisted-pair cables typically terminate with a plastic RJ-11 or RJ-45 plug. There are two main types of twisted-pair cable. In a shielded twisted-pair cable, the wire pairs are coated with a foil shield, which reduces signal noise that might interfere with data transmission. Unshielded twisted-pair cable contains no shielding. It is less expensive than shielded cable, but more susceptible to signal noise.

♦ Coaxial Cable

Coaxial cable is a high-capacity communications cable consisting of a copper wire conductor, a non-conducting insulator, a foil shield, a woven metal outer shielding, and a plastic outer coating. Coaxial cable is typically used to carry cable television signals because its high capacity allows it to carry signals for more than 100 television channels simultaneously. It also provides good capacity for data communications and is used in situations where twisted-pair cable is not adequate to carry the required amount of data. Coaxial cable has a bandwidth that exceeds 100Mbps. Although it has excellent bandwidth, coaxial cable is less durable, more expensive, and more difficult to work with than twisted-pair cable.

♦ Fiber Cable

Fiber-optic cable is a bundle of extremely thin tubes of glass. Each tube, called an optical fiber, is much thinner than a human hair. A fiber-optic cable usually consists of a strong inner support wires;

multiple strands of optical fiber, each covered by a plastic insulator, and a tough outer covering. Unlike twisted-pair and coaxial cables, fiber-optic cables don't conduct or transmit electrical signals. Instead, miniature lasers send pulses of light that represent data through the fibers. Electronics at the receiving end of the fiber convert the light pulses back into electrical signals. Each fiber is a one-way communications channel, which means that at least two fibers are required to provide a two-way communications link.

◆ Radio and Infrared Links

Radio waves provide wireless transmission for mobile communications, such as cellular telephones, and for stationary communications where it is difficult or impossible to install cabling, such as in remote, geographically rugged regions. Radio wave networks operate in frequencies between 1MHz and 3GHz.

Infrared transmissions use a frequency range just below the visible light spectrum to transport data. Infrared transmission is an example of line-of-sight communication, in which the transmitter that sends the signal must have an unobstructed path to the receiver for the transmission to work.

◆ Microwave and Satellite Links

Microwaves are radio waves can be used to provider high-speed transmission of both voice and data. Data is transmitted through the air from one microwave station to another in a manner similar to the way radio signals are transmitted. A disadvantage of microwaves is that they are limited to line-of-sight transmission. This means that microwaves must be transmitted in a straight line and that there can be no obstructions, such as building or mountains, between microwave stations. For this reason, microwave stations are characterized by antennas positioned on top of buildings, towers or mountains.

Communication satellites receive signals from earth, amplify the signals and retransmit the signals back to the earth. Earth stations are communications facilities that use large, dish-shaped antennas to transmit and receive data from satellites.

Keywords

antenna	n.	天线	interfere	v.	干扰，妨碍
bundle	n.	捆，束	miniature	adj.	小型的，小规模的
cellular	adj.	蜂窝的，多孔的	shield	n.	屏蔽，罩
coaxial	adj.	同轴的，共轴的	spectrum	n.	谱，光谱，波谱
conductor	n.	导体，导线	strand	n.	导线束，多心裸电缆
copper	n.	铜，铜制品	susceptible	adj.	易受影响的
durable	adj.	耐用的，坚固的	tough	adj.	强韧的，强健的
infrared	adj.	红外线的，红外区的	unobstruct	v.	非阻塞，无障碍
insulator	n.	绝缘物，绝缘体	woven（weave 的过去分词）	v.	编制，编成

14.3 Reading Material 2：The Colorful Map of Internet

A US-based security consultant is creating a graphical representation of Internet network routes, yielding striking, colorful images that depict those connections.

Barrett Lyon of Network Presence, an information-security consultancy, started the Opte project to map the Internet. "My goal was actually to see if I could do it quickly and with less hardware than other people might need, and to prove it is not this huge, difficult process." Lyon said. Similar efforts, such as the

Lumeta Corp.'s Internet Mapping Project, have operated on a larger scale.

In addition to appealing to general interest, Lyon said, Opte offers benefits such as modeling the Internet and analyzing wasted IP space. For example, he explained, the project could show what happens to the Internet after a natural disaster or important geopolitical event. He said Opte could also show which major companies or organizations have huge blocks of IP addresses that are unused and that could be given to other entities that need them.

Opte tracks Internet routes by using a trace-route program, which records the route a packet takes between two endpoints, indicated by the gateway computer at each hop. The scanrand2 utility performs trace-route searches involving thousands of packets per second.

A single computer—using a 1.7GHz Intel Celeron processor, with 1GB of RAM, working with the Free BSD operating system and other open source applications—runs the Opte project's database, Web pages, image calculations, route collection, and distribution of scanrand2 requests.

Because mapping the Internet from a single computer could miss many Internet sub-networks with which the node may not connect during multiple transmissions, Opte collects information from four computers scattered throughput the Internet.

Opte maps only what used to be called the Internet's Class C networks. The Internet's original routing scheme in the 1970s defined three primary classes of IP address to be assigned to users: classes A (for very large networks), B (for middle-sized networks), and C (for smaller networks). Class D was for multicasts, and Class E was for experimental purposes. Class C represents most Internet networks. Because of this, Lyon said, his project provides a good representation of the Internet.

Keywords

appeal	v.	对……有吸引力	geopolitical	adj.	地理政治论的,地缘政治学的
consultant	n.	顾问	scattered	adj.	不集中的,散漫的
depict	v.	描述	striking	adj.	明显的,显著的,惊人的
disaster	n.	灾难,事故	yield	v.	产生,生出

14.4 动词不定式

动词不定式是非谓语动词的一种，由不定式符号 to 加动词原形构成。之所以称为"不定式"，是因为它的形式不像谓语动词那样受到主语人称和数的限制。但是，动词不定式又具有动词的许多特点：它可以有自己的宾语、状语及宾语补足语。动词不定式和它的宾语、状语及宾语补足语构成不定式短语。动词不定式还有时态和语态的变化，具体信息参见下表。

时态	语态（主动）	语态（被动）	用法
一般式	to do	to be done	表示动作有时与谓语动词表示的动作同时发生，有时发生在谓语动词的动作之后
进行式	to be doing		表示动作正在进行，或者与谓语动词表示的动作同时发生
完成式	to have done	to have been done	表示动作发生在谓语动词表示的动作之前

例：Today we use computers to help us do most of our work.

译文：如今我们使用计算机帮我们做大部分工作。

句中，动词不定式 to help 带有宾语 us 和宾语补足语 do most of our work。

动词不定式通常具有名词性、形容词性和副词性，因此可以充当句子的主语、表语、宾语、定语、状语和补足语。下面分别叙述动词不定式在句中的作用。

1．作主语

动词不定式（短语）作主语，较多地用来表示一个特定的行为或事情，谓语动词需要用第三人称单数，且常用 it 作形式主语。

例 1：To know something about computer is important.

译文：懂得一些计算机的知识很重要。

句中，To know something about computer 是动词不定式短语，在句子中作主语。

不定式短语作主语时，为了句子的平衡，常常把它放在句尾。而用 it 作形式主语代替不定式放在句首。

例 2：It is necessary to learn Visual Basic.

译文：学习 Visual Basic 是很有必要的。

句中，It 是形式主语，而真正的主语是动词不定式 to learn Visual Basic。

例 3：It would perhaps be unwise to forecast undue restrictions on the nature of the ultimate achievement.

译文：在对人类终极目标本质的预测中，如加上过分的限制，可能是不太明智的。

2．作表语

不定式可放在系动词后面作表语。

例 1：To see is to believe.

译文：眼见为实。

句中，to believe 放在 is 后面作表语。

例 2：Our task today is to work out the design.

译文：我们今天的工作是把设计做出来。

3．作宾语

不定式（短语）在某些及物动词后可作宾语。这类及物动词通常有 want、like、wish、hope、begin、decide、forget、ask、learn、help、expect、intend、promise、pledge 等。

例 1：This helps to save coal and reduce the cost of electricity.

译文：这有助于节约用煤以及降低发电成本。

例 2：They decided to do the experiment again.

译文：他们决定再次做这个实验。

当某些动词后面作宾语的不定式必须有自己的补语才能使意思完整时，要用 it 作形式宾语，而将真正的宾语（即不定式）后置。常用这种结构的动词有 think、find、make、consider、feel 等。

例 3：The use of semiconductor devices together with integrated circuits make it possible to develop miniaturized equipment.

译文：半导体装置和集成电路一起使用使得发展微型设备成为可能。

句中，不定式短语 to develop miniaturized equipment 作宾语，it 是形式上的宾语。

4．作定语

动词不定式（短语）作定语时，通常放在它所修饰的名词（或代词）之后。

例 1：He never had the chance to learn computer.

译文：他从来没有学习计算机的机会。

句中，to learn computer 是动词不定式，在句中作定语，修饰和限定 the chance。

有时，动词不定式与它所修饰的名词是逻辑上的动宾关系。

例 2：We usually define energy as the ability to do work.

译文：我们通常将能量定义为做功的能力。

动词不定式作定语除修饰名词外，还可以修饰代词和数词。

例 3：There is something to do.

译文：还有一些事情要做。

5．作状语

不定式作状语可以修饰句中的动词、形容词、副词或全句。主要表示目的、程度、结果、范围、原因等。

例 1：We are glad to hear that you have bought a computer.

译文：听说你买了一台计算机，我们十分高兴。（表示原因）

例 2：To meet our production needs, more and more electric power will be generated.

译文：为了满足生产的需要，将生产越来越多的电力。（表示目的）

例 3：Solar batteries have been used in satellites to produce electricity.

译文：人造卫星上已经用太阳能电池发电了。（表示结果）

应该注意，在"too…to"结构的 too 前面有 not、only、but、never 等含有否定意义的词时，后面的不定式就没有否定意义。

例 4：English is not too difficult to learn.

译文：英语并不难学。

6．作宾语补足语

某些及物动词要求不定式作宾语补足语。宾语补足语是对宾语的补充说明。

例 1：A force may cause an object to move.

译文：力可以使物体移动。

句中，an object 是宾语，不定式 to move 是宾语补足语。

例 2：Conductors allow a large number of electrons to move freely.

译文：导体允许大量的电子自由运动。

当 make、let、have、see、hear、watch、notice、feel 等动词后面用不定式作宾语补足语时，不定式都不带 to。这一点特别重要。

例 3：I often hear people talk about this kind of printer.

译文：我经常听人们谈论这种打印机。

句中，talk about this kind of printer 是个不带 to 的动词不定式短语，在句中作宾语 people 的补足语。

7．作主语补足语

当主动语态的句子变成被动语态时，主动语态句子中的宾语补足语就在被动语态中变成主语补足语。若主动语态中的宾语补足语由动词不定式构成，则该句变为被动语态后它也相应地变为主语补足语。

例 1：He was asked to do the experiment at once.

译文：有人请他马上做实验。

但是，当 make、let、have、see、hear、watch、notice、feel 等动词的句子变为被动语态时，原来在主动语态时作宾语补足语的动词不定式这时也变为主语补足语，此时，动词不定式中的 to 不能省略。

例 2：He was made to finish repairing the printer.

译文：他被迫马上修理打印机。

8．不定式的特殊句型

例 1：I am afraid the box is too heavy for you to carry it.（too…to 太……以至于……）

译文：这箱子太重，恐怕你搬不动。

例 2：Go in quietly so as not to wake the baby.（so as to 以至于，为了）

译文：轻点进去，别惊醒了婴儿。

14.5 Exercises

1. **Translate the following phrases.**

   asynchronous transfer mode
   personal computer
   local area network
   fiber-optic cable
   communication medium
   telephone system
   word processing equipment
   coaxial cable
   metropolitan area network
   high-speed network
   twisted-pair cable
   computer network

2. **Fill in the blanks with appropriate words or phrases.**

   a. hardware resource   b. combined   c. organization   d. a network
   e. 5 and 50km   f. leased line g. connectionless   h. OSI model

   (1) In actual practice, hardware resource sharing and information resource sharing are often _____.

   (2) A group of computers and other devices connected together is called _____.

   (3) At each end of the _____, a router connects to the LAN on one side and a hub within the WAN on the other.

   (4) Two common applications of local area networks are _____ sharing and information resource sharing.

   (5) Protocols can be either _____ or connection-oriented.

(6) A MAN typically covers an area of between _____ diameter.
(7) Protocol stacks are typically based either on the _____ or the TCP/IP model.
(8) Many WANs are built for one particular _____ and are private.

3. **Match the following terms to the appropriate definition.**

   a. MIPS   b. megabyte   c. multimedia
   d. mouse   e. microwave   f. modem

(1) A million bytes. Actually, technically, 1024 kilobytes.
(2) A device that converts digital signals to analog or converts analog to digital, allowing computer data to be carried over normal telephone and cable lines.
(3) Documents that combine text, graphics, sound, movies, or other media.
(4) It is a measure of a computer's processor speed. It represent "peak" execution rates on artificial instruction sequences with few branches.
(5) A nifty device that lets you move about your screen as you like so you can click on links and highlight items.
(6) High frequency radio waves used for telecommunications transmission, usually above 890MHz.

Unit 15 Wireless Network

音频

15.1 Text

In a wireless network, the computers are connected by radio signals instead of wires or cables. Advantages of wireless networks include mobility and no unsightly wires. Disadvantages can include a slower connection than a wired network and interference from other wireless devices. Your computer needs an internal or external wireless network adapter. To see if your computer has a wireless network adapter, do the following:

Open network connections by clicking the "Start" button, clicking "Control Panel", clicking "Network and Internet", clicking "Network and Sharing Center", and then clicking "Manage Network Connections". The adapters installed in your computer are listed.

In the list of available wireless networks, you will see a symbol that shows the wireless signal strength for each network. A strong signal usually means that the wireless network is close or there is no interference. For best performance, connect to the wireless network with the strongest signal. However, if an unsecured network has a stronger signal than a security-enabled one, it is safer for your data if you connect to the security-enabled network.[1] To improve the signal strength, you can move your computer closer to the wireless router or access point, or move the router or access point so it is not close to sources of interference such as brick walls or walls that contain metal support beams.[2]

Public wireless networks are convenient, but if they are not properly secured, connecting to one might be risky. Whenever possible, only connect to wireless networks that require a network security key or have some other form of security, such as a certificate. The information sent over such networks is encrypted, which can help protect your computer from unauthorized access.[3] In the list of available, each network is labeled as either security-enabled or unsecured. If you do connect to a network that is not secure, be aware that someone with the right tools can see everything that you do, including the websites you visit, the documents you work on, and the user names and passwords you use. You should not work on documents or visit websites that contain personal information, such as your bank records, while you are connected to that network.

Wireless vendors face the challenge of supporting increasingly bandwidth hungry applications, such as voice over IP, streaming video and videoconferencing. To dramatically increase throughput, 802.11a proponents had to solve a major challenge of indoor radio frequency. They had to develop a way to resolve the problem of delay spread in the current 2.4GHz, single-carrier and delay-spread system.

Delay spread is caused by the echoing of transmitted radio frequency. As these signals proceed to a certain point, such as a wireless antenna, they often bounce and echo off objects, walls, furniture and floors, and arrive at the antenna at different times due to the different path lengths.[4] A baseband processor, or equalizer, is required to "unravel" the divergent radio frequency signals. The delay spread must be less than the symbol rate, or the rate at which data is encoded for transmission. If not, some of the delayed signal spreads into the next symbol transmission. This can put a ceiling on the maximum bit rate that can be sustained.

With current bit-rate technology, this ceiling tends to be around 10M to 20M bit/sec. The 802.11a standard cleverly solves this challenge through an innovative modulation technique called coded orthogonal frequency division multiplexing (COFDM), which has found earlier application in European digital TV and audio transmission. COFDM breaks the ceiling of the data bit rate by (1) sending data in a massively parallel fashion, and (2) slowing the symbol rate down so each symbol transmission is much longer than the typical delay spread. A guard interval (sometimes called a cyclic prefix) is inserted at the beginning of the symbol transmission to let all delayed signals "settle" before the baseband processor demodulates the data.

Wireless vendors now have a goal to boost wireless throughput beyond 100M bit/sec. While the 802.11a standard currently tops out at 54M bit/sec in 20MHz channels, several firms are developing and proposing high-rate extensions to the 802.11a standard. These proposals generally envision at least doubling throughput to anywhere from 108M to 155M bit/sec.

There is a newly development network for LANs: WLAN (Wireless LAN). Using electromagnetic waves, WLANs transmit and receive data over the air, minimizing the need for wired connections. Thus, WLANs combine data connectivity with user mobility, and through simplified configuration, enable movable LANs. Over the recent several years, WLANs have gained strong popularity in a number of vertical markets, including the health-care, retail, manufacturing, warehousing, and academic arenas. These industries have profited from the productivity gains of using hand-held terminals and notebook computers to transmit real-time information to centralized hosts for processing.

Today WLAN is becoming more widely recognized as a general-purpose connectivity alternative for a broad range of business customers. The following list describes some of many applications made possible through the power and flexibility of wireless LANs.

- ♦ Doctors and nurses in hospitals are more productive because notebook computers with wireless LAN capability deliver patient information instantly.[5]
- ♦ Training sites at corporations and students at universities use wireless connectivity to facilitate access to information, information exchanges, and learning.
- ♦ Senior executives in conference rooms make quicker decisions because they have real-time information at their fingertips.

The increasingly mobile user also becomes a clear candidate for a wireless LAN. Portable access to wireless networks can be achieved using laptop computers and wireless NICs (network interface card). This enables the user to travel to various locations—meeting rooms, hallways, lobbies, cafeterias, classrooms, etc.—and still have access to their networked data. Without wireless access, the user would have to find a network tap.

Wireless Ethernet allows the Ethernet standard to be used with wireless network connections. It is also known as Wi-Fi, though technically the Wi-Fi label can only be used with wireless Ethernet products that are certified by the wireless Ethernet compatibility Alliance. Users of Wi-Fi certified products are assured that their hardware will be compatible with all other Wi-Fi certified hardware. The IEEE 802.11 standard extends the carrier-sensing multiple access principle employed by Ethernet technology to suit the characteristic of wireless communication. The 802.11 standard is intended to support communication between computers located within about 150 meters of each other at speed up to 54 Mbps. Wireless Ethernet is a growing choice for organizations wishing to extend their wired Ethernet network.

WiMAX is a wireless digital communications system, also known as IEEE 802.16, that is intended for wireless "metropolitan area networks". WiMAX can provide broadband wireless access up to 30 miles (50km) for fixed stations, and 3-10 miles (5-15km) for mobile stations. In contrast, the Wi-Fi/802.11 wireless local area network standard is limited in most cases to only 100-300 feet (30-100m).

With WiMAX, Wi-Fi-like data rates are easily supported, but the issue of interference is lessened. WiMAX operates on both licensed and non-licensed frequencies, providing a regulated environment and viable economic model for wireless carriers. WiMAX can be used for wireless networking in much the same way as the more common Wi-Fi protocol. WiMAX is a second-generation protocol that allow higher data rates over longer distances.

Wireless network have many uses. For example, you can send E-mail, receive telephone calls and faxes and read remote documents in your travel. In addition, wireless networks are of great value to fleets of trucks, taxis, buses and repairpersons for keeping in contact with home. Although wireless LAN is easy to install, it also has some disadvantages. Typically they have a lower capacity, which is much slower than wired LAN. The error rates are often much higher too, and the transmissions from different computers can interface with one another.

Keywords

alliance	n.	同盟，联盟	interference	n.	干扰，干涉
arena	n.	活动场所，竞技场所	lessen	v.	减小，减轻
cafeteria	n.	自助餐厅	licensed	adj.	许可的
candidate	n.	候选人	lobby	n.	门廊，门厅
ceiling	n.	天花板，最高限度	mobility	n.	移动性，活动性
divergent	adj.	分歧的，歧义的	orthogonal	adj.	正交的，互相垂直的
dramatically	adv.	戏剧性地，引人注目地	profit	n.	盈余，利润
equalizer	n.	均衡器	repairperson	n.	维修工，维修人员
fingertip	n.	手头，指尖	retail	n.	零售商
furniture	n.	家具，设备	risky	adj.	危险的，冒险的
hallway	n.	门厅，过道	senior	adj.	高级的，上级的
innovative	adj.	革新的，创新的	unravel	v.	拆开，解开
intended	adj.	预期的，打算中的	unsightly	adj.	不好看的，难看的

Notes

[1] However, if an unsecured network has a stronger signal than a security-enabled one, it is safer for your data if you connect to the security-enabled network.

译文：但是，当不安全网络中的信号比启用了安全保护的网络中的信号更强时，为了使数据更安全，最好还是连接到启用了安全保护的网络。

说明：本句有两个由"if"引导的条件状语从句。

[2] To improve the signal strength, you can move your computer closer to the wireless router or access point, or move the router or access point so it is not close to sources of interference such as brick walls or walls that contain metal support beams.

译文：若要提高信号强度，可以将计算机移动到距离无线路由器或访问点更近的位置。或者将路由器或访问点移动到远离干扰源（如砖墙或包含金属支撑梁的墙体）的位置。

说明："To improve the signal strength"作目的状语，有两个并列的主句，"you can move your…"和省略主语的"move the router…"。

[3] The information sent over such networks is encrypted, which can help protect your computer from unauthorized access.

译文：通过此类网络发送的信息是经过加密的信息，这有助于保护计算机免受未经授权的访问。

说明："which"引导的是非限定性定语从句。

[4] As these signals proceed to a certain point, such as a wireless antenna, they often bounce and echo off objects, walls, furniture and floors, and arrive at the antenna at different times due to the different path lengths.

译文：在这些信号到达某一点（如无线天线）的过程中，它们常常因物体、墙壁、家具和地板等而产生反射，由于路径长度不同，信号到达天线的时间也不同。

说明："As these signals proceed to a certain point"作状语，"such as a wireless antenna"是"a certain point"的同位语，主句是并列句结构。

[5] Doctors and nurses in hospitals are more productive because it has acquired notebook computers with wireless LAN capability, which can deliver patient information instantly.

译文：医院的医生和护士工作起来更有效率，因为具备无线局域网连接性能的笔记本电脑及时传递了病人的信息。

说明：本句由"because"引导原因状语从句。

15.2　Reading Material 1：Encryption on the Internet

Modern Web browsers use the secure sockets layer (SSL) protocol for secure transactions like E-commerce purchases and E-bank. SSL works by using a public key for encryption and a different private key for decryption. Because SSL encryption depends so heavily on keys, one normally measures the effectiveness or strength of SSL encryption in terms of key length-number of bits in the key.

The early implementations of SSL in Web browsers, first Netscape 3 and then Microsoft Internet Explorer 3, used a 40-bit SSL encryption standard. Unfortunately, 40-bit encryption proved too easy to decipher or crack in practice. To decipher an SSL communication, one simply needs to generate the correct decoding key.

In cryptography, a common deciphering technique is brute-force decryption; essentially, using a computer to exhaustively calculate and try every possible key one by one. 2-bit encryption, for example, involves four possible key values: 00, 01, 10, and 11. 3-bit encryption involves eight possible values, and so on. Mathematically speaking, 2^n possible values exist for an n-bit key.

Compared to 40-bit encryption, 128-bit encryption offers 88 additional bits of key length. This translates to $2^{88} \times 2^{40}$ possible values required for a brute-force crack. Based on the past history of improvements in computer performance, security experts expect that 128-bit encryption will work well on the Internet for at least the next ten years.

Authentication is an important part of everyday life. The lack of strong authentication has inhibited the development of electronic commerce. It is still necessary for contracts, legal documents and official letters to be produced on paper. Strong authentication is then, a key requirement if the Internet is to be used for electronic commerce. Strong authentication is generally based on modern equivalents of the one time pad. For example tokens are used in place of one-time pads and are stored on smart cards or disks. Digital certificate is an identity card counterpart in the computer society. When a person wants to get a digital certificate, he generates his own key pair, gives the public key as well as some proof of his identification to the Certificate Authority (CA). CA will check the person's identification to assure the identity of the applicant. If the applicant is really the one "who claims to be", CA will issue a digital certificate, with the applicant's name, E-mail address and the applicant's public key, which is also signed digitally with the CA's private key.

It would be nice to have non-forgeable signatures on computer documents, but there are problems with this concept. First, bit streams are easy to copy; the mere presence of such a signature means nothing. Even if a person's signature was made difficult to forge—if it was accomplished by a graphical image of a fingerprint, for example—with today's cut-and-paste software it is all too easy to move a valid signature from one document to another. Second, documents are easy to modify after they are signed, without leaving any evidence of modification.

A digital signature is an electronic signature that can be used to authenticate the identity of the sender of a message or the signer of a document, and possibly to ensure that the original content of the message or document that has been sent is unchanged. A digital signature is a string of bits attached to an electronic document, which could be a word processing file or an E-mail message. This bit string is generated by the signer, and it is based on both the document's data and the person's secret password. Digital signatures are easily transportable, cannot be imitated by someone else, and can be automatically time-stamped. Someone who receives the document can prove that the signer actually signed the document. If the document is altered, the signer can also prove that he did not sign the altered document.

Public-key cryptography can be used for digital signatures, public-key cryptography uses special encryption algorithms with two different keys: a public key that everyone knows, and a private key that only person knows. Public-key algorithms encrypt the contents of an electronic document using both keys. The resulting file is an amalgam of both the public and private keys and the original document's contents.

Keywords

amalgam	n.	混合物	decipher	v.	译解（密码），辨认
authority	n.	权威，官方，职权	fingerprint	n.	指纹
browser	n.	浏览器	forgeable	adj.	伪造的，编造的
certificate	n.	证明书，证书，明证	imitate	v.	模仿，仿效
counterpart	n.	副本	inhibit	v.	抑制，约束
cryptography	n.	加密术，密码学，密码翻译术	mathematically	adv.	数学地，数理地，精确地
			signature	n.	签名，署名，画押

15.3 Reading Material 2: Mobile Networks

The structure of networks and the Internet is changing as more devices connect to it in different ways. At one time it was basically a collection of workstations connected over networks. Networks were then structured in a fixed way, where cabling and the location of networking devices defined the layout of networks. Thus, networks have evolved in a rather random way, as was intended when the Internet was created.

As the Internet has increased its size at an amazing rate, it would have been impossible to create an Internet with multiple connects to billions of nodes, as the routing between nodes would have been too complex. Thus, the Internet now uses layered approach where the network is split into a number of autonomous systems (ASs). At the top level of the Internet, routers simply route between ASs. We now thus have a model which is distributed and decentralized.

The evolution of ASs is only part of the future, as the next major change will come in the connection of other devices, especially mobile ones. It is thus possible, with IPv6 and NAT, to allow many computers to break their physical links to a network, and expand the number of devices which connect to the Internet. It will also be possible for many non-computing devices, such as engine management systems and central heating systems to be assigned network addresses and be connected to the Internet. The mobile devices will typically connect through intermittent and lower-speed connections, which will be pushed towards the peripheral areas of the Internet.

Mobile devices have evolved over several iterations. For mobile phones, the main generations have been:

- First generation (1G). First generation mobile phones had very low transmission rates (typically just a few Kbps).
- Second generation (2G and 2.5G). There are devices improved this to give several hundred Kbps.
- Third generation (3G). These devices give almost workstation network bandwidths (several Mbps), which allows for full multimedia transmissions.

The Internet is thus becoming an exciting and flexible place, where connections depend less on physical connection, and more on mobile ones. The applications of the Internet will thus increase as more devices connect in this way.

A key element of a mobile network is the availability of antennas. One of the most widely established mobile networks is GSM. This network links mobile phones directly on the telephone network. GSM uses a cellular approach where the phone connects to an antenna which gives the strongest signal strength. The phone can then move and can reconnect to other antennas as the signal strength varies.

The advantage of the GSM network is that it is widely available, and covers most of the areas of the world. It has many disadvantages, such as being costly, especially as it is charged for the time of the connection, rather than for the amount of data sent and received.

Keywords

autonomous	*adj.*	自治的，自主的		generation	*n.*	代，时代
cable	*v.*	联网，与……联网		intermittent	*adj.*	间断的，断断续续的
decentralize	*v.*	分散，疏散		mobile	*adj.*	移动的，活动的，运动的
evolve	*v.*	发展，进展，进化				

15.4 状语从句

状语从句通常由从属连词和起连词作用的词组来引导，用来修饰主句中的动词、形容词、副词等。

状语从句可位于主句前，也可位于主句后；前置时，从句后常用逗号与主句隔开；后置时，从句前通常不使用逗号。状语从句在句子中作状语，可表示时间、原因、目的、结果、条件、比较、方式、让步和地点等不同含义。

1．状语从句的分类

1）时间状语从句

引导时间状语从句的连词或词组很多，可根据所表示时间的长短，以及与主句谓语动词行为发生的先后这两点去理解和区别。

这些连词或词组有 when（当……时候）、as（当……时候，随着，一边……一边）、while（在……期间）、before（在……之前）、after（在……之后）、since（自从……以来）、until（till）（直到……才）、as soon as（一……就）、no sooner…than（刚一……就……）、once（一旦）、every time（每次）等。

例 1：It changes speed and direction when it moves.

译文：在运动时它改变速度和方向。

例 2：Check the circuit before you begin the experiment.

译文：检查好线路再开始做实验。

例 3：Electricity has found wide application since it was discovered.

译文：自从发现电以来，它就获得了广泛的应用。

例 4：No sooner had the push button been pressed than the motor began to run.

译文：按钮刚一按下，发动机就启动了。

2）原因状语从句

引导原因状语从句的连词和词组有 because（因为）、as（由于）、since（既然，由于）、now that（既然）、in that（因为）等。其中前 3 个较常用，它们表示原因的正式程度依次为 because > since ≥ as。当原因是显而易见的或已为人们所知，就用 as 或 since。由 because 引导的从句如果放在句首，且前面有逗号,则可以用 for 来代替。但如果不是说明直接原因,而是多种情况加以推断,就只能用 for。

例 1：Electric energy is used most widely mainly because it can be easily produced, controlled, and transmitted.

译文：电能用得最广,主要是因为发电容易,而且控制和输送也方便。

例 2：He is absent today, because he is ill.

译文：他今天缺席是因为他生病了。

3）目的状语从句

目的状语从句由 in order that（为了,以便）、so that（为了,以便）、that（为了）、lest（以免,以防）、for fear that（以免,以防）等引导。

例 1：He handled the instrument with care for fear that it should be damaged.

译文：他小心地操作仪器,生怕把它弄坏。

例 2：You must speak louder so that you can be heard by all.

译文：你必须说大声点以便大家都能听到。

4）结果状语从句

引导结果状语从句的连词有 so that（结果,以致）、so … that（如此……以致）、such … that（这样的……以致）等。注意 so 后接形容词或副词,而 such 后跟名词。so 还可以与表示数量的形容词 many、few、much 及 little 连用,形成固定搭配。

例：This problem is very difficult that it will take us a lot of time to work it out.

译文：这道题很难,我们要用很长时间才能解出。

5）条件状语从句

条件状语从句用来表示前提和条件。通常由以下连词引导。

if（如果）、unless（除非）、provided / providing that（假如）、as long as（只要）、in case（如果）、on condition that（条件是……）、suppose / supposing（假如）等。

例：A physical body will not tend to expand unless it is heated.

译文：除非受热,否则物体不会有膨胀的倾向。

6）比较状语从句

比较状语从句经常是省略句,一般都是省略了重复部分;省略之后不影响句意,反而结构简练。部分比较状语从句还有倒装现象。比较状语从句由下列连词引导。

as … as（像……一样）、than（比）、not so (as) … as（不像……一样）、the more … the more（越……越）、as … so（正如……那样）等。

例 1：Electron tubes are not so light in weight as semiconductor devices.

译文：电子管的重量不如半导体器件那么轻。

例 2：He finished the work earlier than we had expected.

译文：他完成这项工作比我们预计的要早。

7）方式状语从句

方式状语从句通常由 as（如同,就像）、as if (as though)（好像,仿佛）等连词引导。

as 引导的方式状语从句常常是一个省略句。as if 和 as though 两者的意义和用法相同,引出的状语从句常是一个虚拟语气的句子,表示没有把握的推测,或者是一种夸张的比喻。(just) as … so … 引导的方式状语从句通常位于主句后,译为"正如……,就像"。

例 1：The earth itself behaves as though it were an enormous magnet.

译文：地球本身的作用就像一个大磁铁一样。

例 2：They completely ignore these facts as if they never existed.

译文：他们完全不理会这些事实就好像它们不存在一样。

8）让步状语从句

让步状语从句表示在相反的（不利的）条件下，主句行为依然发生了。

引导让步状语从句的有（al）though（虽然）、even if（though）（即使）、as（尽管）、whatever（不管）、however（无论怎样）、no matter（how，what，where，when）不管（怎样，什么，哪里，何时）、whether…or（不论……还是）等。

例 1：It is important to detect such flows, even if they are very slight, before the part is installed.

译文：在安装部件之前，即使变形很轻微，也必须探测出来。

例 2：Much as computer languages differ, they have something in common.

译文：尽管计算机语言之间各不相同，但它们仍有某些共同点。

9）地点状语从句

引导地点状语从句的词常用的有 where（在……地方，哪里）、wherever（在任何地方）、everywhere（每一……地方）等。

例：She found her pen where she had left them.

译文：她的笔是她在原来丢笔的地方找到的。

2．状语从句的翻译方法简介

状语从句的常用翻译方法有顺译法、倒译法、转译法和缩译法。

1）顺译法

一般的句子可以按照原文提供的顺序直接翻译。当表示目的、原因等的状语从句在主句之前出现时，直接按照原句语序翻译。如果这些状语出现在主句之后，可以将它们提前或保持原句顺序，翻译在主句之后，对主句意思起到补充说明的作用。

例 1：Whenever you need any specific information, you can search it by Internet.

译文：每当你需要任何专门的信息时，都可以通过互联网搜索得到。

例 2：The Internet is so powerful that you can get various information through it.

译文：互联网是如此强大，以至于你可以通过它获取各种各样的信息。

2）倒译法

当原文中的时间状语和地点状语在主句后面时，必须倒译；当原文中的原因状语从句、条件状语从句和让步状语从句在主句后面时，一般也可以倒译；另外，当特殊比较从句在主句后面时，必须倒译。

例 1：Many businesses became aware of network when they bought an expensive laser printer and wanted all the PCs to print to it.

译文：当企业购买了一台昂贵的激光打印机，并希望其全部的计算机都能使用该打印机时，他们就想到了网络。

例 2：In the Start menu folder, find the shortcut to the program you want to start each time you start Windows 2000, and drag it to the Startup folder.

译文：在启动菜单文件夹时，找到每次启动 Windows 2000 时想要启动的程序的快捷方式，然后把它拖到启动文件夹中。

3）转译法

当通过对原文的逻辑含义进行分析后，会发现 when、where 不再单纯地表示时间、地点，或者翻译成"当……"或"在……"不合适时，可以考虑这些词也可以表示"如果"的意思。另外，当状语从句比较短，而关联词可以省略时，可以把状语从句翻译成并列成分，这样也使得句子比较紧凑。

例 1：Where the Hz is too small a unit, we may use the MHz.

译文：当用赫兹作单位太小时，我们可以使用兆赫兹。

例 2：Our whole physics universe, when reduced to the simplest terms, is made up of two things: energy and matter.

译文：我们的整个物理世界，如果用简单的话来说，是由能和物质这两样东西组成的。

4）缩译法

有些关联词，如"so…that…"，在很多情况下可以省略翻译，这样使得汉语的译文很简练，对于这样的句子就可以采用"缩译法"进行翻译。

例：Computers work so fast that they can solve a very difficult problem in a few seconds.

译文：计算机工作得如此迅速，一个很难的题目几秒钟内就可以解决。

15.5　Exercises

1. **Translate the following phrases.**

    radio signal

    wireless network

    control panel

    mobile user

    secure sockets layer

    smart card

    digital certificate

    certificate authority

    digital signature

    public-key cryptography

    wireless digital communication system

    coded orthogonal frequency division multiplexing

2. **Fill in the blanks with appropriate words or phrases.**

    a. personal information　　b. radio　　　　c. wired connections　　　d. user
    e. symbol rate　　　　　　f. standard　　　g. business customers　　　h. wireless

(1) Wireless vendors now have a goal to boost _____ throughput beyond 100M bit/sec.

(2) You should not work on documents or visit websites that contain _____, such as your bank records, while you are connected to that network.

(3) Today WLAN are becoming more widely recognized as a general-purpose connectivity alternative for a broad range of _____.

(4) Delay spread is caused by the echoing of transmitted _____ frequency.

(5) Wireless Ethernet allows the Ethernet _____ to be used with wireless network connections.

(6) Using electromagnetic waves, WLANs transmit and receive data over the air, minimizing the need for _____.

(7) The delay spread must be less than the _____, or the rate at which data is encoded for transmission.

(8) The increasingly mobile _____ also becomes a clear candidate for a wireless LAN.

3. Match the following terms to the appropriate definition.

a. offline   b. packet   c. password
d. physical Layer   e. project management software f. plug-in

(1) A code used to gain access to a locked system.

(2) Means you are no longer connected to the Internet, whether through your modem, a wireless service, or any other means.

(3) Special programs you can download off the Internet to do cool things like watch the latest videos or listen to music.

(4) A series of bits containing data and control information, including source and destination node addresses, formatted for transmission from one node to another.

(5) The lowest layer of the OSI model and is implemented by the physical channel.

(6) It is a term covering many type of software, including scheduling, cost control and budget management, resource allocation, collaboration software, communication, quality management and documentation or administration systems, which are used to deal with the complexity of large projects.

Unit 16 Internet of Things

16.1 Text

Internet of Things (IoT) is an integrated part of future Internet. It could be defined as a dynamic global network infrastructure with self configuring capabilities. It is based on standard and interoperable communication protocols where physical and virtual "things" have identities, physical attributes and virtual personalities and use intelligent interfaces and are seamlessly integrated into the information network.[1]

The Internet of Things builds on foundation of existing technologies, such as RFID, and is being enabled by the availability of low cost sensors, the drop in price of data storage, the development of "Big Data" analytics software that can work with trillions of pieces of data, as well implementation of IPv6, which will allow Internet addresses to be assigned to all of these new devices. Funding and research for the Internet of Things is being spearheaded by the European Union and China (where it is known as the Sensing Planet), and in the United States by companies such as IBM's Smarter Planet initiative. Although challenges remain before the Internet of Things is fully realized, it is coming closer and closer to fruition.

The Internet of Things will create a dynamic network of billions or trillions of wireless identifiable "things" communicating with one another and integrating the developments from concepts like pervasive computing, ubiquitous and ambient intelligence. IoT hosts the vision of ubiquitous computing and ambient intelligence and enhances them by requiring a full communication and a complete computing capability among things and integrating the elements of continuous communication, identification and interaction.[2] The Internet of things fuses the digital world and the physical world by bringing different concepts and technical components together; pervasive networks, miniaturization of devices, mobile communication, and new models for business processes.

In the IoT, "Things" are expected to become active participants in business, information and social processes where they are enabled to interact and communicate among themselves. They can also interact with the environment by exchanging data and information "sensed" about the environment, while reacting autonomously to the "real/physical world" events and influencing it by running processes that trigger actions and create services with or without direct human intervention.[3]

Interface in the form of services facilitate interactions with these "smart things" over the Internet, query and change their state and any information associated with them and take into account security and privacy issues.

Applications, services, middleware components, networks, and endpoints will be structurally connected in entirely new ways. Initially there will be commercial and physical challenges to establishing global ubiquitous network connectivity and initially the many connected things and devices may have limited ability to engage in 2-way network connectivity. It is important that the architectural design for the Internet of things support effective two-way caching and data synchronization techniques, as well as network-connected endpoints for virtual representations of the connected things and devices, which can be used for monitoring their location, condition and state, as well as sending requests and instructions to them.[4]

Everything from individuals, groups, communities, objects, products, data, services, processes will be connected by the IoT. Connectivity will become in the IoT a kind of commodity, available to all at a very low cost and not owned by any private entity. In this context, there will be the need to create the right situation-aware development environment for stimulating the creation of services and proper intelligent middleware, to understand and interpret the information, to ensure protection from fraud and malicious attack and to guarantee privacy.[5]

From a technical point of view, the Internet of Things is not the result of a single novel technology; instead, several complementary technical developments provide capabilities that taken together help to bridge the gap between the virtual and physical world. These capabilities include:

- Communication and cooperation: Objects have the ability to network with Internet resources or even with each other, to make use of data and services and update their state. Wireless technologies such as GSM and UMTS, Wi-Fi, Bluetooth, ZigBee and various other wireless networking standards, are of primary relevance here.
- Addressability: Within an Internet of Things, objects can be located and addressed via discovery, look-up or name services, and hence remotely interrogated or configured.
- Identification: Objects are uniquely identifiable. RFID, NFC (near field communication) and optically readable bar codes are examples of technologies with which even passive objects which do not have built-in energy resources can be identified (with the aid of a "mediator" such as an RFID reader or mobile phone). Identification enables objects to be linked to information associated with the particular object and that can be retrieved from a server, provided the mediator is connected to the network.
- Sensing: Objects collect information about their surroundings with sensors, record it, forward it or react directly to it.
- Actuation: Objects contain actuators to manipulate their environment (for example by converting electrical signals into mechanical movement). Such actuators can be used to remotely control real-world processes via the Internet.
- Embedded information processing: Smart objects feature a processor or microcontroller, plus storage capacity. These resources can be used, for example, to process and interpret sensor information, or to give products a "memory" of how they have been used.
- Localization: Smart things are aware of their physical location, or can be located. GPS or the mobile phone network is suitable technologies to achieve this, as well as ultrasound time measurements, UWB (ultra-wide band), radio beacons (e.g. neighboring WLAN base stations or RFID readers with known coordinates) and optical technologies.

◆ User interfaces: Smart objects can communicate with people in an appropriate manner (either directly or indirectly, for example via a smart phone). Innovative interaction paradigms are relevant here, such as tangible user interfaces, flexible polymer-based displays and voice, image or gesture recognition methods.

Most specific applications only need a subset of these capabilities, particularly since implementing all of them is often expensive and requires significant technical effort. Logistics applications, for example, are currently concentrating on the approximate localization (i.e. the position of the last read point) and relatively low-cost identification of objects using RFID or bar codes. Sensor data (e.g. to monitor cool chains) or embedded processors are limited to those logistics applications where such information is essential such as the temperature-controlled transport of vaccines.

Keywords

actuator	n.	激发器	malicious	adj.	有恶意的，存心不良的，预谋的
addressability	n.	寻址能力			
ambient	adj.	周围的	mediator	n.	调解人，斡旋人，中间人
beacon	n.	烽火，灯塔，界标，信号所	microcontroller	n.	微控制器
			miniaturization	n.	小型化，微缩技术
billion	num.	十亿，数以十亿计	participant	n.	参加者，有关系的
commodity	n.	日用品	personality	n.	个性，人品
community	n.	社会，社团，共有	pervasive	adj.	普遍深入的
complementary	adj.	补充的，补足的，互补的	polymer-based	adj.	基于聚合的
			relevance	n.	有关系，适当，实质性
fraud	n.	欺骗，欺诈	seamlessly	adv.	无缝地，无空隙地
fruition	n.	成就，结果	spearhead	n.	前锋，尖端，先锋
identifiable	adj.	可以确认的，可以识别的	stimulate	v.	刺激，激励
			tangible	adj.	有形的，真实的
initiative	adj.	起始的，初步的	trillion	num.	万亿
interoperable	adj.	可共同使用的,可共同操作的	ultrasound	n.	超声波
			vaccine	n. adj.	疫苗，预防疫苗的
interrogate	v.	询问，质问	vision	n.	视觉，景象，美景
localization	n.	局限，地方化			

Notes

[1] It is based on standard and interoperable communication protocols where physical and virtual "things" have identities, physical attributes and virtual personalities and use intelligent interfaces and are seamlessly integrated into the information network.

译文：它基于标准的共同遵守的通信协议。物联网中的物理和虚拟"物"有身份、物理属性和虚拟人格，并使用智能接口无缝整合到信息网络中。

说明：本句的"It"指物联网，"where…"是一个定语从句，修饰和限定 It，最后的两个"and"连接了3个并列谓语，它们的主语是"physical and virtual 'things'"。

[2] IoT hosts the vision of ubiquitous computing and ambient intelligence and enhances them by requiring a full communication and a complete computing capability among things and integrating the elements of continuous communication, identification and interaction.

译文：物联网的远景是泛在计算与环境智能，这通过完全的通信和完备的计算能力以及整合各种通信、识别和交互要素来增强。

说明：本句中的"by requiring a full communication…"是介词短语作方式状语。

[3] They can also interact with the environment by exchanging data and information "sensed" about the environment, while reacting autonomously to the "real/physical world" events and influencing it by running processes that trigger actions and create services with or without direct human intervention.

译文：它们可以互相通信与影响，自治地响应"真实/物理世界"的事件并通过触发行为和创建需要人直接干涉或无须干涉的服务，同时也能够通过交换数据和"感知到的"环境信息与环境交互。

说明：本句由"by"引导的成分作方式状语；"while…"作时间状语；"that trigger…"是一个定语从句，修饰和限定"processes"。

[4] It is important that the architectural design for the Internet of things support effective two-way caching and data synchronization techniques, as well as network-connected endpoints for virtual representations of the connected things and devices, which can be used for monitoring their location, condition and state, as well as sending requests and instructions to them.

译文：因此物联网结构设计应该支持双向缓存和数据同步技术，以及用于虚拟表示连接的物和设备状况的网络端点，这十分重要。这些网络端点可以用来监控物和设备的位置、条件和状态，也可以向它们发送请求和指令。

说明：本句中的"It"是形式主语，"which can be used for…"是一个非限定性定语从句，对"network-connected endpoints"作进一步的补充说明。

[5] In this context, there will be the need to create the right situation-aware development environment for stimulating the creation of services and proper intelligent middleware, to understand and interpret the information, to ensure protection from fraud and malicious attack and to guarantee privacy.

译文：在这种情况下，需要建立恰当的有情景意识的开发环境来激发创建服务和适当智能中间件、了解和解释信息、确保防止欺骗和恶意攻击并保护隐私。

说明：本句的 4 个动词不定式短语"to create the right…, to understand…, to ensure protection… and to guarantee privacy"作定语，修饰和限定"the need"，"for stimulating the creation…"是一个介词短语，作定语，修饰"development environment"。

16.2　Reading Material 1：The Future of IoT

No discussion of the future Internet would be complete without mentioning the Internet of Things (IoT), also sometimes referred to as the Industrial Internet. Internet technology is spreading beyond the desktop, laptop, and tablet computer, and beyond the smart phone, to consumer electronics, electrical appliance, cars, medical devices, utility systems, machines of all types, even clothing—just about everything can be equipped with sensors that collect data and connect to the Internet, enabling the data to be analyzed with data analytics software.

The future network of networks will be laid out as public /private infrastructures and dynamically extended and improved by edge points created by the "things" connecting to one another. In fact, in the IoT communications will take place not only between people but also between people and their environment.

Communication will be seen more among terminals and data centers than among nodes as in current networks. Growth of storage capacity at lower and lower costs will result in the local availability of most information required by people or objects. This, coupled with the enhanced processing capabilities and always-on connectivity, will make terminals gain a main role in communications.

Terminals will be able to create a local communication network and may serve as a bridge between communication networks thus extending, particularly in urban environments, the overall infrastructure capacity. This will likely determine a different view of network architectures. The future Internet will exhibit high levels of heterogeneity, as totally different things, in terms of functionality, technology and application fields, are expected to belong to the same communication environment.

The IoT will bring tangible business benefits, such as the high-resolution management of assets and products, improved life-cycle management, and better collaboration between enterprises. Many of these benefits are achieved through the use of unique identification for individual things together with search and discovery services, enabling each thing to interact individually, building up an individual life history of its activities and interactions over time.

The Internet of things allows people and things to be connected anytime, anyplace, with anything and anyone, ideally using any path/network and any service. This implies addressing elements such as convergence, content, collections, computing, communication, and connectivity in the context where there is seamless interconnection between people and things and /or between things and things.

In the vision of Internet of things, it is foreseeable that any "thing" will have at least one unique way of identification, creating an addressable continuum of "things" such as computers, sensors, people, actuators, refrigerators, TVs, vehicles, mobile phones, clothes, food, medicines, books, passports, luggage, etc. Having the capability of addressing and communicating with each other and verifying their identities, all these "things" will be able to exchange information and, if necessary, be deterministic. It is also desirable that some "things" have multiple virtual addresses and identities to participate in different contexts and situations under different "personalities".

Many "things" will be able to have communications capabilities embedded within them and will be able to create a local communication network in an ambient environment together with other "things". These ad-hoc networks will connect with other communication networks locally and globally and the functionalities of the "things" will be influenced by the communications capabilities and by the context. "Things" could retrieve reference information and start to utilize new communication means based on their environment.

Keywords

appliance	n.	器械，装置，设备	heterogeneity	n.	异质，异种
continuum	n.	连续体，闭联集	mention	v.	说到，提起，提及
convergence	n.	聚合，汇合，收敛	seamless interconnection		无缝连接
deterministic	adj.	确定性的	tangible	adj.	实在的，确实的，有实质的
foreseeable	adj.	有先见之明，可预见到的	urban	adj.	城市的，在城市里的

16.3　Reading Material 2：RFID Technology

RFID (radio frequency identification) is a system in which an object is uniquely identified by transmitting its identity (a unique serial number) through radio waves. RFID system works well in harsh or dirty environment, without the need for line of sight, whereas other automatic identification technologies like bar codes fails to operate in such environment. Devoid of human errors, RFID provides an easy way to collect information about a product, place, time or transaction quickly.

Wireless technology development and progress is changing people's lifestyles, and RFID technology is a typical example. RFID is a method of identifying unique items using radio waves. When a consumer in the store select a full car of goods to be checked out, the amount of goods purchased will be displayed on the screen at once in seconds without going through any barcode scanning so consumers can reduce the waiting time. Also the Taipei MRT Easy Card also has RFID technology, which combines the ticket, electronic wallet, credit cards and other functions, so that passengers can take a bus from the MRT to the car park by just using a card. We can see from the above example that the short-range RFID technology is getting mature and applied to daily life. Embedded wireless products integrated with simple features and easy carry not only make people quick and convenient, but also means that the wireless significant leap forward in the era of communication technology has come.

The market by using RFID technology has developed a "non-contact sensing" identification card. In an ideal environment, the identification card and reader do not need contact, and card reader can read and write data stored in tags so to interpret the information and identification represented. The so-called radio frequency identification system referred to as RFID is a "non-contact" type of automatic identification technology.

RFID technology is becoming a part of daily life. It has been widely used in many applications like healthcare, patrolling, baggage, toll road, storage management, access control, track management, transportation control, production automation, combined ticket, automatic identification and so on.

Many experts see RFID as an enabler for ubiquitous computing-the integration of computation into the environment: any device, anytime, and anywhere. RFID enabled mobile phones could be a first step in this direction. The combination of mobile phones and RFID technologies promises great potential in the market for mobile telecommunication services.

Keywords

barcode	*n.*	条形码	mature	*adj.*	成熟的，慎重的
harsh	*adj.*	严厉的，苛刻的，粗糙的，荒芜的	non-contact	*n.*	无触点，非接触
			patrol	*v.*	巡视，巡逻，侦查
interpret	*v.*	说明，解释，把……理解为	Radio Frequency Identification		无线射频识别技术
leap	*v.*	跳跃，跃起	radio wave		无线电波

16.4　定语从句

定语从句在句子中起定语作用，修饰一个名词或代词，有时也可修饰一个句子。被定语从句修饰的名词、词组或代词称先行词，定语从句通常跟在先行词的后面。

例：This is the software that I would like to buy.

译文：这就是我想买的那个软件。

"that I would like to buy"是定语从句，"the software"是先行词。

通常，定语从句都由关系代词 that、which、who、whom、whose 和关系副词 when、where、why、how 引导。关系代词和关系副词往往放在先行词和定语从句之间，起联系作用，同时还代替先行词在句中担任一定的语法成分，如主语、宾语、定语和状语等。

例：The man who will give us a lecture is a famous professor.

译文：将要给我们讲演的人是位著名的教授。

该句中，"who will give us a lecture"是由关系代词"who"引导的定语从句，修饰先行词"the man"，"who"在从句中作主语。

定语从句根据其与先行词的密切程度可分为限定性定语从句和非限定性定语从句。

1．限定性定语从句

限定性定语从句与先行词关系密切，是整个句子不可缺少的部分，没有它，句子的意思就不完整或不明确。这种定语从句与主句之间不用逗号隔开，译成汉语时，一般先译定语从句，再译先行词。

限定性定语从句如果修饰人，一般用关系代词 who，有时也用 that。若关系代词在句子中作主语，则 who 用得较多，且不可省略；若关系代词在句子中作宾语，就应当使用宾格 whom 或 that，但在大多数情况下都可省略。若表示所属，就应用 whose。

限定性定语从句如果修饰物，用 that 较多，也可用 which。它们在句中既可作主语，也可作宾语。若作宾语，则大多可省略。

例：Those who agree with me please put up your hands.

译文：同意我的观点的人请举手。

who agree with me 是定语从句，修饰 Those。who 既是引导词，又在句中作主语，who 不能省略。

例：PCTOOLS are tools whose functions are very advanced.

译文：PCTOOLS 是功能很先进的工具。

因为 functions 和 tools 之间是所属关系，故用所有格 whose。

例：Mouse is an instrument which operators often use.

译文：鼠标是操作员经常使用的一种工具。

which 引导的定语从句修饰 an instrument。因为 which 在从句中作 use 的宾语，故可省略。

下面各例也是限定性定语从句。

例 1：That is the reason why I am not in favor of the plan.

译文：这就是我不赞成该计划的原因。

例 2：Potential energy is the energy that a body has by virtue of its position.

译文：势能是指物体由于自身的位置而具有的能量。

例 3：I want to buy a car which color is blue.

译文：我想买一辆蓝色的车。

2．非限定性定语从句

非限定性定语从句与先行词的关系比较松散，从句只对先行词附加说明，如果缺少，不会影响句子的主要意思。从句与主句之间常用逗号隔开，译成汉语时，从句常单独译成一句。

非限定性定语从句在修饰人时用 who、whom 或 whose，修饰物时用 which，修饰地点和时间时用 where 和 when 引导。关系代词 that 和关系副词 why 不能引导非限定性定语从句。

例 1：We do experiments with a computer, which helps to do many things.

译文：我们利用计算机做实验，计算机可帮助做许多工作。

Which 引导的非限定性定语从句是对先行词 a computer 的说明。

例 2：The meeting will be put off till next week, when we shall have made all the preparations.

译文：会议将推迟到下周，那时我们将做好一切准备。

例 3：They'll fly to America, where they plan to stay for 10 days.

译文：他们将飞往美国，计划在那逗留 10 天。

例 4：Mechanical energy is changed into electric energy, which in turn is changed into mechanical energy.

译文：机械能转变为电能，而电能又转变为机械能。

16.5　Exercises

1. **Translate the following phrases.**

   Internet of Things

   pervasive computing

   ubiquitous computing

   ambient intelligence

   mechanical movement

   intelligent interfaces

   information network

   mobile communication

   2-way network connectivity

   data synchronization techniques

   embedded information processing

   physical location

2. **Fill in the blanks with appropriate words or phrases.**

   a. future　　　b. people　　　c. technical　　　d. dynamic
   e. processor　　f. commodity　　g. middleware　　h. sensors

   (1) Internet of Things (IoT) is an integrated part of _____ Internet.

   (2) IoT could be defined as a _____ global network infrastructure with self configuring capabilities.

   (3) Applications, services, _____ components, networks, and endpoints will be structurally connected in entirely new ways.

   (4) Connectivity will become in the IoT a kind of _____, available to all at a very low cost and not owned by any private entity.

(5) From a _____ point of view, the Internet of Things is not the result of a single novel technology.

(6) Objects collect information about their surroundings with _____, record it, forward it or react directly to it.

(7) Smart objects feature a _____ or microcontroller, plus storage capacity.

(8) Smart objects can communicate with _____ in an appropriate manner (either directly or indirectly, for example via a smart phone).

3. **Match the following terms to the appropriate definition.**

   a. laser printer   b. LCD   c. online
   d. login   e. LED   f. newsgroup

(1) Ongoing discussions that are spread around by E-mail among people on the Internet who share a common interest.

(2) Simply means that you are connected to the Internet.

(3) The act of connecting to a computer system by giving your credentials. It usually includes your "username" and "password".

(4) It is a semiconductor device that emits incoherent narrow-spectrum light when electrically biased in the forward direction. The color of the emitted light depends on the chemical composition of the semi-conducting material used, and can be near-ultraviolet, visible or infrared.

(5) It is a thin, flat display device made up of any number of color or monochrome pixels arrayed in front of a light source or reflector. It is prized by engineers because it uses vary small amounts of electric power, and is therefore suitable for use in battery-powered electronic devices.

(6) It uses laser beams in the printing process. This type of non-impact printer can attain the unbelievable speeds of approximately 20,000 lines per minute.

Unit 17 Computer Virus

17.1 Text

A virus is a program that reproduces its own code by attaching itself to other executable files, so the virus code is executed when the infected executable file is executed. Here, the program (COM or EXE file) refers to an executable file. To attach might mean physically adding to the end of a file, inserting into the middle of a file, or simply placing a pointer to a different location on the disk somewhere where the virus can find it.

Most viruses place self-replicating codes in other programs, so that when those other programs are executed, even more programs are "infected" with the self-replicating codes. These self-replicating codes, when caused by some event, may do a potentially harmful act to your computer.[1]

Categories of Computer Viruses

A macro virus infects a set of instructions called a "macro". A macro is essentially a miniature program that usually contains legitimate instructions to automate document and worksheet production. Experts estimate that more than 2000 viruses exist. Nevertheless, most virus damage is caused by fewer than 10 viruses. Of these viruses, macro viruses account for more than 90 percent of virus attacks.

A modern Trojan horse is a computer program that appears to perform one function while actually doing something else. A Trojan horse is not a virus, but it might carry a virus. Instead of carrying a virus, a Trojan horse can be designed to destroy data or steal passwords.

Time bombs and logic bombs are two computer viruses which lurk in your computer system without your knowledge. A time bomb is a computer program that stays in your system undetected until it is triggered by a certain event in time, such as when the computer system clock reaches a certain date. A time bomb is usually carried by a virus or Trojan horse. A logic bomb is a computer program that is triggered by the appearance or disappearance of specific data.

A worm is a program designed to enter a computer system—usually a network—through security "holes". Like a virus, a worm reproduces itself. Unlike a virus, a worm doesn't need to be attached to a document or executable program to reproduce.

The Spread of Computer Virus

Viruses can be disguised as attachments of funny images, greeting cards, or audio and video files.

Viruses can spread through download on the Internet. They can be hidden in illicit software and other files or programs you might download. And it will spread when you run the files or programs. Some viruses will spread on the Internet at certain time, that may be one day at certain time. Of course, when you use E-mail on the Internet, you must be careful with the uncertain mails. If you are not sure who emailed the letter, you should not open it, as it may be with a virus.

Viruses can also spread through the USB port when you use USB disk, MP3 and so on. Today, there are many viruses designed for the USB. When you insert the USB device, the antivirus software will not scan the USB device automatically, and the USB device will be the virus source to the computer.[2] So we must be careful when we use the MP3 and other movable stored devices.

The Symptoms of Virus Infection

Viruses remain free to proliferate only as long as they exist undetected. Accordingly, the most common viruses give off no symptoms of their infection. Antivirus tools are necessary to identify these infections. However, many viruses are flawed and do provide some tip-offs to their infection. Here are some indications to watch for:

- Changes in the length of programs
- Changes in the file date or time stamp
- Longer program load time
- Slower system operation
- Reduced memory or disk space
- Bad sectors on your floppy
- Unusual error messages
- Unusual screen activity
- Failed program execution
- Unexpected writes to a drive

The Harm of Virus

The main harms caused by viruses are as following.

- Format the whole disk or the sector of the disk.
- Delete important information on the hard disk.
- Delete data files and execute files on the floppy disk, hard disk and Internet.
- Occupy the disk space.
- Modify and damage the data of the file.
- Reduce the speed of the CPU.
- Disturb the operation by the user and damage the common display of the computer.

Preventive Methods

There are two common methods used to detect viruses. First, and by far the most common method of virus detection is using antivirus software. The disadvantage of this detection method is that users are only protected from viruses that predate their last virus definition update. The second method is to use a heuristic algorithm to find viruses based on common behaviors. This method has the ability to detect viruses that antivirus security firms have not yet created a signature for.

Many users install antivirus software that can detect and eliminate known viruses after the computer downloads or runs the executable files. They work by examining the content of the computer's memory and the files stored on fixed or removable drives, and comparing those files against a database of viruses. Users must update their software regularly to patch security holes. Antivirus software also needs to be regularly updated in order to gain knowledge about the latest threats.

Recovery Methods

Once a computer has been compromised by a virus, it is usually unsafe to continue using the same computer without completely reinstalling the operating system.[3] However, there are a number of recovery options that exist after a computer has a virus.

♦ Data Recovery

We can prevent the damage done by viruses by making regular backups of data (and the operating system) on different media, that are either kept unconnected to the system, or not accessible for other reasons, such as using different file systems.[4] In this way, if data is lost through a virus, one can start again using the backup.

♦ Virus Removal

One possibility is a tool known as system restore, which restores the registry and critical system files to a previous checkpoint. Often, a virus will cause a system to hang, and a subsequent hard reboot will render a system restore point from the same day.

♦ OS Reinstallation

As a last ditch effort, if virus is on your system and antivirus software can not clean it, then reinstalling the OS may be required.[5] To do this, the hard drive is completely erased and the OS is installed from media (eg. CDROM). Do not forget, important files should first be backed up before you reinstall the OS.

Keywords

antivirus	n.	防病毒（软件），抗病毒（软件）		predate	v.	在日期上先于
				preventive	n.	预防措施，预防法
attach	v.	附上，加上		proliferate	v.	繁殖，扩散
backup	n.	备份		registry	n.	注册表
compromised	v.	危害，危及		reinstallation	n.	重装
disturb	v.	干扰，打扰		removal	n.	移位，移除
flaw	n.	缺陷，缺点		render	v.	给予，反映
heuristic	adj.	启发式的，探索式的		restore	v.	恢复，复兴
illicit	adj.	违法的，违禁的		symptom	n.	征兆，症状
infect	v.	传染，散布病毒		trigger	v.	触发，激起
legitimate	adj.	合法的，正常的		worksheet	n.	工作单，记工单，操作单
lurk	v. n.	潜伏，潜在				

Notes

[1] These self-replicating codes, when caused by some event, may do a potentially harmful act to your computer.

译文：当这些自我复制代码被一些事件触发时，或许会做出一些对计算机有潜在危害的行为。

说明：本句的"when caused by some event"是插入语，作时间状语。

[2] When you insert the USB device, the antivirus software will not scan the USB device automatically, and the USB device will be the virus source to the computer.

译文：当插入 USB 设备时，反病毒软件不能自动扫描 USB 设备，该 USB 设备就会成为计算机的病毒源了。

说明："When you insert the USB device"是时间状语，后跟两个并列的句子。

[3] Once a computer has been compromised by a virus, it is usually unsafe to continue using the same computer without completely reinstalling the operating system.

译文：一旦计算机已经受病毒危害，在未完全重装操作系统的情况下使用同一台计算机通常是不安全的。

说明："Once a computer has been compromised by a virus"作条件状语，"it"是形式主语，真正的主语是不定式结构。

[4] We can prevent the damage done by viruses by making regular backups of data (and the Operating System) on different media, that are either kept unconnected to the system, or not accessible for other reasons, such as using different file systems.

译文：我们可以通过定期在不同媒介上备份数据（和操作系统）的方法来避免病毒所造成的危害，它们与系统不连接，或者由于其他缘由而不可访问，如使用不同的文件系统。

说明："by making regular…"作状语，"that are either …, or not …"作定语，修饰"different media"，"such as using different file systems"作"other reasons"的同位语。

[5] As a last ditch effort, if virus is on your system and antivirus software can not clean it, then reinstalling the OS may be required.

译文：作为最后的努力，如果你的系统上有病毒而反病毒软件又不能清除它，那么就可能需要重装操作系统了。

说明：本句由"if"引导条件状语从句，主语是分词短语"reinstalling the OS"。

17.2　Reading Material 1：Computer Virus Timeline

The computer virus is an outcome of the computer overgrowth in the 1980s. The cause of the term "computer virus" is the likeness between the biological virus and the evil program infected with computers. A computer virus is a program designed to replicate and spread on its own, generally with the victim being obvious to its existence.

1981: Apple Viruses 1, 2 and 3 are some of the first viruses "in the wild" or in the public domain. Found on the Apple Ⅱ operating system, the viruses spread through Texas A&M via pirated computer games.

1983: Fred Cohen, while working on his dissertation, formally defined a computer virus as a computer program that can affect other computer programs by modifying them in such a way as to include a (possibly evolved) copy of itself.

1986: Two programmers named Basit and Amjad replaced the executable code in the boot sector of a floppy disk with their own code designed to infect each 360KB floppy accessed on any drive.

1987: The Lehigh virus, one of the first viruses, infects command.com files.

1988: One of the most common viruses, Jerusalem, was unleashed. Activated every Friday, the virus affects both .exe and .com files and deletes any programs run on that day.

1990: Symantec launched Norton Antivirus, one of the first antivirus programs developed by a large company.

1992: The dark avenger mutation engine (DAME) was created. It is a toolkit that turns ordinary viruses into polymorphic viruses. The virus creation laboratory (VCL) was also made available. It is the first actual virus creation kit.

1998: Currently harmless and yet to be found in the wild, Strange Brew is the first virus to infect Java files. The virus modifies CLASS files to contain a copy of itself within the middle of the file's code and to begin execution from the virus section.

1999: The Melissa virus, W97M/Melissa, executes a macro in a document attached to an E-mail, which forwards the document to 50 people in the user's Outlook address book. The virus also infects other Word documents and subsequently mails them out as attachments. Melissa spread faster than any previous virus, infecting an estimated 1 million PCs.

2000: The Love Bug, also known as the ILOVEYOU virus, sends itself out via Outlook, much like Melissa. The virus comes as a VBS attachment and deletes files, including MP3 and JPG. It also sends usernames and passwords to the virus's author.

2002: The LFM-926 virus appeared in early January, displaying the message "Loading.Flash.Movie" as it insects Shockwave Flash (.swf) files. Celebrities named viruses continue with the "Shakira", "Britney Spears" and "Jennifer Lopez" viruses emerging. The Klez worm, an example of the increasing trend of worms that spread through E-mail, overwrite files, create hidden copies of the originals, and attempt to disable common antivirus products. The Bugbear worm also made it first appearance in September. It is a complex worm with many methods of infecting systems.

2003: In January the relatively benign "Slammer" (Sapphire) worm because the fastest spreading worm to date, infecting 75,000 computers in approximately ten minutes, doubling its numbers every 8.5 seconds in its first minute of infection. The Sobig worm becomes the first one to join the spam community. Infected computer systems have the potential to become spam relay points.

2004: The fast-spreading Sasser computer worm affected an estimated one million computers running Windows in May. Victims include businesses, such as British Airways, banks, and government offices, including Britain's Coast Guard. The worm does not cause irreparable harm to computers or data, but it does slow computers and cause some to quit or reboot without explanation.

Keywords

benign	adj. 良性的	overgrowth	n.	生长过快，繁茂
biological	adj. 生物学上的	pirate	n.	盗版，盗印
celebrity	n. 名人，知名之士	spam	n.	电子垃圾
dissertation	n. 论文，学位论文	timeline	n.	年鉴
evil	adj. 有害的，邪恶的	toolkit	n.	工具箱
evolve	v. 进化，进展，发展	unleash	v.	释放，解放
irreparable	adj. 不能挽回的，不能恢复的	victim	n.	牺牲者，受害者

17.3　Reading Material 2：Using Antivirus Software

If you detect the symptoms of a virus on your computer system, you should immediately take steps to stop the virus from spreading. If you are connected to a network, alert the network administrator that you found a virus on your workstation. The network administrator can then take action to prevent the virus from spreading throughout the network. If you think that a computer has invaded your personal computer, you can find and remove it by using antivirus software. Even if you don't detect any symptoms of a virus, it is not a bad idea to use your antivirus software every month or two, just in case a virus is stealthily lurking on your hard disk.

Before you can use antivirus software, it must be installed on your computer. Because new virus appear every week, it is important that your antivirus software be kept up-to-date so that it contains the information necessary to hunt for new viruses as well as old ones. Typically, the companies that publish antivirus software provide updates that you can easily download from the Internet.

You run your antivirus software just as you would any program-by using the Windows Start button, then selecting the program from the Program menu. Typically, the antivirus program will scan your computer's memory and hard disk. If it detects a virus, you will see a message indicating the name of the virus. Then you can decide whether you want to remove it immediately.

Some antivirus software stays active and watches for viruses whenever your computer is on. When operating in this mode, you don't have to remember to periodically initiate a virus scan because your antivirus software automatically checks any new programs that you install, any macros that you receive, and any files that you download from the Internet.

Keywords

alert	v.	使警戒，使警惕	initiate	v.	启动，开始
indicate	v.	指示，表示	stealthily	adv.	偷偷地，秘密地

17.4　Terms

1. von Neumann Architecture

An early computer created by Hungarian mathematician John von Neumann. It included three components used by most computers today: a CPU; a primary fast-access memory; and a secondary slow-access storage area, like a hard drive. The machines stored instructions as binary values and executed instructions sequentially—the processor fetched instructions one at a time and processed them.

2. RS-232-C

RS is the acronym for "Recommended Standard". It is an accepted industry standard for serial communications connections. Adopted by the Electrical Industries Association, this RS defines the specific lines and signal characteristics used by serial communications controllers to standardize the transmission of serial data between devices. The letter C denotes that the current version of the standard is the third in a series.

3. BIOS

BIOS is the acronym of basic input/output system. It is the built-in software that determines what a computer can do without accessing programs from a disk. The BIOS is typically placed in a ROM chip that comes with the computer. This ensures that the BIOS will always be available and will not be damaged by disk failures.

4. CRT

CRT is the acronym for "cathode ray tube". It is the display device used in most computer displays, video monitors, televisions and oscilloscopes. The CRT developed from Philo Farnsworth's work was used in all television sets until the late 20th century and the advent of plasma screens, LCDs, DLP, OLED displays, and other technologies.

5. Word Size

A computer word is the amount of data that a CPU can manipulate at one time. Data size is often measured in bits or bytes. Most newer CPU chips are designed for 64 bit words.

6. Algorithm

An algorithm is a formula or set of steps for solving a particular problem. To be an algorithm, a set of rules must be unambiguous and have a clear stopping point. Algorithms can be expressed in any language, from natural languages to programming languages.

7. RAM

Random access memory, commonly known by its acronym RAM, is a type of computer storage whose contents can be accessed in any order. This is in contrast to sequential memory devices such as magnetic tapes, discs and drums, in which the mechanical movement of the storage medium forces the computer to access data in a fixed order.

8. ROM

ROM is the acronym for "read only memory". It is used as a storage medium in computers. Because it cannot easily be written to, its main uses lie in the distribution of software that is very closely related to hardware, and not likely to need frequent upgrading.

17.5 Exercises

1. **Translate the following phrases.**

    computer virus
    self-replicating
    Trojan horse
    time bomb
    logic bomb
    heuristic algorithm
    removable drive
    security hole

antivirus software

network administrator

virus scan

username and password

2. **Fill in the blanks with appropriate words or phrases.**

   a. other programs   b. USB port   c. function   d. attachments

   e. instructions   f. executable program   g. E-mail   h. free

   (1) Viruses can be disguised as _____ of funny images, greeting cards, or audio and video files.

   (2) Most viruses place self-replicating codes in _____.

   (3) A macro virus infects a set of _____ called a "macro".

   (4) Viruses can also spread through the _____ when you use USB disk.

   (5) Unlike a virus, a worm doesn't need to be attached to a document or _____ to reproduce.

   (6) A modern Trojan horse is a computer program that appears to perform one _____ while actually doing something else.

   (7) Viruses remain _____ to proliferate only as long as they exist undetected.

   (8) When you use _____ on the Internet, you must be careful with the uncertain mails.

3. **Match the following terms to the appropriate definition.**

   a. primary memory   b. platform   c. pattern recognition

   d. protocol   e. router   f. resistor

   (1) It is a category of computer storage, and it is used to store data that is in active use.

   (2) In computing and technology, it refers to a framework on which applications may be run. The term may occasionally refer to a type of computer game.

   (3) An agreed-upon format for transmitting data between two devices. It can be implemented either in hardware or in software. From a user's point of view, the only interesting aspect about the term is that your computer or device must support the right ones if you want to communicate with other computers.

   (4) It is an important field of computer science, particularly visual and sound patterns. It is central to optical character recognition (OCR), voice recognition, and handwriting recognition.

   (5) An electronic component that is deliberately designed to have a specific amount of resistance.

   (6) A device that forwards data packets along networks. It is connected to at least two networks, commonly two LANs or WANs or a LAN and its ISP network. It uses headers and forwarding tables to determine the best path for forwarding the packets, and they use protocols such as ICMP to communicate with each other and configure the best route between any two hosts.

Unit 18 Office Automation

18.1 Text

Office automation (OA) is the application of computer and communications technology to improve the productivity of clerical and managerial office workers. In the mid-1950s, the term was used as a synonym for almost any form of data processing, referring to the ways in which bookkeeping tasks were automated. After some years of disuse, the term was revived in the mid-1970s to describe the interactive use of word and text processing systems, which would later be combined with powerful computer tools, thereby leading to a so-called "integrated electronic office of the future".

Personal computer-based office automation software has become an indispensable part of electronic management in many countries. Word processing programs have replaced typewriters; spreadsheet programs have replaced ledger books; database programs have replaced paper-based electoral rolls, inventories and staff lists; personal organizer programs have replaced paper diaries; and so on.

Office automation encompasses six major technologies:
- Data processing—Information in numeric form usually calculated by a computer.
- Word processing—Information in text form—word and numbers.
- Graphics—Information that may be in the form of numbers and words, then keyed into a computer and displayed on a screen in a graph, chart, table, or other visual form that makes it easier to understand.
- Image—Information in the form of pictures. Here an actual picture or photograph is taken, entered into the computer, and shown on a screen.
- Voice—The processing of information in the form of spoken words.
- Networking—The linking together electronically of computer and other office equipment for processing data, words, graphics, image, and voice.

Initially, systems sold by major manufacturers were aimed at clerical and secretarial personnel. These were mainly developed to do word processing and record processing (maintenance of small sequential files, such as names and addresses, which are ultimately sorted and merged into letters). More recently, attention has also been focused on systems, which directly support the principals (managers and professional workers). Such systems emphasize the managerial communications function.

Today's organizations have a wide variety of office automation hardware and software components at their disposal. The list includes telephone and computer systems, electronic mail, word processing, desktop publishing, database management systems, two-way cable TV, office-to-office satellite broadcasting, on-line database services, and voice recognition and synthesis. Each of these components is intended to automate a task or function that is presently performed manually. But experts agree that the key to attaining office automation lies in integration—incorporating all the components into a whole system such that information can be processed and communicated with maximum technical assistance and minimum human intervention.

The fast pace of modern business requires critical information quickly. At the same time, government demands and business bureaucracy require extensive amounts of paperwork. As a result, modern business offices are reexamining traditional methods of doing office work to find better ways to capture and communicate information when and where it is needed.[1] They seek the most efficient method to generate, record, process, file, and communicate or distribute information.[2] Modern technology offers office automation as the foundation of an economical solution. Present and future office system aim to develop integrated information processing networks that bring together everything a firm needs to conduct its daily business effectively.

(1) Word Processing

Word processing refers to the methods and procedures involved in using a computer to create, edit, and print documents. Most standard word processing features are supported, including footnotes and mail-merge but no tables or columns. The interface uses customizable toolbars, and the editing screen is a zoom-able draft mode that optionally displays headers, footnotes, and footers. An un-editable print preview displays a full-page or facing-page view. Fonts, keyboard layouts, and input direction change when you select a new language, and keyboard layouts can be customized.[3]

Word processors vary considerably, but all word processors support some basic features. Word processors that support only the basic features (and maybe a few others) are called text editors. Most word processors, however, support additional features that enable you to manipulate and format documents in more sophisticated ways. These more advanced word processors are sometimes called full-featured word processors. Full-featured word processors usually support the following features:

- Insert and delete text
- Cut, paste and copy
- Page size and margins
- Search and replace
- Word wrap
- Print
- File management
- Font specifications
- Footnotes and cross-references
- Graphics
- Headers, footers and page numbering
- Layout
- Macros

- Merge
- Spell checker
- Tables of contents and indexes
- Thesaurus
- Windows
- WYSIWYG (what you see is what you get)

(2) Desktop Publishing

One important application that exemplifies the relationship of personal computing to office automation is desktop publishing, a term that was coined in 1985 by Paul Brainard, founder of Aldus Corporation.[4] In its simplest form, desktop publishing entails the use of personal computers to prepare and print a wide variety of typeset-or near-typeset-quality documents. The process involves composing the text, manipulating graphics, making up or composing the document, and publishing the finished product on a laser printer or typesetter.[5]

(3) Video Conferencing

Video conferencing involves the linking of remote sites by one-way or two-way television. If meeting rooms or conference rooms in offices can be equipped with the necessary audiovisual facilities, travel time and money can be saved by holding a teleconference instead of a face-to-face conference. Many businesses are experimenting with sales and board meetings through video conferencing. The cost is still high, especially if the video conference involves a direct two-way satellite channel connection.

(4) Videotext

Videotext is a form of electronic publishing that consists of a database connected to terminals or personal computers. It stores and displays information in blocks of information called frames or pages. Videotext offers a simple user interface by displaying pages of information that contain text and graphics and providing an easy way for an untrained user to access one or more databases of information.

Offices using videotext can electronically publish standard operating procedure manuals, sales literature, in-house newsletters, company wide phone lists, and other information that is subject to periodic change. Some videotext systems permit users to conduct electronic transactions. For example, an office worker might view an inventory database of office supplies, and then electronically place an order for various supplies.

Keywords

bureaucracy	n.	官僚机构，官僚	founder	n.	创立人，创始人
clerical	adj.	职员的，办公室人员的	indispensable	adj.	必需的，重要的
coin	v.	制造	intervention	n.	干预，干涉
disposal	n.	配置，安排	ledger	n.	总账，分类账
disuse	n.	废弃，废止	literature	n.	文献，著作
entail	v.	使产生，引起，带来	managerial	adj.	管理人的，经理的
exemplify	v.	示范，表明	manually	adv.	手工地，人工地
footer	n.	页脚	margin	n.	边距，边缘
footnote	n.	脚注	newsletter	n.	业务通信，时事通信

secretarial	*adj.*	秘书的	thesaurus	*n.*	词库，词汇集
synonym	*n.*	同义词	typesetter	*n.*	排字工人，排字机

Notes

[1] As a result, modern business offices are reexamining traditional methods of doing office work to find better ways to capture and communicate information when and where it is needed.

译文：因此，现代商业办公室都在对传统的办公方法进行研究与调整，以找到可以随时随地获取信息和传递信息的较好的途径。

说明：本句的"to find better…"作目的状语。"doing office work"作定语，修饰"traditional methods"。

[2] They seek the most efficient method to generate, record, process, file, and communicate or distribute information.

译文：这些部门正在寻求用来产生、记录、处理、归档、交流或发布信息的最有效方法。

说明：本句的不定式结构作定语，修饰宾语。

[3] Fonts, keyboard layouts, and input direction change when you select a new language, and keyboard layouts can be customized.

译文：当选择一种新语言时，字体、键盘布局和输入方向均可变化，并且键盘布局可以定制。

说明：主句是两个并列句，"Fonts, keyboard layouts, and input direction"是第一分句的主语，"when you select a new language"作时间状语。

[4] One important application that exemplifies the relationship of personal computing to office automation is desktop publishing, a term that was coined in 1985 by Paul Brainard, founder of Aldus Corporation.

译文：桌面印刷系统是一种重要的应用软件，它表明了个人计算机与办公自动化之间的关系。该术语是1985年由ALDUS公司创始人保罗·布赖纳德提出的。

说明：本句的"that exemplifies…"作定语，修饰主语，"a term"是表语"desktop publishing"的同位语，"that was coined…"作定语，修饰"a term"。

[5] The process involves composing the text, manipulating graphics, making up or composing the document, and publishing the finished product on a laser printer or typesetter.

译文：此工序包括排字、绘图、形成文件，最后在激光打印机或排字机上出版。

说明：本句中的宾语是4个分词短语，即"manipulating …, making…"。

18.2　Reading Material 1：Excel

　　Microsoft Excel is a spreadsheet program that allows you to organize data, complete calculations, make decisions, graph data, develop professional looking reports, publish organized data on the Web, and access real-time data from Web sites. Using Microsoft Excel, you can create hyperlinks within a worksheet to access other Office documents on the network, an organization's intranet, or the Internet. You also can save worksheets as static and dynamic Web pages that can be viewed using your browser. Static Web pages cannot be changed by the person viewing them. Dynamic Web pages give the person viewing them many Excel capabilities in their browser. In addition, you can create and run queries to retrieve information from a Web page directly into worksheet.

Electronic spreadsheet can be used in all walks of life, for example, accountants use electronic spreadsheets to check financial statements and compile payrolls; in commerce, electronic spreadsheets are used to prepare budget and perform comparison on quoted prices; teachers use electronic spreadsheets to record marks of courses students get in examinations; scientists use electronic spreadsheets to analyze experimental data. Housewives use electronic spreadsheets to keep track of family expenditure. So, you can see how extensively and flexibly electronic spreadsheets are used. More importantly, electronic spreadsheets can relieve us from repetitive and tedious computations and allow us to concentrate our mind on analysis and assessment of the computed results, to bring our professional potential into full play.

An Excel file is a workpad, which may contain several worksheets. In the worksheet, at the intersection of each row and each column, there is a small square area called the "unit square". Single click the unit square, a blank frame occurs around it, and this unit square is activated. Each unit square can be represented by row number and column letter. In Excel, integral data include character style, value style, date style and logic style, in addition, annotation information and formulas can also be input in unit square. Selection of operands in the base on which revision and edition can be performed. After being selected, the operand is highlighted.

♦ Selection of Contents in a Unit Square

At the beginning of the content to be selected press down left key of the mouse and keep pressing it, then drag it to the end of the content to be selected.

♦ Section of Unit Square

Single click a unit square, this square is activated (selected); single click the row head or column head, this row/column is respectively selected.

♦ Selection of Continual Unit Squares

Put the mouse at the upper left corner and then drag it to lower right corner of an area, to select the continual unit squares within this area.

♦ Selection of Non-continual Unit Squares

Press Ctrl key and single click the unit squares.

♦ Selection of all Contents of the Current Worksheet

Press down Ctrl+A keys or single click "All Select" button at the upper left corner of the unite square.

In Excel, formulas and functions can be used to perform statistics or computations of data in electronic spreadsheets. When data change, the results of computations will automatically upgrade. All formulas begin with the sign =. It contains arithmetic operation signs, text operation signs, comparative operation sign and quoting operation sign—totally four categories of signs. Functions are built-in formulas pre-defined in Excel, including SUM, AVERAGE, COUNT, and MAX, etc.

Various statistic charts in electronic spreadsheets represent the data in unit squares, so as to make the data easy and intuitive to understand, meanwhile, when the data change, the charts automatically follow the change.

In Excel, charts include "imbedded chart" and "independent chart". The former is placed and displayed and printed together with worksheets; the latter is an independently generated worksheet, and it is printed separated from the original datasheet. In Excel, there are 2-dimensional chart and 3-dimentational charts. They are gradually generated by using the "chart direction" button in tool bar or "picture" command in "insert" menu. The user can edit the generated charts. There are zigzag chart, bar chart, pie chart and others available for selection.

Excel has not only abilities to perform simple data management and computation but also to set up database. To use datasheet, increase or delete record, and do sequencing sieving, classifying and summing up data.

Keywords

accountant	n.	会计员，出纳	payroll	n.	职工名册，工资表
assessment	n.	评价，评定	professional	adj.	专业的，职业的
expenditure	n.	开销，费用	quote	v.	引用，引证
formula	n.	公式，定则	zigzag	adj.	Z字形的，锯齿形的，折线的
highlight	v.	强调，使显著			
intersection	n.	交叉点，交叉线			

18.3 Reading Material 2：Telephone

A telephone is one of the simplest devices you have in your house. It is so simple because the telephone connection to your house has not changed in nearly a century. If you have an antique phone from 1920s, you could connect it to the wall jack in your house and it would work fine.

As you see, telephone only contains three parts and they are all simple.

(1) A Switch

The function of a switch is to connect and disconnect the phone from the network. This switch is generally called the hook switch. It connects when you lift the handset.

(2) A Speaker

This is generally a little 50-cent, 8-ohm speaker of some sort.

(3) A Microphone

In the past, telephone microphones have been as simple as carbon granules compressed between two thin metal plates. Sound waves from your voice compress and decompress the granules, changing the resistance of the granules and modulating the current flowing through the microphone.

The only problem with the phone is that when you talk, you will hear your voice through the speaker. Most people find that annoying, so any "real" phone contains a device called a duplex coil or something functionally equivalent to block the sound of your own voice from reaching your ear. A modern telephone also includes a bell so it can ring and a touch-tone keypad and frequency generator.

In a modern phone there is an electronic microphone, amplifier and circuit to replace the carbon granules and loading coil. The mechanical bell is often replaced by a speaker and a circuit to generate a pleasant ringing tone.

The telephone network starts in your house. A pair of copper wires runs from a box at the road to a box (often called an entrance bridge) at your house. From there, the pair of wires is connected to each phone jack in your house.

Alone the road runs a thick cable packed with 100 or more copper pairs. Depending on where you are located, this thick cable will run directly to the phone company's switch in your area or it will run to a box about the size of a refrigerator that acts as a digital concentrator.

Keywords

antique	*adj.*	旧式的，过时的	handset	*n.*	送受话器
carbon	*n.*	碳，炭精电极	hook	*n.*	钩
compress	*v.*	压缩，浓缩	jack	*n.*	插口，插座
duplex	*adj.*	双的，二重的	refrigerator	*n.*	电冰箱
granule	*n.*	颗粒，细粒，粒状斑点			

18.4 Terms

1. CGA

It is a video adapter board introduced by IBM in 1981, and the acronym is Color Graphics Adapter. The CGA is capable of several character and graphics modes, including character modes of 40 or 80 horizontal characters by 25 vertical lines with 16 colors, and graphics modes of 640 horizontal pixels by 200 vertical pixels with 2 colors, or 320 horizontal pixels by 200 vertical pixels with 4 colors.

2. Clock Speed

Usually when referring to a computer, the term "clock rate" is used to refer to the speed of the CPU. The clock rate is the fundamental rate in cycles per second, measured in hertz, at which a computer performs its most basic operations such as adding two numbers or transferring a value from one processor register to another.

3. SCSI

It is a standard high-speed parallel interface, and its meaning is small computer system interface. A SCSI interface is used to connect microcomputers to SCSI peripheral devices, such as many hard disks and printers, and to other computers and Local Area Networks.

4. Cache

Cache is a special high-speed storage mechanism. It can either a reserved section of main memory or an independent high-speed storage device. Two types of caching are commonly used in personal computers: memory caching and disk caching. When data is found in the cache, it is called a cache hit, and the effectiveness of a cache is judged by its hit rate.

5. Motherboard

A motherboard, also known as a main-board, logic board or system board, and sometimes abbreviated as mobo, is the central or primary circuit board making up a complex electronic system, such as a computer.

6. Partitioning

Partitioning a hard drive enable you to logically divide the physical capacity of a single drive into separate areas called partitions. You can then treat each of the partition as an independent disk drive, such as a C drive and a D drive.

7. Driver

A driver is a program that controls a device. Every device, whether it is a printer, disk drive, or keyboard, must have a driver program. Many drivers, such as the keyboard driver, come with the operating system. A driver acts like a translator between the device and programs that use the device. Each device has its own set of specialized commands that only its driver knows.

8. Disk Cache

When disk caching is being used, during any disk access the computer system also fetches program or data contents located in neighboring disk areas, such as the entire track, and transports them to a dedicated part of RAM.

18.5 Exercises

1. **Translate the following phrases.**

   office automation
   computer and communication technology
   word processing program
   desktop publishing
   on-line database service
   integrated information processing network
   video conferencing
   face-to-face conference
   electronic publishing
   periodic change
   real-time data
   spreadsheet program

2. **Fill in the blanks with appropriate words or phrases.**

   a. office automation   b. management   c. basic features   d. footnotes
   e. organizations   f. modern business   g. word processors   h. manufacturers

   (1) Personal computer-based office automation software has become an indispensable part of electronic _____ in many countries.
   (2) Most standard word processing features are supported, including _____ and mail-merge but no tables or columns.
   (3) Modern technology offers _____ as the foundation of an economical solution.
   (4) Word processors that support only the _____ (and maybe a few others) are called text editors.
   (5) The fast pace of _____ requires critical information quickly.
   (6) Word processors vary considerably, but all _____ support some basic features.
   (7) Initially, systems sold by major _____ were aimed at clerical and secretarial personnel.
   (8) Today's _____ have a wide variety of office automation hardware and software components at their disposal.

3. **Match the following terms to the appropriate definition.**

   a. resolution　　　　b. RGB　　　　　c. response time
   d. router　　　　　　e. receiver　　　f. repeater

   (1) The length of time between the occurrence of an event and the response of an instrument or circuit to that event.

   (2) A system for representing the colors to be used on a computer display.

   (3) Hosts that are connected to more than one network and route messages between them.

   (4) A measure of picture resolving capabilities of a television system determined primarily by bandwidth, scan rates and aspect ratio. Relates to fineness and details perceived.

   (5) A network device that repeats signals from one cable onto one or more other cables, which restoring signal timing and waveforms.

   (6) A electronic device which can convert electromagnetic waves into either visual or aural signals, or both.

Unit 19 Virtual Reality

19.1 Text

音频

Virtual reality (VR) is a new technology, which emerged after the 1980s. But within several years, it had pervaded into various domains—science, technology, engineering, medicine, culture, entertainment, and its potential in application is just conspicuous. One definition of VR is a wide-field presentation of computer-generated, multi-sensory information that tracks a user in real time. This definition is used at the Electronic Visualization Laboratory in Chicago (EVL). In other words, VR means that the user is surrounded by a computer-generated image that changes depending on the movements—i.e. shifts in perspective—of the user.

With present achievements, using computer hardware and software and advanced sensors, researchers can generate a 3-dimensional artificial virtual environment, where you can walk around, watch in every direction and touch every object in the environment.[1] Everything in the environment is harmoniously combined with other objects so realistically that you may feel you are in a physical environment, but really you are roaming in a virtual world.

An important aspect of any immersive system is tracking. To create the illusion of a real surrounding, and not just a computer image, the image must change when the user's position (and thus perspective) inside the virtual world changes.[2] To do this, the system has to keep track of where the user's eyes are and in what direction they are looking. This will allow the perspective shown in the images to change according to the user's movements, which gives the user the same impression of objects as in a real surrounding. This is inherently difficult, and even more so in the fully immersive environment of the six-surface cube system.

HMD and BOOM

Head-mounted displays (HMDs), which typically also include earphones for the auditory channel as well as devices for measuring the position and orientation of the user, have been the primary VR visual device for much of the 1990s.[3]

In HMDs, projectors feed real-time images to small screens attached inside a kind of helmet that the user wares. In the beginning, HMDs often showed mono pictures, and the user's head movements were tracked. In modern HMDs, stereo pictures are standard, since the technique has become cheaper. Typically, HMDs either have poor resolution or are too heavy. Because only one person at a time can use an HMD, discussion of images is much more difficult.

To increase image quality, Fakespace invented the BOOM (Binocular Omni-Orientation Monitor). Very small monitors are mounted on a mechanical arm, and users look into the monitors like they would look into a pair of binoculars. Tracking occurs when the user moves the arm, which changes the perspective. When a user releases the BOOM, another person can look at the same thing from the same perspective, which is an advantage over HMDs.[4] Since real monitors are used, the resolution is good.

HMDs and BOOMs are similar devices in that the user is fully immersed in the virtual environment and does not see his/her actual surroundings. The BOOM solves several of the limitations of the HMD (e.g., resolution, weight, field of view), but at the expense of reducing the sense of immersion by requiring the user to stand or sit in a fixed position.

CAVE

The concept of a room with graphics projected from behind the walls was invented at EVL in 1992. The images on the walls are usually in stereo to give a depth cue. The main advantage over ordinary graphics systems is that the users are surrounded by the projected images, which means that the images are the users' main field of vision. This is usually called a "CAVE", CAVE Automatic Virtual Environment. The first CAVE was created by the faculty, staff, and students at EVL.

Since then this back-projection method of VR has gained a strong following. It has obvious benefits; it is easy for several people to be in the room simultaneously and therefore see images together; and it is easy to mix real and virtual objects in the same environment. Also, because users see for example their own hands and feet as part of the virtual world, they get a heightened sense of being inside that world.

Various CAVE-like environments exist all over the world today. Most of these have up to four projection surfaces; images are then usually projected on three walls and the floor. Adding projection on the ceiling gives a fuller sense of being enclosed in the virtual world. Projection on all six surfaces of a room allows users to turn around and look in all directions. Thus, their perception and experience are never limited, which is necessary for full immersion. An immersive visualization environment can be used for everything from science to art, and from industrial simulations to education.

While HMDs require that users interact in virtual spaces (they can not see each other in their "real" environment), the immersive room offers the significant advantage of permitting user interaction, discussion, and analysis in the real world. However, the computational cost of generating scenes within an immersive room is very high. Two images must be generated at high refresh rates for each wall in the immersive room. In addition, each wall requires a high-quality projector, and since back projection is used, a large allocation of space is required for projection length.[5] Costing over one-half million dollars, immersive rooms exist only in a handful of large research organizations and corporations.

The VR Responsive Workbench

The VR Responsive Workbench operates by projecting a computer-generated, stereoscopic image off a mirror and then onto a table surface that is viewed by a group of users around the table. Using stereoscopic shuttered glasses, users observe a 3D image displayed above the tabletop. By tracking the group leader's head and hand movements using magnetic sensors, the Workbench permits changing the view angle and interacting with the 3D scene. Other group members observe the scene as manipulated by the group leader, facilitating easy communication between observers about the scene and defining future actions by the group leader. Interaction is performed using speech recognition, a pinch glove for gesture recognition, and simulated laser pointer.

VRML

The virtual reality modeling language (VRML) is a file format for describing interactive 3D objects and worlds. VRML is designed to be used on the Internet, Intranets, and local client systems. VRML is also intended to be a universal interchange format for integrated 3D graphics and multimedia. VRML may be used in a variety of application areas such as engineering and scientific visualization, multimedia presentations, entertainment and educational titles, web pages, and shared virtual worlds.

VRML is capable of representing static and animated dynamic 3D and multimedia objects with hyperlinks to other media such as text, sounds, movies, and images. VRML browsers, as well as authoring tools for the creation of VRML files, are widely available for many different platforms. VRML supports an extensibility model that allows new dynamic 3D objects to be defined and a registration process that allows application communities to develop interoperable extensions to the base standard. There are mappings between VRML objects and commonly used 3D application programming interface (API) features.

The VRML specification defines a file format that integrates 3D graphics and multimedia. Conceptually, each VRML file is a 3D time-based space that contains graphic and aural objects that can be dynamically modified through a variety of mechanisms. VRML defines a primary set of objects and mechanisms that encourage composition, encapsulation, and extension.

Keywords

aural	adj.	耳朵的，听觉的	inherently	adv.	内在地，固有地
binocular	n.	双筒望远镜	mono	adj.	平面的，单声道的
composition	n.	合成，结构，构造	orientation	n.	方向，定向
conspicuous	adj.	显著的，触目的	pervade	v.	普及，蔓延
earphone	n.	耳机，耳塞	projector	n.	投射器，投影机
faculty	n.	教职员	realistically	adv.	现实主义地，现实地
harmoniously	adv.	和谐地，融洽地	registration	n.	记录，注册，登记费
heighten	v.	加深，加强	roam	v.	漫游，漫步
helmet	n.	头盔，防护帽	stereo	adj.	立体的，立体声的
illusion	n.	幻觉，幻影	tabletop	n.	桌面上
immersive	adj.	浸入式的，沉浸的	workbench	n.	工作台

Notes

[1] With present achievements, using computer hardware and software and advanced sensors, researchers can generate a 3-dimensional artificial virtual environment, where you can walk around, watch in every direction and touch every object in the environment.

译文：借助现有的成果，利用计算机硬件和软件及先进的传感器，研究人员能够创造三维人工虚拟环境，你可在其中漫步、四下观望和触摸环境中的每个物体。

说明："With present …advanced sensors"作状语，"where"引导的是地点状语。

[2] To create the illusion of a real surrounding, and not just a computer image, the image must change when the user's position (and thus perspective) inside the virtual world changes.

译文：为了创造一种真实环境的幻象，而不仅仅是计算机图景，这些图景必须随着虚拟世界中的用户位置的变化（由此产生的视角变化）而变化。

说明：本句的主句是"the image must change"，"To create…"作目的状语，"when…"作时间状语。

[3] Head-mounted displays (HMDs), which typically also include earphones for the auditory channel as well as devices for measuring the position and orientation of the user, have been the primary VR visual device for much of the 1990s.

译文：典型的头盔式显示器（HMD）也包括听觉通道用的耳机，以及测量用户的位置与方向的一些设备，在20世纪90年代的大多数日子里，它们一直是主要的VR视觉设备。

说明：这是一个长句，主语是"Head-mounted displays"，"which typically also…"是非限定定语从句，修饰主语，"for much of the 1990s"是时间状语。

[4] When a user releases the BOOM, another person can look at the same thing from the same perspective, which is an advantage over HMDs.

译文：当一个用户不用这个BOOM后，另一个用户可以从相同的视角观察到相同的事物，这一点BOOM要优于HMD。

说明："When a user releases the BOOM,"是时间状语，"which is an advantage over HMDs"作补语。

[5] In addition, each wall requires a high-quality projector, and since back projection is used, a large allocation of space is required for projection length.

译文：另外，每面墙需要高质量的投影仪，并且因为使用了背投，为了投影长度需要配给大的空间。

说明：这是并列句，前句用主动语态，后句用被动语态，"since back projection is used"作原因状语。

19.2 Reading Material 1：Computers and Journalism

Technology has benefited print journalism, too. For decades, typesetters transferred reporters' handwritten stores into neatly set columns of type. Today, reporters use computers and word processing software to tap out their stories and run a preliminary check of spelling and grammar.

Stories are submitted via computer network to editors, who use the same software to edit stories to fit time and space constraints. The typesetting process has been replaced by desktop publishing software and computer to plate (CTP) technology. Digital pages produced with desktop publishing software are sent to a raster image processor (RIP), which converts the pages into dots that form words and images. After a page has been RIPed, a plate-setter uses lasers to etch the dots onto a physical plate, which is then mounted on the printing press to produce printed pages. CTP is much faster and more flexible than typesetting, so publishers can make last-minute changes to accommodate late-breaking stories.

Personal computers have also added a new dimension to the news-gathering process. Reporters were once limited to personal interviews, observation, and fact gathering at libraries, but can now make extensive use of Internet resources and E-mail. Web sites and online databases provide background information on all sorts of topics. Other resources include newsgroups and chat rooms, where reporters can monitor public opinion on current events and identify potential sources.

Most major networks maintain interactive Web sites that offer online polls and bulletin boards designed to collect viewers' opinions. Although online poll respondents are not a representative sample of the population, they can help news organizations gauge viewer opinions and determine whether news coverage is comprehensive and effective.

E-mail has changed the way reporters communicate with colleagues and sources. It is often the only practical method for contacting people in remote locations or distant time zones, and it's useful with reluctant sources. "Vetting" e-mail sources can be difficult, however, so reporters tend not to rely on these sources without substantial corroboration.

For broadcast journalism, digital communications play a major role in today's "live on the scene" television reporting. Most news organizations maintain remote production vans, sometimes called "satellite news gathering (SNG) trucks," that travel to the site of breaking news, raise their antennas, and begin to broadcast. These complete mobile production facilities include camera control units, audio and video recording equipment, and satellite or microwave transmitters.

On the scene reporting no longer requires a truck full of equipment, however. Audiovisual editing units and video cameras have gone digital, making them easier to use and sized to fit in a suitcase. A new breed of "backpack journalists" carries mini-DV cameras, notebook computers, and satellite phones.

Backpack journalists connect their mini-cams to notebook computers with a FireWire cable, transfer their video footage to the hard disk, and then edit the footage using consumer-level video editing software. The resulting video files, compressed for transmission over a satellite phone, are sent to newsroom technicians, who decompress and then broadcast them—all in a matter of seconds.

One disadvantage of backpack journalists' use of mini-cams and compression is that the video quality usually isn't as crisp as images filmed with studio cameras. News organizations with high standards were hesitant to use this lower quality video, but have found that viewers would rather see a low-quality image now than a high-quality image later. To many viewer, a few rough edges just make the footage seem more convincing, more "you are there".

Computers, the Internet, and communications technology make it possible to instantly broadcast live reports across the globe, but live reporting is not without controversy. A reporter who arrives at the scene of a disaster with microphone in hand has little time for reflection, vetting, and crosschecking, so grievous errors, libelous images, or distasteful video footage sometimes find their way into news reports. Technology has given journalists a powerful arsenal of tools for gathering and reporting the news, but has also increased their accountability for accurate, socially responsible reporting.

Keywords

accommodate	v.	适应，调节，供应	grievous	adj.	悲痛的，严重的，难忍的
accountability	n.	有责任，有义务	handwritten	adj.	手写的，用笔写的
arsenal	n.	军械库，武器库	interview	v. n.	面谈，协商，访问
audiovisual	n.	视听	journalism	n.	新闻业
controversy	n.	争论，辩论	libelous	adj.	诽谤的，用语言中伤他人的
corroboration	n.	加强，坚固，坚定	observation	n.	观察，观测
crisp	adj.	易碎的，脆的，有力的	online poll		在线民意测验
distasteful	adj.	令人讨厌的，令人不愉快的	preliminary	adj.	预备的，初步的
			raster image processor (RIP)		光栅图像处理器
gauge	v. n.	量计，标准规格，范围，估计的方法	reluctant	adj.	难处理的，难加工的，厌恶的，勉强的

19.3　Reading Material 2：Virtual Reality in Industry

Computers can be wonderful amplifiers of human capability, and computerization can be a soul-deadening juggernaut of alienation. Is virtual reality the ultimate form of alienation, or can it be used as a tool to diminish some of the alienation that already has occurred as the result of computerization? Consider those cases where the nature of the computer interface, not the use of computer itself, has removed people from the true nature of their work. The economic advantages of computerization will ensure the continued introduction of computer-mediated or computer-controlled systems in the workplace.

Industry—the coordination of matter, energy, and human effort in the production of the ten thousand objects and energies that seem to be vital to modern life—is an area where the right kind of cyberspace application could produce high leverage. And tools tend to evolve very quickly when they are found to produce leverage in an expensive and competitive enterprise. Design—which one might regard as the "front end" of the industrial system—is a discipline in which 3D visual immersion, telepresence, and the ability to manipulate actual design databases by reaching out and touching virtual objects could produce enough competitive advantage to drive virtual reality as a CAD interface. There are areas in the "back end" of industrial production, the molten-metal and caustic-chemical part of the system, where similar points of leverage can be found.

The control of industrial processes, for example, is a little-seen but economically virtual area in which the introduction of computers has had strange effects, and in which a virtual reality interface has high payoff potential.

Process control is somewhere between a science and an art, involving a delicate, vital, and often dangerous minuet of humans and machines. Although those who are not involved with it do not pay much attention to the way process control is managed, the results are essential to many parts of daily life, from the pulp miles that produce the newsprint in our morning papers to the petroleum refineries that fuel our automobiles and the smelters that produce the steel in the freeways we travel, the buildings we enter, the bridges we cross. Process control used to be performed by experts who felt the heat of the smelter on their face, looked at the color of the liquid in the vats, reached right in and grabbed a handful of wood pulp and squeezed it between their fingers. Over the past twenty years, however, every major industry in which process control is a major concern has switched to automated, computer-controlled systems. And that has removed the humans from direct sensory contact, striped them of their ability to directly interact with the process.

Keywords

alienation	n.	异化，转让，疏远		molten	adj.	熔融的，融化了的
caustic	n.	苛性碱，腐蚀性		payoff	n.	出乎意料的事情
computerization	n.	电子计算机计算		petroleum	n.	石油
deadening	n. v.	隔音材料，隔离		refinery	n.	精炼厂
discipline	n. v.	训练，锻炼		smelter	n.	熔炼，冶炼厂
juggernaut	n.	可怕，不可抗拒的力量		soul	n.	心灵，灵魂
minuet	n.	小步舞曲		telepresence	n.	远程监控

19.4 Terms

1. Instruction

An instruction is a form of communicated information that is both command and explanation for how an action, behavior, method, or task is to be begun, completed, conducted, or executed.

2. SRAM

SRAM is the acronym of static random access memory. In a static RAM, binary values are stored using traditional flip-flop logic-gate configurations. A SRAM will hold its data as long as power is supplied to do.

3. DRAM

DRAM is the acronym of dynamic random access memory. It is made with cells that store data as charge on capacitors. The presence or absence of charge in a capacitor is interpreted as a binary 1 or 0.

4. DVD

It was initially developed to store the full contents of a standard two-hour movie, but is now also used to store computer data and software. The acronym DVD for digital video disc refer to a relatively new high-capacity optical storage format that can hold from 4.7GB to 17GB, depending on the number of recording layers and disc sides being used.

5. LCD

LCD is the acronym for "liquid crystal display". It is a thin, flat display device made up of any number of color or monochrome pixels arrayed in front of a light source or reflector. It is prized by engineers because it uses very small amounts of electric power, and is therefore suitable for use in battery-powered electronic devices.

6. Wizard

A wizard—also called a coach, assistant, tutor, expert, and advisor, is a tool to guide you through a step-by-step procedure for completing a standard task, such as prepare a letter, invoice, report, fax, and the like. Wizard do this either by providing directed tutorials or by performing the difficult phases of a task automatically, using the input supplied by the user via some type of dialog box.

7. JavaScript

It is the most popular scripting language on the Internet, and works in all major browsers, such as Internet Explorer, Firefox, Netscape and Mozilla. JavaScript was designed to add interactivity to HTML pages and used in millions of Web pages to improve the design, validate forms, detect browsers and create cookies.

8. Virtual

The term virtual is popular among computer scientists and is used in a wide variety of situations. In general, it distinguishes something that is merely conceptual from something that has physical reality.

19.5 Exercises

1. **Translate the following phrases.**
   virtual reality
   physical environment

immersive system

six-surface cube system

head-mounted display

real-time image

mechanical arm

field of view

back-projection method

magnetic sensor

virtual reality modeling language

electronic visualization laboratory

2. Fill in the blanks with appropriate words or phrases.

   a. aspect   b. HMD   c. students   d. industrial simulations
   e. a file format   f. 1980s   g. 1992   h. environments

(1) Virtual reality is a new technology, which emerged after the _____.

(2) An important _____ of any immersive system is tracking.

(3) Because only one person at a time can use an _____, discussion of images is much more difficult.

(4) The first CAVE was created by the faculty, staff, and _____ at EVL.

(5) The concept of a room with graphics projected from behind the walls was invented at EVL in _____.

(6) Various CAVE-like _____ exist all over the world today.

(7) An immersive visualization environment can be used for everything from science to art, and from _____ to education.

(8) The VRML specification defines _____ that integrates 3D graphics and multimedia.

3. Match the following terms to the appropriate definition.

   a. ring   b. software   c. sector
   d. system board   e. speech recognition   f. search engine

(1) A network topology in which the nodes are connected in a closed loop. Data is transmitted from node to node around the loop, always in the same direction.

(2) It is a sub-division of a track of a magnetic hard disk or optical disc. It stores a fixed amount of data.

(3) It is the central or primary circuit board making up a complex electronic system, such as a computer.

(4) A computer program that is made up of certain instructions or codes that tell your hardware, or computer, what to do.

(5) Uses a robot or computer to search for words or topics on a certain website, or all over the Web.

(6) The field of computer science that deals with designing computer systems that can recognize spoken words.

Unit 20 Artificial Intelligence

音频

20.1 Text

Definitions of artificial intelligence vary along two main dimensions. One is concerned with thought processes and reasoning, the other addresses behavior. Also, some definitions measure success in terms of human performance, whereas the others measure against an ideal concept of intelligence, which we will call rationality. A system is rational if it does the right thing. All this gives us four possible goals to pursue in artificial intelligence: systems that think like humans, systems that think rationally, systems that act like humans and systems that act rationally.

The Goals of Artificial Intelligence

♦ Acting Humanly

The Turing Test, proposed by Alan Turing, was designed to provide a satisfactory operational definition of intelligence. Turing defined intelligent behavior as the ability to achieve human-level performance in all cognitive tasks, sufficient to fool an interrogator. Roughly speaking, the test he proposed is that the computer should be interrogated by a human via a teletype, and passes the test if the interrogator cannot tell if there is a computer or a human at the other end.[1] Programming a computer to pass the test provides plenty to work on. The computer would need to possess the following capabilities: natural language processing to enable it to communicate successfully in English (or some other human language), knowledge representation to store information provided before or during the interrogation, automated reasoning to use the stored information to answer questions and to draw new conclusions and machine learning to adapt to new circumstances and to detect and extrapolate patterns.

♦ Thinking Humanly

If we are going to say that a given program thinks like a human, we must have some way of determining how humans think. We need to get inside the actual working of human minds. There are two ways to do this: through introspection—trying to catch our own thoughts as they go by—or through psychological experiments. Once we have a sufficiently precise theory of the mind, it becomes possible to express the theory as a computer program. If the program's input/output and timing behavior matches human behavior, that is evidence that some of the program's mechanisms may also be operating in humans.[2] The interdisciplinary field of cognitive science brings together computer models from artificial intelligence and experimental techniques from psychology to try to construct precise and testable theories of the working of the human mind.

♦ Thinking Rationally

The development of formal logic in the late nineteenth and early twentieth centuries provided a precise notation for statements about all kinds of things in the world and the relations between them. By 1965, programs existed that could, given enough time and memory, take a description of a problem in logical notation and find the solution to the problem (if one exists).[3]

There are two main obstacles to this approach. First, it is not easy to take informal knowledge and state it in the formal terms required by logical notation, particularly when the knowledge is less than 100% certain. Second, there is a big difference between being able to solve a problem "in principle" and doing so in practice. Even problems with just a few dozen facts can exhaust the computational resources of any computer unless it has some guidance as to which reasoning steps to try first.

♦ Acting Rationally

Acting rationally means acting so as to achieve one's goals, given one's beliefs. An agent is just something that perceives and acts. In this approach, artificial intelligence is viewed as the study and construction of rational agents. In the "laws of thought" approach to artificial intelligence, the whole emphasis was on correct inferences. Making correct inferences is sometimes part of being a rational agent, because one way to act rationally is to reason logically to the conclusion that a given action will achieve one's goals, and then to act on that conclusion.[4] On the other hand, correct inference is not all of rationality, because there are often situations where there is no provably correct thing to do, yet something must still be done.

The Research Subjects of AI

The basic subjects of AI include acquisition of knowledge, representation of knowledge and application of knowledge. In application domain, the research is now focusing on these subjects:

♦ Solutions to Hard Problems

Here hard problems refer to those difficult problems that they have no algorithm or whose algorithms cannot be executed in computer programs.[5] For example, route planning, electricity dispatching, stock market analysis, robot action planning, etc., and in games some problems are difficult, for example, the Tower of Hanol, peasant crossing a river, eight-numbers problems, eight-queen problem, tourist salesman problem.

♦ Automatic Translation

Automatic translation means computer programs doing translation between two different languages. Machine translation is not simply "checking up dictionaries" and translating "word-by-word". True translation should be based on understanding of semantics and grammar rules.

♦ Intelligent Control and Intelligent Management

Intelligent control means AI technology used in control system to solve problems like complexity, incompleteness, unclearness, or uncertainty. With AI technology introduced into management system, information management system, office automation system, and decision-making supportive system, their functions and technologies can be integrated based on expert system, knowledge engineering, pattern recognition and artificial nervous element network, to form a new generation of computerized management system.

♦ Intelligent Decision-making

With AI combined with decision-making process, expert system as an intelligent component, together with model library, method library, database and knowledge base, most intelligent decision-making systems emerge in this way.

♦ Intelligent Simulation

Simulation refers to dynamic modeling experiment. Based on three forms of knowledge—descriptive knowledge, objective knowledge and processing knowledge, it produces another form of knowledge—conclusive knowledge. AI technology is used in the whole process of simulation including building models, practical operation and analysis of results, to direct and improve simulation models.

♦ Machine Learning

The application of AI techniques to software engineering has produced some encouraging results. Of the many AI techniques, machine learning methods have found their way into the software development in the past twenty years. Machine learning deals with the issue of how to build computer programs that improve their performance at some task through experience. It is dedicated to creating and compiling verifiable knowledge related to the design and construction of artifacts. Machine learning algorithms have been utilized in many different problem domains. Some typical applications are: data mining problems where large databases contain valuable implicit regularities that can be discovered automatically; poorly understood domains where there is lack of knowledge needed to develop effective algorithms; or domains where programs must dynamically adapt to changing conditions.

Keywords

artifact	n.	人工制品，人为现象	introspection	n.	内省，反省
cognitive	adj.	认识的，有认识力的	mechanism	n.	机制，结构
dispatch	v.	调度，调遣	obstacle	n.	障碍，阻碍
evidence	n.	证据，根据	provably	adv.	可证明地，可证实地
exhaust	v.	耗尽，用尽	pursue	v.	寻求，追求
extrapolate	v.	推断，判定，推论	rationality	n.	合理性
guidance	n.	指引，指导	regularity	n.	有规则，规律性
interdisciplinary	n.	跨学科	verifiable	adj.	可证实的，可检验的
interrogator	n.	讯问者，审问者			

Notes

[1] Roughly speaking, the test he proposed is that the computer should be interrogated by a human via a teletype, and passes the test if the interrogator cannot tell if there is a computer or a human at the other end.

译文：粗略地说，他提出的试验是计算机要由一个人通过电传打字机来向其提出问题，如果这个人不知道另一端是人还是计算机，那么该试验就算通过了。

说明：本句的"that"引导的是表语从句，"if the interrogator…"是条件状语从句。

[2] If the program's input/output and timing behavior matches human behavior, that is evidence that some of the program's mechanisms may also be operating in humans.

译文：如果程序的输入/输出和时机选择的行为与人类的行为相当，那就证明某些程序的机制对人类也可能有效。

说明："If the program's…"作条件状语，"that is evidence"是主句，"that some of the program's…"是主语从句。

[3] By 1965, programs existed that could, given enough time and memory, take a description of a problem in logical notation and find the solution to the problem (if one exists).

译文：到1965年，出现了这样的程序：如果给定足够的时间和内存，它们就能用逻辑符号来描述一个问题并能求出问题的解（如果解存在）。

说明：本句的插入语"given enough time and memory"是条件状语，"By 1965"是时间状语。

[4] Making correct inferences is sometimes part of being a rational agent, because one way to act rationally is to reason logically to the conclusion that a given action will achieve one's goals, and then to act on that conclusion.

译文：做出正确的推理有时可以作为一个合理代理的一部分，这是因为一种方法能够合理地实施也就能够对给定行为达至某人目标的结论进行逻辑推理，然后再按照该结论去行动。

说明："because one way…"作原因状语。

[5] Here hard problems refer to those difficult problems that they have no algorithm or whose algorithms cannot be executed in computer programs.

译文：这里的难题是指那些根本没有算法或算法在计算机程序中无法执行的难题。

说明：本句中修饰宾语的定语从句有两个，即"that they have no algorithm"和"whose algorithms cannot be executed in computer programs"。

20.2 Reading Material 1：Expert System

The reliance on the knowledge of a human domain expert for the system's problem solving strategies is a major feature of expert systems. An expert system is a set of programs that manipulate encoded knowledge to solve problems in a specialized domain that normally requires human expertise. An expert system's knowledge is obtained from expert sources and coded in a form suitable for the system to use in its inference or reasoning processes.

The expert knowledge must be obtained from specialists or other sources of expertise, such as texts, journal articles, and data bases. This type of knowledge usually requires much training and experience in some specialized field such as medicine, geology, system configuration, or engineering design. Once a sufficient body of expert knowledge has been acquired, it must be encoded in some form, loaded into a knowledge base, then tested, and refined continually throughout the life of the system.

Expert systems differ from conventional computer systems in several important ways.

1. Expert systems use knowledge rather than data to control the solution process. Much of the knowledge used is heuristic in nature rather than algorithmic.

2. The knowledge is encoded and maintained as an entity separate from the control program. As such,

it is not compiled together with the control program itself. This permits the incremental addition and modification of the knowledge base without recompilation of the control programs.

3. Expert systems are capable of explaining how a particular conclusion was reached, and why requested information is needed during a consultation.

4. Expert systems use symbolic representations for knowledge and perform their inference through symbolic computations that closely resemble manipulations of natural language.

5. Expert systems often reason with meta-knowledge; that is, they reason with knowledge about themselves, and their own knowledge limits and capabilities.

The reasoning of an expert system should be open to inspection, providing information about the state of its problem solving and explanations of the choices and decisions that the program is making. Explanations are important for a human expert, such as a doctor or an engineer, if he or she is to accept the recommendations from a computer. Indeed, few human experts will accept advice from another human, let alone a machine, without understanding the justifications for it.

The exploratory nature of AI and expert system programming requires that programs be easily prototyped, tested, and changed. AI programming languages and environments are designed to support this iterative development methodology. In a pure production system, for example, the modification of a single rule has no global syntactic side effects. Rules may be added or removed without requiring further changes to the large program. Expert system designers often comment that easy modification of the knowledge base is a major factor in producing a successful program.

A further feature of expert systems is their use of heuristic problem-solving methods. As expert system designers have discovered, informal "tricks of the trade" and "rules of thumb" are an essential complement to the standard theory presented in textbooks and classes. Sometimes these rules augment theoretical knowledge in understandable ways; often they are simply shortcuts that have, empirically, been shown to work.

It is interesting to note that most expert systems have been written for relatively specialized, expert level domains. These domains are generally well studied and have clearly defined problem-solving strategies. Problems that depend on a more loosely defined notion of "common sense" are much more difficult to solve by these means. In spite of the promise of expert systems, it would be a mistake to overestimate the ability of this technology. Current deficiencies include:

- Difficulty in capturing "deep" knowledge of the problem domain.
- Lack of robustness and flexibility.
- Inability to provide deep explanations.
- Difficulties in verification.
- Little learning from experience.

In spite of these limitations, expert systems have proved their value in a number of important applications.

Keywords

empirically   *adv.* 以经验为根据地   exploratory   *n.* 探索性

inability	n.	无能，无能为力	refined	adj.	精炼的，精制的
incremental	adj.	增量的	resemble	v.	像，相似
inspection	n.	检验，审查	shortcut	n.	捷径
justification	n.	证明为正当，正当的理由	specialist	n.	专家
recommendation	n.	推荐，介绍	syntactic	n.	句法学
recompilation	n.	重新编译			

20.3　Reading Material 2：Genetic Algorithm

If we are solving a problem, we are usually looking for some solution which will be the best among others. The space of all feasible solutions is called search space. Each point in the search space represents one possible solution. Each possible solution can be "marked" by its value (or fitness) for the problem. With genetic algorithm (GA), we look for the best solution among a number of possible solutions—represented by one point in the search space.

Looking for a solution is then equal to looking for some extreme value (minimum or maximum) in the search space. At times the search space may be well defined, but usually we know only a few points in the search space. In the process of using GA, the process of finding solutions generates other points (possible solutions) as evolution proceeds.

The problem is that the search can be very complicated. One may not know where to look for a solution or where to start. There are many methods one can use for finding a suitable solution, but these methods do not necessarily provide the best solution. Some of these methods are hill climbing, simulated annealing and the genetic algorithm. The solutions found by these methods are often considered as good solutions, because it is not often possible to prove what the optimum is.

Genetic Algorithms are a part of evolutionary computing, which is a rapidly growing area of artificial intelligence. Evolutionary computing was introduced in the 1960s by I. Rechenberg in his work "Evolution strategies". His idea was then developed by other researchers. Genetic algorithms were invented by John Holland and developed by him and his students and colleagues. Genetic algorithms are inspired by Darwin's theory of evolution. Solution to a problem solved by genetic algorithms uses an evolutionary process.

Algorithm begins with a set of solutions (represented by chromosomes) called population. Solutions from one population are taken and used to form a new population. This is motivated by a hope, that the new population will be better than the old one. Solutions which are then selected to form new solutions (offspring) are selected according to their fitness—the more suitable they are the more chances they have to reproduce. This is repeated until some condition is satisfied.

Outline of the basic genetic algorithm:

Step 1　Generate random population of n chromosomes (suitable solutions for the problem).

Step 2　Evaluate the fitness $f(x)$ of each chromosome x in the population.

Step 3　Create a new population by repeating following steps until the new population is complete.

(1) Select two parent chromosomes from a population according to their fitness (the better fitness, the

bigger chance to be selected).

(2) With a crossover probability cross over the parents to form new offspring. If no crossover was performed, offspring is the exact copy of parents.

(3) With a mutation probability mutate new offspring at each locus (position in chromosome).

(4) Place new offspring in the new population.

Step 4　Use new generated population for a further run of the algorithm.

Step 5　If the end condition is satisfied, stop, and return the best solution in current population.

Step 6　Go to Step 2.

Keywords

anneal	v.	使退火	fitness	n.	适应值
chromosome	n.	染色体	genetic	adj.	遗传的
colleague	n.	同事，同行	locus	n.	座位，位置
crossover	n.	杂交	mutation	n.	变异
evolutionary	n.	进化	offspring	n.	后代，子孙
extreme	adj.	极端的，非常的	optimum	n.	最适条件，最适度
feasible	adj.	可行的，可用的	population	n.	种群，群体

20.4　Terms

1. Structured Programming

It frequently employs a top-down design model, in which developers map out the overall program structure into separate subsections. A defined function or set of similar functions is coded in a separate module or sub-module, which means that codes can be loaded into memory more efficiently and that modules can be re-used in other programs.

2. Connection-oriented

In telecommunications, connection-oriented describes a means of transmitting data in which the devices at the end points use a preliminary protocol to establish an end-to-end connection before any data is sent. Connection-oriented protocol service is sometimes called a "reliable" network service, because it guarantees that data will arrive in the proper sequence.

3. Window

The principle component of the GUI is the Window. A window is a rectangular area of information that is displayed on the screen. These Windows can contain programs and documents, as well as menus, dialog boxes, icons, and a variety of other types of data.

4. Procedural Language

In procedural language, the basic programming element is the procedure (a named sequence of statements, such as a routine, subroutine, or function). The most widely used high-level languages, such as C, Pascal and FORTRAN, are all procedural languages.

5. Virtual Circuit

A virtual circuit is a circuit or path between points in a network that appears to be a discrete, physical path but is actually a managed pool of circuit resources from which specific circuits are allocated as needed to meet traffic requirement.

6. Menu

A menu is a set of options, usually text based, from which the user can choose to initiate a desired action in a program. At the top of many windows is a menu bar showing the main menu categories. Drop-down menus display on the screen when the user selects an items on the menu bar.

7. Desktop Database

The desktop database is a pair of invisible files in Mac OS classic that contain the location information of all the files on your computer's disk drives. These files are called "Desktop DB". Rebuilding the desktop database is also sometimes referred to as simply "rebuilding the desktop".

8. Thread

In most traditional operating systems, each process has an address space and a single thread of control. Thread is sometimes called lightweight process. In many respects, threads are like little mini-processes. Each thread runs strictly sequentially and has its own program counter and stack to keep track of where it is.

20.5 Exercises

1. **Translate the following phrases.**

    extrapolate pattern
    psychological experiment
    cognitive science
    logical notation
    computational resource
    correct inference
    tourist salesman problem
    automatic translation
    intelligent control
    decision-making supportive system
    expert system
    knowledge engineering

2. **Fill in the blanks with appropriate words or phrases.**

    a. translation   b. problem domains   c. experiment   d. software engineering
    e. one's goals   f. humans   g. something   h. computer programs

    (1) If we are going to say that a given program thinks like a human, we must have some way of determining how _____ think.

    (2) Acting rationally means acting so as to achieve _____, given one's beliefs.

    (3) An agent is just _____ that perceives and acts.

(4) Automatic translation means _____ doing translation between two different languages.
(5) Machine _____ is not simply "checking up dictionaries" and translating "word-by-word".
(6) Simulation refers to dynamic modeling _____.
(7) The application of AI techniques to _____ has produced some encouraging results.
(8) Machine learning algorithms have been utilized in many different _____.

3. **Match the following terms to the appropriate definition.**

   a. script   b. SSL   c. security certificate
   d. spider   e. token   f. token ring

(1) A chunk of information (often stored as a text file) that is used by the SSL protocol to establish a secure connection.
(2) A protocol designed by Netscape Communications to enable encrypted, authenticated communications across the Internet.
(3) A program that automatically fetches Web pages. They are used to feed pages to search engines.
(4) Developed by IBM, this 4 or 16 Mbps network uses a ring topology and a token-passing access method.
(5) A program or sequence of instructions that is interpreted or carried out by another program rather than by the computer processor.
(6) The character sequence or frame, passed in sequence from node to node, to indicate that the node controlling it has the right to transmit for a given amount of time.

Unit 21 Electronic Commerce

21.1 Text

Electronic-Commerce Basics

The term electronic-commerce typically refers to business transactions that are conducted electronically over a computer network. It encompasses all aspects of business and marketing processes enabled by Internet and Web technologies. E-Commerce wares include many kinds of physical products, digital products, and services. Physical products offered at E-Commerce sites include such goods as clothing, shoes, skateboards, and cars. Most of these products can be shipped to buyers through the postal service, a parcel delivery service, or a trucking company.

With the wide use of computer, the maturity and the wide adoption of Internet, the permeation of credit cards, the establishment of secure transaction agreement and the support and promotion by governments, the development of E-Commerce is becoming prosperous, with people starting to use electronic means as the media of doing business.[1]

E-Commerce is not a kind of pure technique itself, but an application process of Internet and information technologies in doing business. In such process, the tools used for information exchanging, transmitting, processing and storing based on paper media are replaced by the tools based on electronic media.[2]

Increasingly, E-Commerce goods include digital products, such as news, music, video, databases, software, and all types of knowledge-based items. The unique feature of these products is that they can be transformed into bits and delivered over the Web. Consumers can get them immediately upon completing their orders, and no one pays shipping costs. E-Commerce merchants also peddle services, such as online medical consultation, distant education, or custom sewing. Some of these services can be carried out by computers. Others require human agents. Services can be delivered electronically, as in the case of a distance education course.

Many E-Commerce activities are classified as B2C (business-to-consumer) in which individual consumers purchase goods and services from online merchants. C2C (consumer-to-consumer) is another popular e-commerce model, in which consumers sell to each other at popular online auctions. B2B (business-to-business) E-Commerce involves one enterprise buying goods or services from another enterprise. B2G (business-to-government) E-Commerce aims to help businesses sell to governments.

E-Commerce Site

E-Commerce seems simple from the perspective of a shopper who connects to an online store, browses the electronic catalog, selects merchandise, and then pays for it. Behind the scenes, an E-Commerce site uses several technologies to display merchandise, keep track of shoppers' selections, collect payment data, attempt to protect customers' privacy, and prevent credit card numbers from falling into the wrong hands.

An E-Commerce site's URL, such as www.amazon.com, acts as the entry to the online store. A Web page at this location—sometimes referred to as an electronic storefront—welcomes customers and provides links to various parts of the site. The goods and services for sale appear in a customer's browser window.

An E-Commerce site usually includes some mechanism for customers to select merchandise and then pay for it. Customer orders might be processes manually in a small business. Most high-volume e-commerce businesses, however, use as much automation as possible; their order-processing systems automatically update inventories, and then print packing slips and mailing labels.

If you have done any shopping online, you have probably used an online shopping cart. Most shopping cars work because they use cookies to store information about your activities on a Web site. Cookies work with shopping carts in one of two ways, depending on the E-Commerce site. An E-Commerce site might use cookies as a storage bin for all the items you load into your shopping cart. Some E-Commerce sites use cookies simply as a way to uniquely identify each shopper. These sites generated a unique ID number that is stored along with your item selections in a server-side database.

Online Auction

An online auction is the electronic equivalent to old-fashioned yard sales, rummage sales, and auctions. Hosted by an Internet auction site, such as eBay, sellers post their merchandise or services and buyers bid for the posted items. The auction site does not own or hold the merchandise; rather, it merely facilitates the online bidding process and the transactions between buyers and sellers.

You can expect to bid on new, used, closeout, over-stoke, or refurbished items at an online auction. Different types of services are also available. The merchandise covers a very broad spectrum of categories, such as antiques, collectibles, books and comics, coins, stamps, entertainment memorabilia, art, home and garden, cameras and camcorders, computers, cars and boats, sporting goods, apparel, jewelry, musical instruments, toys, and even theater tickets, special travel deals, and real estate.

The services up for auction are just as varied as the merchandise. For example, you can bid on tutoring lessons, printer repair, interior design, custom-made clothing, shopping assistance, printing and personalization, graphic design, and Web and computer services.

At an online auction site, specially designed computer software takes the place of an auctioneer. Server databases store information about auction items, and customer ratings of sellers. Server-side scripts accept bids and notify winners. Not only are auction sites automated; software tools for bidders are also hot. For example, sniper bots can make last-minute bids and auction alert tools can notify you when items on your watch list come up for bid.

Benefits of Electronic-Commerce

As computer network facilitates information exchange in a speedy and inexpensive way, Internet now penetrates into almost every corner of the world. Small and medium sized enterprises (SMEs) can forge global relationships with their trading partners everywhere in the world. High-speed network makes geographical distance insignificant. Businesses can sell goods to customers outside traditional markets, explore new markets and realize business opportunities more easily. SMEs who can't afford establishing overseas offices and strongholds can now increase their exposure to every corner of the world.

Paper-based buying and selling often requires large clerical staffs. Envelopes need to be prepared, and errors need to be corrected. Organizations can achieve major clerical staff reduction by automating billing and bill payment. Significant staff saving can also be achieved in many other ways through the use of electronic methods. For example, sales and support staff can be reduced when customers are able to obtain information online rather than via telephone inquiries.[3] Staff for printing and mailing catalogues can also be eliminated when paper catalogues are converted to online versions. Orders entered online rather than via a telephone conversation require less staff time while generating fewer errors.

Internet provides companies with many markets in the cyber world and numerous chances for product promotion. Besides, relationships with buyers can also be enhanced. By the use of multimedia capabilities, corporate image, product and service brand names can be established effectively through the Internet.[4] Detailed and accurate sales data can help to reduce stock level and thus the operating cost. Detailed client information such as mode of consumption, personal preferences and purchasing power, etc. can help businesses to set their marketing strategies more effectively.

Many customers are turning to online financial markets trading because they have more control over their own funds as well as because these systems are so easy to use compared with placing telephone orders.[5] Many other firms now support stock, bond, futures, and options trading with direct dial-in to their computers, offering deep discounts for those who trades prefer electronic trading because orders can be entered at any time, 24 hours a day, with confirmations arriving almost immediately. Investors can also check their account status at any time and do not have to wait for monthly statements.

Keywords

antique	n.	古董，古物
apparel	n.	衣服，服饰
auction	n.	拍卖
auctioneer	n.	拍卖商
automation	n.	自动化
bidder	n.	（拍卖时的）出价人
bond	n.	债券
camcorder	n.	（便携式）摄像机
closeout	n.	清仓销售的
collectible	n.	收藏品
comic	n.	连环漫画（册）
cookie	n.	指一种临时存储网络用户信息的结构
custom	adj.	定制的
discount	n.	折扣，贴现，贴现率
establishment	n.	设置，制定
fund	n.	资金
future	n.	期货
investor	n.	投资者
jewelry	n.	珠宝，首饰
maturity	n.	成熟，完成
merchandise	n.	商品
peddle	v.	（沿街）叫卖，兜售
penetrate	v.	渗透入，进入
permeation	n.	普及
prosperous	adj.	繁荣的，良好的
refurbish	v.	整修，把……翻新
skateboard	n.	滑板
sniper	n.	狙击手
spectrum	n.	频谱，范围
storefront	n.	店面，铺面
stronghold	n.	根据地，中心点
trucking	n.	货车，运输业
tutor	v.	辅导
ware	n.	商品，货物

Notes

[1] With the wide use of computer, the maturity and the wide adoption of Internet, the permeation of credit cards, the establishment of secure transaction agreement and the support and promotion by governments, the development of E-Commerce is becoming prosperous, with people starting to use electronic means as the media of doing business.

译文：随着计算机的广泛使用、互联网的完备和普及、信用卡的普及、安全交易协议的建立和政府的支持和鼓励，以及人们开始以电子手段作为商业媒介，电子商务发展得越来越繁荣。

说明：这是一个长句，"With the wide use…by governments"作状语，主句是"the development of E-Commerce is becoming prosperous"，"with people starting to…"也是状语。

[2] In such process, the tools used for information exchanging, transmitting, processing and storing based on paper media are replaced by the tools based on electronic media.

译文：在这个运用过程中，以纸质为基础的用于信息交流、传递、处理及存储的工具被以电子媒体为基础的工具所取代。

说明：本句中的"used for information exchanging, transmitting, processing and storing based on paper media"作定语，修饰主语。

[3] For example, sales and support staff can be reduced when customers are able to obtain information online rather than via telephone inquiries.

译文：例如，当客户在线获得联机的信息而不是通过电话询问时就可以减少销售和维护人员。

说明：本句中主语是"sales and support staff"，"when"引导时间状语从句。

[4] By the use of multimedia capabilities, corporate image, product and service brand names can be established effectively through the Internet.

译文：使用多媒体设备，互联网能很有效地建立起公司形象、产品和服务品牌。

说明："By the use of multimedia capabilities"作状语，本句的主语"corporate image, product and service brand names"比较长，谓语用被动语态形式。

[5] Many customers are turning to online financial markets trading because they have more control over their own funds as well as because these systems are so easy to use compared with placing telephone orders.

译文：许多客户转向在线式金融市场交易，因为在这里他们对自己的资金有更大的控制权，还因为这些系统与电话订单相比更加容易使用。

说明：本句有两个由"because"引导的原因状语从句，修饰主句。

21.2　Reading Material 1：The Introduction of E-Commerce

Electronic commerce is defined as the use of computer applications communicating over networks to allow buyers and sellers to complete a transaction or part of a transaction. E-Commerce is doing business through electronic media. It means using simple, fast and low-cost electronic communications to transact, without face to face meeting between the two parties of the transaction. Now, it is mainly done through Internet and electronic data interchange (EDI). E-Commerce was first developed in the 1960s.

Businesses can gather information on products, buyers and competitors through Internet so as to increase their own competitiveness. Businesses can maintain their competitive advantage by establishing close contact with their customers and consumers at anytime through Internet by providing the latest information on products and services round the clock. On the other hand, data can be updated at anytime, eliminating the problem of outdated information.

The difference between Traditional Business and E-Commerce

In traditional commerce surroundings, a company should spend quite a lot of time and effort to investigate and decide what and how to produce or sell, then, advertise in mediums such as TV, radio, bulletin board on street and so on, contact the buyer through phone, fax, mail or in face to face. After looking at the goods, the buyer bargains, places the order or signs the agreement with the seller. The seller prepares the goods and finally sends the goods to the buyer. The buyer pays for the deal in cash or by check. So, as you can see, in traditional commerce surroundings, people spend more time and efforts to do business. Sometimes mistakes and delay will happen.

In E-Commerce surroundings, people use Internet to do business. The contact between the buyer and the seller is no longer face to face, but through Internet, on which both the buyers and the sellers have far more chances to know each other, not only one to one, but many to many. No geographical constrains, no time limited, all of them can find satisfied partners. The buyer uses browser to seek out the product list on the home page of the website or on-line shops based on the wanted domain names, and see the packing, outlook, style, color and price of products, sometimes, even can bargain on-line. If he is satisfied with the goods, he can place an order by click the keyboard. The seller immediately receives the message. If he accepts the order, he can use Internet to arrange goods production, distribution as well as payment. All data are transmitted just in time and business operation becomes more efficient.

Classification of the EC Field by the Nature of the Transactions

A common classification of EC is by the nature of transaction. The following types are distinguished:

Business-to-business (B2B). Most of EC today is of this type. It is the electronic market transactions between organizations.

Business-to-consumer (B2C). There are retailing transactions with individual shoppers. Businesses provide consumers with online shopping through Internet, allowing consumers to shop and pay their bills online. This type of offering saves time for both retailers and consumers.

Consumer-to-consumer (C2C). In this category, consumer sells directly to consumers. Examples are individuals selling in classified ads and selling residential property, cars, and so on. Advertising personal services on the Internet and selling the knowledge and expertise is another example of C2C. Service auction sites allow individuals to put items up for auctions finally, many individuals are using Intranets and other organizational internal networks to advertise items for sale or services.

Consumer-to-business (C2B). This category includes individuals who sell products or services to organizations, as well as individuals who seek sellers, interact with them, and conclude a transaction.

 Government-to-citizen (G2C). Among various kinds of services provided by governments, many of them can be done through electronic media. Providing public services electronically not only provides citizens timesaving and high-quality services, but also improves efficiency and cost effectiveness.

Government-to-Business (G2B). This mode of trading often describes the way in which government purchases goods and services through electronic media such as Internet.

Keywords

bargain	v.	议价，约定	expertise	n.	专门知识，专门技能
bulletin	n.	公告，布告，告示	investigate	v.	调查，研究
citizen	n.	市民，平民，公民	purchase	v.	买，购买
competitiveness	n.	竞争能力，竞争力	retailer	n.	零售商，零售店
competitor	n.	竞争者，敌手	timesaving	v.	省时
effectiveness	n.	效益，有效			

21.3 Reading Material 2: E-Government

By definition, E-Government is simply the use of information and communications technology, such as the Internet, to improve the processes of government. Thus, E-Government is in principle nothing new. Governments were among the first users of computers. But the global proliferation of the Internet, which effectively integrates information and communications technology on the basis of open standards, combined with the movement to reform public administration known as New Public Management, has for good reason generated a new wave of interest in the topic.

E-Government promises to make government more efficient, responsive, transparent and legitimate and is also creating a rapidly growing market of goods and services, with a variety of new business opportunities.

To some, E-Government might seem to be little more than an effort to expand the market of E-Commerce from business to government. Surely there is some truth in this. E-Commerce is marketing and sales via the Internet. Since governmental institutions take part in marketing and sales activities, both as buyers and sellers, it is not inconsistent to speak of E-Government applications of E-Commerce. Governments do after all conduct business.

But E-Commerce is not at the heart of E-Government. The core task of government is governance, the job of regulating society, not marketing and sales. In modern democracies, responsibility and power for regulation is divided up and shared among the legislative, executive and judicial branches of government. Simplifying somewhat, the legislature is responsible for making policy in the form of laws, the executive for implementing the policy and law enforcement, and the judiciary for resolving legal conflicts. E-government is about improving the work of all of these branches of government, not just public administration in the narrow sense.

E-Government gives New Public Management fresh blood. Not only does information and communications technology provide the infrastructure and software tools needed for a loosely coupled network of governmental units to collaborate effectively, the infiltration of this technology into government agencies tends to lead naturally to institutional reform, since it is difficult to maintain strictly hierarchical channels of communication and control when every civil servant can collaborate efficiently and directly with anyone else via the Internet.

E-Government is not only or even primarily about reforming the work processes within and among governmental institutions, but is rather about improving its services to and collaboration with citizens, the business and professional community, and nonprofit and nongovernmental organizations such as associations, trade unions, political parties, churches, and public interest groups.

Keywords

church	n.	教会，教派	judiciary	n.	法院系统，司法部
governance	n.	管理，统治	legislative	n.	立法机关
infiltration	n.	渗入，渗透	legitimate	adj.	合法的，正统的
institution	n.	机关，机构	proliferation	n.	激增，扩散

21.4 Terms

1. WYSIWYG

WYSIWYG is the acronym for " what you see is what you get". A WYSIWYG editor or program is one that allows a developer to see what the end result will look like while the interface or document is being created. A WYSIWYG editor can be contrasted with more traditional editors that require the developer to enter descriptive codes and do not permit an immediate way to see the results of the markup.

2. LAN

LAN is the acronym of local area network. A group of computers and other devices dispersed over a relatively limited area and connected by a communication link that enables any device to interact with any other on the network. LANs commonly include microcomputers and shared resources such as laser printers and large hard disks.

3. WAN

WAN is the acronym of wide area network. A WAN is a communications network that connects geographically separated areas. The size of the network can be extended by adding a new switch and another communication line.

4. ISP

ISP is the acronym for "Internet service provider". It is a company that provides access to the Internet. For a monthly fee, the service provider gives you a software package, username, password and access phone number. Equipped with a modem, you can then log on to the Internet and browse the WWW and Usenet, and send and receive e-mail.

5. TCP/IP

TCP/IP is the acronym of Transmission Control Protocol/ Internet Protocol. It is built into the UNIX system and has become the defacto standard for data transmission over networks, including the Internet.

6. Online Community

An online community or virtual community is a group of people that primarily or initially communicates or interacts via the Internet. There are several motivations that lead people to contribute to online communities. Various online media, such as Wikis, Blogs, Chat rooms, Internet forums, Electronic mailing lists, are becoming ever greater knowledge-sharing resources. Many of these communities are highly cooperative and establish their own unique culture.

7. URL

URL is the acronym of uniform resource locator. An URL specifies the protocol to be used in accessing the resource, the name of the server on which the resource resides, and, optionally, the path to a resource.

8. Gateway

A gateway is a device that connects networks using different communications protocols so that information can be passed from one to the other. A gateway both transfers information and converts it to a form compatible with the protocols used by the receiving network. In enterprises, the gateway is the computer that routes the traffic from a workstation to the outside network that is serving the Web pages. In homes, the gateway is the ISP that connects the user to the Internet.

21.5 Exercises

1. **Translate the following phrases.**

    packing slip
    rummage sale
    real estate
    electronic commerce
    sniper bot
    server-side
    medical consultation
    mailing label
    interior design
    old-fashioned
    over-stoke
    yard sale

2. **Fill in the blanks with appropriate words or phrases.**

    a. shopping carts   b. computer network   c. online   d. governments
    e. enterprise   f. knowledge-based   g. auction   h. pay for

    (1) The term electronic-commerce typically refers to business transactions that are conducted electronically over a _____.
    (2) B2G (business-to-government) E-Commerce aims to help businesses sell to _____.
    (3) B2B (business-to-business) E-Commerce involves one enterprise buying goods or services from another _____.
    (4) E-Commerce goods include digital products, such as news, music, video, databases, software, and all types of _____ items.
    (5) Cookies work with _____ in one of two ways, depending on the E-Commerce site.
    (6) If you have done any shopping _____, you have probably used an online shopping cart.
    (7) An online _____ is the electronic equivalent to old-fashioned yard sales, rummage sales,

and auctions.

(8) An E-Commerce site usually includes some mechanism for customers to select merchandise and then _____ it.

3. **Match the following terms to the appropriate definition.**

    a. transistor   b. telnet   c. transmission media

    d. touch screen   e. transport layer   f. unzip

(1) A type of display screen that has a touch-sensitive transparent panel covering the screen. Instead of using a pointing device such as a mouse or light pen, you can use your finger to point directly to objects on the screen.

(2) The command and program used to login from one Internet site to another.

(3) A semiconductor device consisting of three or four layers used for switching or amplification at frequencies ranging from direct-current to ultrahigh.

(4) Materials or substances capable of carrying one or more signals in a communications channel.

(5) Equivalent to the OSI model layer 4 and part of 5. At this level, TCP establishes a logical connection with the receiving computer and determines the size of the segments to be sent.

(6) To restore a compressed file to its original form. Refer to "uncompress".

Unit 22　Electronic Payment System

22.1　Text

The Internet Payment

Payment is a very important step in the E-Commerce procedure. The Internet payment means conducting fund transfer, money receiving and disbursing electronically. Internet payment, sometimes called electronic payment, is developed for the demand of E-Commerce applications.

The former of the electronic payment is electronic remittance conducted by bank. First, a client deposits money in bank A, when he needs to pay his partner in another city, the client entrusts bank A to transfer the sum of money from his bank account to bank B in that city, in which the partner has an account. After bank B confirms the partner's identity, then transfers the money to his account. In early time, those data exchange between banks are done by phone or telegraphy. In recent years, they use EDI system replace the phone or telegraph system. This financial EDI system is called EFT (electronic funds transfer) system. EFT system greatly saved the data exchanging time and reduced the operating cost.

Now, many banks and software developing companies are bent on developing a new kind of electronic payment mechanism that is the digital fund system. Apparently, the digital funds are only a series of digital numbers, but those numbers are not created casually or freely. Behind the digital money, real money supporting it, otherwise, there will be a mass. It has the ability of being used universally as circulation medium. That is, it must be accepted by many banks, merchants or customers, otherwise, it can only be used in small circles and not suitable for the feature of E-Commerce applications.

There are many kinds of digital money, or electronic token, which mainly could be classed into two categories: real-time payment mechanism, such as digital cash, electronic wallet and smart card, and after time payment mechanism, such as electronic check (net check) or electronic credit card (Web credit card).

The advanced encryption and authentication system can make digital money be used more safely and privately than paper money, and at the same time, keep the feature of "anonymity of payer" of paper money. The digital cash system is constructed based on digital signature and cryptograph techniques. This system mostly takes use of the public/private key pairs methods.

First, the client deposits money in the bank, and opens an account. The bank will give the client the digital cash software used for client-side and a bank's public key. The bank uses sever-side digital cash software and private key to encode the message. The client can decode such message by using client-side digital cash software and the bank's public key. When client needs some digital money to pay someone, he initiates that sum of digital money he needs by using the digital cash software, then, he encodes it and transfers the message to someone.[1] Someone encodes this message, his own account message and depositing orders and sends them to the bank. The bank decodes these messages with its private key and confirms the authenticity of both the client and someone. Then transfer the money from client's account to someone's account.

Since the digital cash is only a digital number, so it needs to be verified as authentic. Banks must have the ability to track whether the e-cash is duplicated or reused, without linking the personal shopping behavior to the person who bought that cash from the bank, to keep the feature of "anonymity of payer" of real money.[2] In the whole checking process, the seller can't capture any private data of his client. He can verify the digital cash sent by the client with the bank's public key.

Another mechanism for Internet payment is e-check. Its former is the financial EDI system used in Value Added Networks. Some electronic check software can be packed in smart card. Thus, clients can bring it very conveniently and can write checks at any computer site. Some electronic-check software can provide e-wallets for their customers, which also enable the customers to write checks at anywhere even they have no bank account.

Electronic Bank

An electronic bank is not a real bank, it is a virtual bank, in which, the bankers, software designers, and hardware producers and ISPs work together. Each of them fulfils its own work in its profession and coordinates perfectly with each other. In an electronic bank, all the bank service contents are put on the home pages of the server for clients to choose.[3] Clients don't need to store personal financial software, they use their browser to get access to the bank's website and ask for services. All the services are conducted on the bank server. The bank interacts with its clients through website. The clients may use a portable smart card and conduct their bank businesses anywhere at any time, such as pay a bill, buy some digital money, fill out an electronic bank and so on. They don't need to buy PC, they can use any Internet connected computer to make his bank payments.

These new types of bank server software, which adopt up the modern information technologies, can provide convenient, real-time, secure and reliable bank services.[4] They can also provide other relevant services, such as stock quotation, tax and polices consultation, exchange and interest rates information. The more convenient the software are, the more users there may be, the more profits the company will get and better services they provide, thus, the banks and the clients are in win/win relationship, clients get cheaper and convenient services. Banks greatly reduce the operating cost. For on, Internet, the cost of running a website is much lower than building a bank branch, and on Internet, the cost for serving one person is the same as that for serving 10 thousand persons. With the rapid development of electronic banks, software developing companies and ISPs will get more and more businesses, they are all in win/win relationship with electronic banks.

Electronic Funds Transfer

Electronic funds transfer (EFT) is the automatic transfer of "funds" from one institution's computer to another's. EFT is a subset of EDI, and refers to the transfer of value electronically from buyer to seller.

EFT has been the major implementation by the banks to eliminate paper processing. Some EFT solutions enlisted in the fight against paperwork volumes: screen-based cash management systems, Swift (international), Bits (domestic), direct entry (magnetic media), automatic teller machine (ATM), EFTPOS (card-based, international and domestic), home banking, phone banding, and bill payment.

Credit Cards

Credit cards, (and debit cards that share their networks) are the preferred method of payment among online consumers. Reasons for credit cards popularity as a payment method: Easy to use, no technology hurdles, and almost everyone has one; Consumers trust card companies to conduct secure transactions; Most merchants accept them. Parties involved in the credit card transactions are customer, customer's credit card Issuer (Issuing Bank), merchant, merchant's bank, and company processing credit card transactions over the Internet.[5]

Steps involved in the credit card payment process.

(1) The customer selects the product or service via shopping cart or catalogue on your website.

(2) The order is totaled, and a request is made to purchase.

(3) The customer enters the sensitive data such as Credit card number, shipping address etc.. on the given form.

(4) The data is then submitted, collected and passed securely through web service.

(5) Web server payment systems securely contacts the processor associated with the merchant's Merchant Account, and verifies that the credit car number is valid.

(6) If the previous step passes correct, then the purchase price will be reserved from the "Issuing Bank" of the consumer's credit card, and allocated to the merchant's Merchant Account.

Keywords

account	n.	账户，账目	enlist	v.	参加，协助，支持
anonymity	n.	匿名	hurdle	n.	障碍，困难
apparently	adv.	明显地，显而易见地	issuer	n.	发行者
casually	adv.	偶然，临时	quotation	n.	行情，行市
circulation	n.	流通，通货	remittance	n.	汇款，汇款额
consultation	n.	协商，咨询	telegraphy	n.	电报
deposit	n.	储蓄，存款	universally	adv.	普遍地，全面地

Notes

[1] When client needs some digital money to pay someone, he initiates that sum of digital money he needs by using the digital cash software, then, he encodes it and transfers the message to someone.

译文：当客户需要将一些数字现金支付给某人时，就用数字现金软件生成他所需的数字现金，给它加密并传给某人。

说明："When…"是时间状语，"he initiates…"和"he encodes it and transfers the…"是并列句。

[2] Banks must have the ability to track whether the e-cash is duplicated or reused, without linking the personal shopping behavior to the person who bought that cash from the bank, to keep the feature of "anonymity of payer" of real money.

译文：银行必须具备能追踪电子现金是否被复制或重复使用的能力，而又不能将个人购买行为与从银行购买现金的人联系起来，以保持真钱币的"匿名"特色。

说明："to track…"作定语，修饰宾语"ability"，"without…"是条件状语，"to keep…"是目的状语。

[3] In an electronic bank, all the bank service contents are put on the home pages of the server for clients to choose.

译文：在电子银行，银行服务内容放在服务器的主页上供客户选择。

说明："In an electronic bank"作状语，主句用被动语态，表示客观。

[4] These new types of bank server software, which adopt up the modern information technologies, can provide convenient, real-time, secure and reliable bank services.

译文：这些新型银行服务软件采用了现代信息技术，能提供方便、实时、安全和可靠的银行服务。

说明：本句中"which"引导的是非限定性定语从句，修饰主语，"convenient, real-time, secure and reliable"修饰宾语"bank services"。

[5] Parties involved in the credit card transactions are customer, customer's credit card Issuer (Issuing Bank), merchant, merchant's bank, and company processing credit card transactions over the Internet.

译文：在信用卡交易中的参与者有顾客、顾客的信用卡发行人（发行的银行）、商人、商人的银行、在Internet上处理信用卡事务的公司。

说明："involved in the credit card transactions"作定语，修饰主语，本句的表语比较长，"customer, customer's credit card Issuer (Issuing Bank), merchant, merchant's bank, and company processing credit card transactions over the Internet."都是表语。

22.2　Reading Material 1：E-Commerce Security

E-Commerce security is the top concern when setting up an online store. SSL certificates, encryption, public and private keys …, it can all be very confusing. If you don't bother with security at all or don't understand it well enough to take the necessary steps to secure your site, your website and your business could be at risk.

♦ Establish Security Control Mechanisms

The level of security is also a consideration. You should protect your system against hacking and virus attack. Firewalls, intrusion detection systems, virus scanning software can be used. Besides, some security measures such as keeping your user IDs and passwords secret, changing your password regularly, etc.. should also be adopted. Higher level of security is expected for payment transactions. If you want to obtain customers' personal information online, secure transfer and storage of data should be ensured. A data privacy statement should also be published.

♦ SSL—Secure Socket Layers

SSL (secure socket layer) technology creates a connection between two devices, such as a computer and web server, to transmit data securely. An SSL certificate is absolutely necessary for any E-Commerce site or any site accepting sensitive information. Not only do you have to have an SSL certificate, but you have to use it.

♦ Securing E-mail

If you receive your orders from your web store via E-mail, having an SSL certificate may not help.

E-mail is sent in plain text, and it is possible to intercept E-mail messages in transit and read their contents.

There are ways to secure E-mail. PGP is an encryption technology that encrypts E-mails as they are sent and decrypts them when received. For this technology to work, both the sender and recipient must have certificates installed on their computers that can identify and decrypt the messages. The technology has progressed over the years, but it is still not easy enough to setup and use to be of use to the masses.

Another way to receive secure E-mails through your website is to have the E-mail delivered to an E-mail account on the same server or domain on which the secure order was placed. Since they will not be routed through the public Internet, they will remain secure until you attempt to download them from your E-mail client.

- Update Your Software on a Regular Basis

Updating your software may seem trivial, but patches and software updates are often released to fix security holes that hackers can use to gain access to your system. Patch your system regularly.

- Be Careful of the Scripts and Programs You Install on Your Server

Free scripts and software sound great, but sometimes they can be poorly written and have potential security holes that can allow a hacker to gain access to your system. Make sure the software you install has been thoroughly tested and there are some forms of supports available should you have a problem.

- Use Secure FTP to Transfer Files

FTP and Telnet send passwords in clear text. Hackers can get this type of information and gain access to your entire site. Use Secure FTP and SSH to access your site, the entire sessions are encrypted just as with SSL to ensure that no one is able to intercept and read your passwords or other sensitive information.

Shopping online opens up a whole world of goods and services. The WWW has expanded the international marketplace in a way never before possible, giving consumers' unlimited choices. But shopping electronically opens up a whole world of questions, especially when you are dealing with vendors in other countries.

Mastering some tips is necessary to help you when you go shopping online.

- Know Whom You Are Dealing with

Do some homework to make sure a company is legitimate before doing business with it. Identify the company's name, its physical address, including the company where it is based, and an E-mail address or telephone number, so you can contact the company with questions or problems. And consider dealing only with vendors that clearly state their policies.

- Understand the Terms, Conditions and Costs

Find out up front what you are getting for your money, and what you are not. Get a full, itemized list of costs involved in the sale, with a clear designation of the currency involved, terms of delivery or performance, and terms, conditions and methods of payment. Look for information about restrictions, limitations or conditions of the purchase.

- Protect Yourself When Paying Online

Look for information posted online that describes the company's security policies, and checks whether the browser is secure and encrypts your personal and financial information during online transmission. That makes the information less vulnerable to hackers.

◆ Look Out for Your Privacy

All businesses require information about you to process an order. Some use it to tell customers about products, services or promotions, but others share or sell the information to other vendors. Go shopping only from online vendors that respect your privacy. Look for the vendor's privacy policy on the website. The policy statement should reveal what personal identifying information is collected about you and how it will be used, and give you the opportunity to refuse having your information sold or shared with other vendors. It also should tell you whether you can correct or delete information the company already has about you.

Keywords

bother	n. v.	烦恼，操心，担忧	regularly	adv.	定期地，经常地
confuse	v.	混淆，混乱	restriction	n.	限制，限定，束缚
designation	n.	指出，指明	reveal	v.	显露，揭露，透露
intercept	v.	拦截，截击	thoroughly	adv.	彻底地，充分地
intrusion	n.	入侵，闯入	trivial	adj.	琐碎的，轻微的
itemize	v.	详细列举	vulnerable	adj.	脆弱的，有弱点的
progress	v.	发展，进展，前进			

22.3 Reading Material 2: Using Credit Cards for Online Purchases

Credit cards are by far the most popular form of online payments for consumers. Credit cards are widely accepted by merchants around the world and provide assurances for both the consumer and the merchant. A consumer is protected by an automatic 30-day period in which he can dispute an online credit card purchase. A merchant has a high degree of confidence that a credit card can be safely accepted from an unseen purchaser.

Online purchases require an extra degree of security not required in offline purchases. Because online transactions pass from the consumer to the merchant over the Internet, which is an open and highly vulnerable network, merchants must work to protect sensitive information, such as credit card numbers. Additional software, such as the secure sockets layer (SSL) protocol, protects transactions in transit from being viewed by unauthorized parties.

Anyone who has made a purchase on the Web knows that using a credit card to pay for purchases is easy. Most people have concluded that conducting credit card transactions over the Internet, when accompanied by built-in safeguards such as secure servers, is as safe as presenting a credit card to a merchant in person.

Currently, online shoppers use credit and charge cards for a majority of their Internet purchases. A credit card is a card that has a preset spending limit based on the user's credit limit. A user can pay the balance of the credit card or a minimum amount for each billing period. Credit card issuers charge interest on any unpaid balances.

Credit cards have several features that make them an attractive and popular choice both with consumers and merchants in online and offline transactions. For merchants, credit cards provide fraud protection. When a merchant accepts credit cards for online payment—called card not present because the merchant's location and purchaser's location are different—they can authenticate and authorize purchases using a credit card processing network.

Perhaps the biggest advantage of using credit cards is their worldwide acceptance. You can pay for goods with credit cards anywhere in the world, and the currency conversion, if needed, is automatic. For online transactions, credit cards are particularly advantageous. When a consumer reaches the electronic checkout, she enters the credit card's number and her shipping and billing information in the appropriate fields to complete the transaction. The consumer does not need any special hardware or software to complete the transaction.

Credit card service companies charge merchants per-transaction fees and monthly processing fees. These fees can add up, but online and offline merchants view them as a cost of doing business. The consumer pays no direct fees for using credit cards, but the prices of goods and services are slightly higher than they would be in an environment free of credit cards altogether.

Keywords

dispute	v.	阻止，反对	purchaser	n.	买主，购买人
offline	adj.	离线的	safeguard	n.	防护措施，保护措施
merchant	n.	商人，贸易商	slightly	adv.	稍微地，略微地
preset	v.	预先设置	worldwide	adv.	世界范围

22.4 Terms

1. Intranet

An Intranet is simply the application of Internet technology within an internal or closed user-group. Intranets are company specific and do not have to have a physical connection to the Internet.

2. Cyberspace

Cyberspace is a metaphor for describing the non-physical terrain created by computer systems. Online systems, for example, create a cyberspace within which people can communicate with one another, do research, or simply window shop. Like physical space, cyberspace contains objects and different modes of transportation and delivery.

3. HTML

HTML is the acronym of Hypertext Markup Language. HTML is the set of markup symbols or codes inserted in a file intended for display on WWW browser page. It allows users to produce Web pages that include text, graphics, and pointers to other Web pages. A markup language describes how documents are to be formatted. It thus contains explicit commands for formatting.

4. Switch

The switching elements in most wide area networks are specialized computers used to connect two or more transmission lines. When data arrive on an incoming line, the switching element must choose an outgoing line to forward them on.

5. Telnet

Telnet is a user command and an underlying TCP/IP protocol for accessing remote computers. Through Telnet, an administrator or another user can access someone else's computer remotely. With Telnet, you log on as a regular user with whatever privileges you may have been granted to the specific application and data on that computer.

6. Hub

In a network, a device joining communication lines at a central location, providing a common connection to all devices on the network. The term is an analogy to the hub of a wheel.

7. HTTP

HTTP is the acronym for "hypertext transfer protocol". It is the set of rules for transferring files on the WWW. As soon as a Web user opens their Web browser, the user is indirectly making use of HTTP. HTTP is an application protocol that runs on top of the TCP/IP suite of protocols.

8. Router

A router is an intermediary device on a communications network that expedites message delivery. On a single network linking many computers through a mesh of possible connections, a router receives transmitted messages and forwards them to their correct destinations over the most efficient available route.

22.5 Exercises

1. **Translate the following phrases.**

    electronic remittance
    electronic funds transfer
    electronic payment mechanism
    digital fund system
    circulation medium
    digital cash
    electronic wallet
    value added network
    electronic check software
    win/win relationship
    automatic teller machine
    shopping cart

2. **Fill in the blanks with appropriate words or phrases.**

    a. money receiving   b. paper   c. the transfer   d. real bank
    e. authentication system   f. number   g. computer   h. data

    (1) The Internet payment means conducting fund transfer, _____ and disbursing electronically.

(2) EFT is a subset of EDI, and refers to _____ of value electronically from buyer to seller.

(3) The advanced encryption and _____ can make digital money be used more safely and privately than paper money.

(4) Since the digital cash is only a digital _____, so it needs to be verified as authentic.

(5) In the whole checking process, the seller can't capture any private _____ of his client.

(6) An electronic bank is not a _____, it is a virtual bank.

(7) They don't need to buy PC, they can use any Internet connected _____ to make his bank payments.

(8) EFT has been the major implementation by the banks to eliminate _____ processing.

3. **Match the following terms to the appropriate definition.**

   a. uploading   b. usenet   c. user ID
   d. user interface   e. validation rules   f. virus

(1) Some people write software which can be transferred from one computer to the other and do things which can be harmless, annoying, or very damaging to your computer.

(2) Process of transferring documents, graphics, and other objects from a computer to a server on the Internet.

(3) A collection of notes on various subjects that are posted to servers on a world-wide network. Each subject collection of posted notes is known as a newsgroup.

(4) It is a set of commands or menus through which a user communicates with a program.

(5) Check that analyzes entered data to help ensure that it is correct. It is also called validity check.

(6) Unique combination of characters, such as letters of the alphabet and numbers, that identifies a specific user.

Electronic Payment System

Unit 23 Electronic Marketing

音频

23.1 Text

By using the Internet, manufacturers can directly contact customers without using intermediaries. The manufacturer's direct marketing can be realized as long as they sell established brands and their home site is well known. If a manufacturer's site does not have high visibility, just opening a home page and passively waiting for customers' access may not contribute greatly to sales. Therefore, it is necessary for companies to heavily advertise their Web sites' address. Any cost-effective advertisement method can be employed for this purpose. One example is to link the site to well known electronic directories, and most manufacturers use the directory service of intermediaries. These intermediary sites are called electronic shopping malls.

At least during the initial electronic stage, established distributors like department stores and discount stores were not the major players in electronic retailing. The traditional distributors used their home pages and electronic catalogs to attract customers to the physical stores. Therefore, we need to study the competition structure of electronic distributors, brokers, and online department stores.

Initially, the main concern for electronic marketing involved securing technologies necessary to implement Internet-based marketing, such as powerful search capability and secure electronic payment. However, today the main concern of management is shifting to how to utilize the opportunity of Internet-based marketing to enhance competitiveness in harmony with existing marketing channels.

Increasingly, companies are classifying customers into groups and creating targeted messages for each group. The sizes of these targeted groups can be smaller when companies are using the Web in some cases, just one customer at a time can be targeted.[1] New research into the behavior of Web site visitors has even suggested ways in which Web sites can respond to visitors who arrive at a site with different needs at different times.

Most companies use the term "marketing mix" to describe the combination of elements that they use to achieve their goals for selling and promoting their products and services. When a company decides which elements it will use, it calls that particular marketing mix its marketing strategy. Companies—even those in the same industry—try to create unique presences in their markets. A company's marketing strategy is an important tool that works with its Web presence to get the company's message across to both its current and prospective customers.[2]

Most marketing classes organize the essential issues of marketing into the four Ps of marketing: product, price, promotion, and place. Product is the physical item or service that a company is selling. The intrinsic characteristics of the product are important, but customers' perceptions of the product, called the product's brand, can be as important as the actual characteristics of the product[3].

The price element of the marketing mix is the amount the customer pays for the product. In recent years, marketing experts have argued that companies should think of price in a broader sense, that is, the total of al financial costs that the customer pays to obtain the product. This total cost is subtracted from the benefits that a customer derives from the product to yield an estimate of the customer value obtained in the transaction. The Web can create new opportunities for creative pricing and price negotiations through online auctions, reverse auctions, and group buying strategies. These Web-based opportunities are helping companies find new ways to create increased customer value.

Promotion includes any means of spreading the word about the product. On the Internet, new possibilities abound for communicating with existing and potential customers. Companies are using the Internet to engage in meaningful dialogs with their customers using E-mail and other means.

For years, marketing managers dreamed of a world in which instant deliveries would give all customers exactly what they wanted when they wanted it.[4] The issue of place is the need to have products or services available in many different locations. The problem of getting the right products to the right places at the best time to sell them has plagued companies since commerce began. Although the Internet does not solve all of these logistics and distribution problems, it can certainly help.

- Product-based Marketing Strategies

Managers at many companies think of their businesses in terms of the products and services they sell. This is a logical way to think of a business because companies spend a great deal of effort, time, and money to design and create those products and services. If you ask managers to describe what their companies are selling, they usually provide you with a detailed list of the physical objects they sell or use to create a service.[5] When customers are likely to buy items from particular product categories, or are likely to think of their needs in terms of product categories, this type of product-based organization makes sense.

- Customer-based Marketing Strategies

The communication structures on the Web can become much more complex than those in traditional mass media outlets such as broadcast and print advertising. When a company takes its business to the Web, it can create a Web site that is flexible enough to meet the needs of many different users. Instead of thinking of their Web sites as collections of products, companies can build their sites to meet the specific needs of various types of customers. A good first step in building a customer-based marketing strategy is to identify groups of customers who share common characteristics. The use of customer-based marketing approaches was pioneered on B2B sites. B2B sellers were more aware of the need to customize product and service offerings to match their customers' needs than were the operators of B2C Web sites. In recent years, B2C sites have increasingly added customer-based marketing elements to their Web sites.

Advertising is an attempt to disseminate information in order to effect a buyer-seller transaction. The Internet redefined the meaning of advertising. The Internet has enabled consumers to interact directly with advertisers and advertisements. The Internet has provided the sponsors with two-way communication and E-mail capabilities, as well as allowing the sponsors to target specific groups on which they want to spend their advertising dollars, which is more accurate than traditional telemarketing. Finally, the Internet enables a truly one-to-one advertisement.

Some of the major methods used for advertisements:

◆ Banners

A banner contains a short text or graphical message to promote produce. Advertisers go to great lengths to design a banner that catches consumers' attention. A major advantage of using banners is the ability to customize them to the target audience. One can decide which market segments to focus on. Banners can even be customized to one-to-one targeted advertisement. There are several different forms of placing banner advertising on the Internet on others' Web sites. The most common forms are: Banner Swapping, Banner Exchanges, and Paid Advertising.

◆ Spot Leasing

Search engines often provide space (spot) in their home page for any individual business to lease. The duration of the lease depends upon the contract agreement between the Web site host and the lessee.

◆ E-mail

E-mail is emerging as a marketing channel that affords cost-effective implementation and better, quicker response rates than other advertising channels. A list of E-mail addresses can be a very powerful tool because you are targeting a group of people you know something about.

◆ URL (Universal Resource Locators)

The major advantage of using URL as an advertising tool is that it is free. Anyone can submit its URL to a search engine and be listed. Also, by using URL the targeted audience can be locked and unwanted viewers can be filtered because of the keyword function.

Keywords

auction	n.	拍卖，标售	lessee	n.	租户，承租人
audience	n.	观众	meaningful	adj.	意味深长的
brand	n.	品牌，商标	negotiation	n.	谈判，协商
broker	n.	经纪人，代理人	passively	adv.	被动地，消极地
competitiveness	n.	竞争力	pioneer	v.	提倡，开辟
disseminate	v.	传播，普及	plague	v.	麻烦，困扰，折磨
harmony	n.	调和，与……协调一致	promotion	n.	促进，振兴
intermediary	adj.	中介的，媒介的	prospective	adj.	将来的，未来的
intrinsic	adj.	内在的，本来的	sponsor	n.	倡议者，主办者

Notes

[1] The sizes of these targeted groups can be smaller when companies are using the Web in some cases, just one customer at a time can be targeted.

译文：在有些情况下，当公司采用网络时，这些目标群体的规模可能更小，有时候甚至可能一个客户就是一个目标群体。

说明：本句由"when"引导时间状语从句。

[2] A company's marketing strategy is an important tool that works with its Web presence to get the company's message across to both its current and prospective customers.

译文：公司的营销策略是一个与现有客户和潜在客户能获得公司信息的网上展示相结合的重要工具。

说明：本句中"that"引导的定语修饰表语。

[3] The intrinsic characteristics of the product are important, but customers' perceptions of the product, called the product's brand, can be as important as the actual characteristics of the product.

译文：产品的本质特性很重要，但是客户对产品的感知和产品的实际特性一样重要，客户对产品的认知度被称为"品牌"。

说明：这是一个并列句，"called the product's brand"是插入语。

[4] For years, marketing managers dreamed of a world in which instant deliveries would give all customers exactly what they wanted when they wanted it.

译文：多年来，营销经理都梦想着能存在这样的一个世界：当客户有所需求时，公司能够在最短的时间内满足他们的需求。

说明："in which"引导的是定语从句，"when they wanted it"是时间状语。

[5] If you ask managers to describe what their companies are selling, they usually provide you with a detailed list of the physical objects they sell or use to create a service.

译文：如果你要求经理描述一下他们的公司正在卖的产品，通常会提供给你一本关于他们正在销售或用之制造服务的实物的详细目录。

说明："If"引导的是条件状语，"of the physical objects…"作定语。

23.2 Reading Material 1: Logistics and Supply-Chain Management

As goods move, so must information. To move the right goods to the right place at the right time in the right condition with the right documents, the answers to all the "right" questions must be known. Where should the truck driver pick up this shipment? Who should receive this package? How much inventory of an item is in stock and how much should be produced? Where is this shipment now? Information permeates the logistics system.

Peak logistics efficiency and effectiveness demand a superior integrated logistics information system (ILIS). Without ready access to accurate information, integrated logistics operations lose both efficiency and effectiveness. Integrated logistics will not sustain a strategic, competitive edge. Priority applications of ILIS are inventory status, tracing and expediting, pickup and delivery, order convenience, order accuracy, balancing inbound and outbound traffic opportunities, and order processing. The quality of the information flowing through the ILIS is of utmost importance. The saying "garbage in-garbage out" applies to any information system. Three concerns stand out on quality information: (1) getting the right information, (2) keeping the information accurate, and (3) communicating the information effectively.

An integrated logistics information system can be defined as: the involvement of people, equipment, and procedures required to gather, sort, analyze, evaluate, and then distribute needed information to the appropriate decision-makers in a timely and accurate manner so they can make quality logistics decisions.

An ILIS gathers information from all possible sources to assist the integrated logistics manager in making decisions. It also interfaces with marketing, financial, and manufacturing information systems. All of this information is then funneled to top level management to help formulate strategic decisions. The ILIS has four primary components: the order processing system, research and intelligence system, decision support system, reports and outputs system.

Many global manufacturing companies are involved in implementing new information systems and technology for supply-chain management. Initial applications include financial systems, production planning, distribution and inventory management systems. Most will take four-to-five years to implement and cost millions of dollars in direct expenses.

Supply-chain management improvements are directed at making the supply-chain relationship faster and more consistent, and lowering the cost of working capital by using inventory as the buffer of the last rather than the first resort. In hyper-competition, the focus is on creating value primarily by improving information use and quality in customer data, after-sales service and order fulfillment, and only secondarily by defining more consistent information for upstream processes.

Most companies are beginning to face "hyper-competition", where firms position themselves against one another in an aggressive fashion, as opposed to moderate competition where firms are positioned "around" each other. With moderate competition, barriers are used to limit new entrants and sustainable is possible so long as industry leaders cooperate to restrain competitive behavior. However, hyper-competitive firms are constantly seeking to disrupt the competitive advantage of industry leaders and create new opportunities.

The operational focus of supply-chain management projects may also be at issue in moderate versus hyper-competitive markets. In the former, investing in upstream projects (new financial systems, production planning or inventory management systems) may offer substantial benefits, including consistent information sharing and improved cross-functional cooperation. In hyper-competitive conditions, the focus needs to be on process and information systems with high return-on-investment and added customer value. The operational focus will shift to the demand side and emphasize customer interaction, account management, after-sales service and order processing. To sustain competitive advantage in hyper-competition, a firm may seek to eliminate the need for detailed management reporting and controls, or market forecasts and production plans. Instead, a firm can substitute real-time, online product movement information from its dealers and retailers or simplify controls and management reporting by delivering the organization and empowering employees to improve process quality continuously.

Keywords

barrier	n.	栅栏，障碍	logistics	n.	物流，物流学，后勤学
disrupt	v.	打断，使分裂，破坏	inbound	adj.	入境的，入站的
empower	v.	授权，准许	integrated logistics information system		集成物流信息系统
evaluate	v.	对……评估，对……作评价			
expedite	v.	加快，派遣，迅速的	inventory	n.	存货，报表，清单
fulfillment	n.	履行（条约、义务），完成	involvement	n.	包含，含有，牵连的事务
funnel	v.	汇集，使汇集	moderate	adj.	温和的，适度的

outbound	adj.	外出的，出口的	shipment	n.	装货，载货
peak	v.	减少，缩小	supply-chain		供应链
priority	n.	优先权，优先，先	upstream	adj.	溯流而上的
return-on-investment		投资回报			

23.3 Reading Material 2：Customer Relationship Management

An important element of success doing business on the Web is identifying your customers and connecting with them. The people who visit electronic commerce Web sites are either customers or potential customers. You will learn how a Web site can help firms identify potential customers, encourage existing customers to visit the site, and reach out to new customers.

Businesses use two general ways of identifying and reaching customers: personal contact and mass media. In personal contact, the firm's employees individually search, qualify, and contact potential customers. This personal contact approach to identifying and reaching customers is sometimes called prospecting. Companies that use mass approach prepare advertising and promotional materials about the company and its products or services. They then deliver these messages to potential customers by broadcasting them on television or radio, printing them in newspapers or magazines, posting them on highway billboards, or mailing them.

Many businesses use a combination of mass media and personal contact to identify and reach customers. For example, an insurance company might use mass media to create and maintain the public's general awareness of its insurance products and reputation, while its salespersons might use prospecting technique to identify specific potential customers. Once an individual becomes a customer, the insurance company maintains contact through a combination of personal contact and mailings.

The nature of the Web, with its two-way communication features and traceable connection technology, allows firms to gather much more information about customer behavior and preferences than they could use micro marketing approaches. For the first time, companies could measure large numbers of things that were happening as customers and potential customers gathered information and made purchase decisions. The idea of technology-enabled customer relationship management became possible when promoting and selling via the Web. Technology-enabled customer relationship management occurs when a firm obtains detailed information about a customer's behavior, preferences, needs, and buying patterns, and uses that information to set prices, negotiate terms, tailor promotions, add product features, and otherwise customize its entire relationship with that customer.

Many business researchers distinguish between commerce in the physical world, or marketplace, and commerce in the information world, which they termed the marketspace. In the information world's marketspace, digital products and services are delivered through electronic communication channels, such as the Internet.

For years, businesses have valued information as a way of increasing sales, controlling costs, or both. However, few companies have considered how information itself might be a source of value. In the marketspace, firms can use information to create new value for customers.

Keywords

awareness	n.	知道，晓得	prospect	n.	潜在客户
mass	n.	大众，集团，民众	reputation	n.	声誉，名望
promotional	adj.	宣传，推广	tailor	v.	裁定

23.4 Terms

1. Client/Server

Client/Server describes the relationship between two computer programs in which one program, the client, makes a service request from another program, the server, which fulfills the request. In a network, the client/server model provides a convenient way to interconnect programs that are distributed efficiently across different locations.

2. CAI

CAI is the acronym for "computer aided instruction". It refers to instruction or remediation presented on a computer. Computer programs are interactive and can illustrate a concept through attractive animation, sound, and demonstration. They allow students to progress at their own pace and work individually or problem solve in a group.

3. Event

Event is an action or occurrence detected by a program. Events can be user actions, such as clicking a mouse button or pressing a key, or system occurrences, such as running out of memory. Most modern applications, particularly those that run in Windows environments, are said to be event-driven, because they are designed to respond to events.

4. VPN

VPN is the acronym for "virtual private network". VPN is a network that uses a public telecommunication infrastructure, such as the Internet, to provide remote offices or individual users with secure access to their organization's network. A virtual private network can be contrasted with an expensive system of owned or leased lines that can only be used by one organization.

5. ERP

ERP is the acronym for "enterprise resource planning". The term ERP originally referred to how a large organization planned to use organizational wide resources. Now ERP is a way to integrate the data and processes of an organization into one single system. Usually ERP systems will have many components including hardware and software, in order to achieve integration, most ERP systems use a unified database to store data for various functions found throughout the organization.

6. Open Source

Generically, open source refers to a program in which the source code is available to the general public for use or modification from its original design free of charge. Open source code is typically created as a collaborative effort in which programmers improve upon the code and share the changes within the community.

7. Grid

Grid is an infrastructure that gathers a large diversity of distributed physical resources and software, such as supercomputers, parallel machines, clusters of PCs, massive storage systems, sensors, and special devices. Grid facilitates controlled sharing of resources across organizational boundaries.

8. Usenet

Usenet is a collection of user-submitted notes or messages on various subjects that are posted to servers on a worldwide network. Each subject collection of posted notes is known as a newsgroup. There are thousands of newsgroups and it is possible for you to form a new one. Most newsgroups are hosted on Internet-connected servers, but they can also be hosted from servers that are not part of the Internet.

23.5 Exercises

1. **Translate the following phrases.**

    electronic marketing

    electronic shopping mall

    discount store

    electronic retailing

    electronic catalog

    competition structure

    secure electronic payment

    customer-based marketing strategy

    two-way communication

    traditional telemarketing

    search engine

    electronic data interchange

2. **Fill in the blanks with appropriate words or phrases.**

    a. manufacturers   b. disseminate information   c. company   d. price element
    e. graphical message   f. product   g. targeted messages   h. businesses

    (1) Promotion includes any means of spreading the word about the _____.
    (2) By using the Internet, _____ can directly contact customers without using intermediaries.
    (3) Increasingly, companies are classifying customers into groups and creating _____ for each group.
    (4) A banner contains a short text or _____ to promote produce.
    (5) Managers at many companies think of their _____ in terms of the products and services they sell.
    (6) Product is the physical item or service that a _____ is selling.
    (7) The _____ of the marketing mix is the amount the customer pays for the product.
    (8) Advertising is an attempt to _____ in order to effect a buyer-seller transaction.

Electronic Marketing

3. **Match the following terms to the appropriate definition.**

   a. viewable size   b. WAN   c. world wide web
   d. worksheet   e. worm   f. wireless network

(1) A computer network which usually spans larger geographic area, such as cities, counties, states, nations and planets.

(2) Diagonal measurement of the actual viewing area provided by the screen in a CRT monitor.

(3) An interface or system that connects you to other computers all throughout the world.

(4) It is a piece of paper, often preprinted in a way designed to help organize material for learning or clear understanding. The term is also extended to designate a single two-dimensional array of data within a computerized spreadsheet program. Rows and columns used to organize data in a spreadsheet.

(5) A program or algorithm that replicates itself over a computer network and usually performs malicious actions, such as using up the computer's resources and possibly shutting the system down.

(6) It is the term used to describe any computer network where there is no physical wired connection between sender and receiver, but rather the network is connected by radio waves or microwaves to maintain communications.

Unit 24　Mobile Commerce

24.1　Text

音频

With the introduction of the world wide web, electronic commerce has revolutionized traditional commerce and boosted sales and exchanges of merchandise and information. Recently, the emergence of wireless and mobile networks has made possible the admission of electronic commerce to a new application and research subject: mobile commerce, which is defined as any transaction with a monetary value that is conducted via a mobile telecommunications network.[1] A somewhat looser approach would be to characterize mobile commerce as the emerging set of applications and services people can access from their Internet-enabled mobile devices. Mobile commerce is an effective and convenient way to deliver electronic commerce to consumers from anywhere and at anytime. Realizing the advantages to be gained from mobile commerce, many major companies have begun to offer mobile commerce options for their customers.[2]

Commerce is the exchange or buying and selling of commodities on a large scale involving transportation from place to place. It is boosted by the convenience and ubiquity conveyed by mobile commerce technology. There are many examples showing how mobile commerce helps commerce. For example, consumers can now pay for the products in a vending machine or a parking fee by using their cellular phones; mobile users can check their bank accounts and perform account balance transfers without needing to go a bank or access an ATM; etc.

Just-in-time delivery is critical for the success of today's businesses. Mobile commerce allows a business to keep track of its mobile inventory and make time definite deliveries, thus improving customer service, reducing inventory, and enhancing a company's competitive edge.[3] Most major delivery services, such as UPS and FedEx, have already applied these technologies to their business operations worldwide.

Traffic is the movement (as of vehicles or pedestrians) through an area or along a route. The passengers in vehicles or pedestrians are mobile objects, the ideal clients of mobile commerce. Also, traffic control is usually a major headache for many metropolitan areas. Using the technology of mobile commerce can easily improve traffic in many ways. For example, it is expected that a mobile handheld device will have the capabilities of a GPS (global positioning system), e.g., determining the driver's position, giving directions, and advising on the current status of traffic in the area; a traffic control center could monitor and control the traffic according to the signals sent from mobile devices in vehicles.

Travel expenses can be costly for a business. Mobile commerce could help reduce operational costs by providing mobile travel management services to business travelers. It can deliver a compelling and memorable experience to customers at the point of need by using the mobile channels to locate a desired hotel nearby, purchase tickets, make transportation arrangements, and so on. It also extends the reach of relationship-oriented companies beyond their current channels and helps the mobile users to identify, attract, serve, and retain valuable customers.

It is a short number of steps from owning a smart phone or tablet, to searching for products and services, browsing, and then purchasing. The resulting mobile commerce is growing at over 50% a year, significantly faster than desktop E-Commerce at 12% a year. The high rate of growth for mobile commerce will not, of course, continue forever.

A study of the top 400 mobile firms by sales indicates that 73% of mobile commerce is for retail goods, 25% for travel, and 2% for ticket sales. Increasingly, consumers are using their mobile devices to search for people, places, and things—like restaurants and deals on products they saw in a retail store. The rapid switch of consumers from desktop platforms to mobile devices is driving a surge in mobile marketing expenditure.

Compared to an electronic commerce system, a mobile commerce system is much more complicated because components related to mobile computing have to be included.[4] The following outline gives a brief description of a typical procedure that is initiated by a request submitted by a mobile user.

- Mobile commerce applications: A content provider implements an application by providing two sets of programs: client-side programs, such as a user interface on a micro-browser, and server-side programs, such as database accesses and updating.
- Mobile stations: Mobile stations present user interfaces to the end users, who specify their requests on the interfaces.[5] The mobile stations then relay user requests to the other components and display the processing results later using the interfaces.
- Mobile middleware: The major purpose of mobile middleware is to seamlessly and transparently map Internet contents to mobile stations that support a wide variety of operating systems, markup languages, micro-browsers, and protocols. Most mobile middleware also encrypts the communication in order to provide some level of security for transactions.
- Wireless networks: Mobile commerce is possible mainly because of the availability of wireless networks. User requests are delivered to either the closest wireless access point (in a wireless local area network environment) or a base station (in a cellular network environment).
- Wired networks: This component is optional for a mobile commerce system.
- Host computers: This component is similar to the one used in electronic commerce. User requests are generally acted upon in this component.

Without trust and security, there is no mobile commerce period. How could a content provider or payment provider hope to attract customers if it cannot give them a sense of security as they connect to paying services or make purchases from their mobile devices? Consumers need to feel comfortable that they will not be charged for services they have not used, that their payment details will not find their way into the wrong hands, and that there are adequate mechanisms in place to help resolve possible disputes.

As we review different aspects of mobile security, it is important to keep in mind that security always requires an overall approach. A system is only as secure as its weakest component, and securing networking transmission is only one part of the equation. Technically speaking, there are a number of different dimensions to network security, each corresponding to a different class of threat or vulnerability. Protecting against one is no guarantee that you will not be vulnerable to another.

Keywords

adequate	*adj.*	适当的，足够的，恰当的	metropolitan	*adj.*	首都的，主要城市的
admission	*n.*	允许进入，许可，承认	monetary	*adj.*	货币的，金融的，财政上的
boost	*v.*	促进，提高，宣传，升	optional	*adj.*	可选择的，随意的，任意的
cellular	*adj.*	蜂窝的，多孔的	pedestrian	*adj.*	徒步的，步行的
compel	*v.*	强迫，胁迫	purchase	*v.*	买，购买，购得物
costly	*adj.*	昂贵的，奢华的	surge	*n. v.*	大浪，波涛，高涨，起伏
dispute	*v.*	辩论，争论	transparently	*adv.*	透明地，透彻地
expenditure	*n.*	支出，花费，开销	ubiquity	*n.*	泛在，无所不在
looser	*adj.*	松的，不精确的	vulnerability	*n.*	脆弱性，攻击，弱点
memorable	*adj.*	可记忆的，难忘的	vulnerable	*adj.*	易受攻击的，易受伤的
merchandise	*n.*	商品，商业，推销，买卖			

Notes

[1] Recently, the emergence of wireless and mobile networks has made possible the admission of electronic commerce to a new application and research subject: mobile commerce, which is defined as any transaction with a monetary value that is conducted via a mobile telecommunications network.

译文：最近，无线网络和移动网络的出现使得电子商务进入了新的应用及科研主题：移动商务。移动商务可以定义为通过移动的无线电通信网络进行的涉及货币价值的任何交易。

说明：本句的"wireless and mobile networks"作主语的定语，"which is defined as…"作非限定性定语从句。

[2] Realizing the advantages to be gained from mobile commerce, many major companies have begun to offer mobile commerce options for their customers.

译文：很多大公司已经意识到移动商务带来的好处，开始为客户提供移动商务选择。

说明：本句的分词结构"Realizing the advantages…"作状语。

[3] Mobile commerce allows a business to keep track of its mobile inventory and make time definite deliveries, thus improving customer service, reducing inventory, and enhancing a company's competitive edge.

译文：移动商务让企业可以跟踪其移动库存并适时地投递，这样可以改善客户服务、减少库存并增强公司的竞争力。

说明：本句的"to keep track of…"作宾语补足语，"thus improving customer…"是目的状语。

[4] Compared to an electronic commerce system, a mobile commerce system is much more complicated because components related to mobile computing have to be included.

译文：与电子商务系统相比，移动商务系统要复杂得多，因为它必须包含与移动计算相关的成分。

说明：本句的"Compared to an electronic commerce system"是条件状语，由"because"引导的是原因状语从句。

[5] Mobile stations present user interfaces to the end users, who specify their requests on the interfaces.

译文：移动站向终端用户显示用户界面，这些用户在界面上细化他们的要求。

说明：本句的"who specify their…"是非限定性定语从句，修饰"the end users"。

24.2 Reading Material 1：O2O

What do Groupon, Open Table, Restaurant.com and SpaFinder all have in common? They grease the wheels of online-to-offline commerce. Groupon's growth has been nothing short of extraordinary, but it is merely a small subset of an even larger category which we'd like to call online-to-offline commerce, or O2O commerce, in the view of other commerce terms like B2C, B2B, and C2C.

The key to O2O is that it finds consumers online and brings them into real-world stores. It is a combination of payment model and foot traffic generator for merchants (as well as a "discovery" mechanism for consumers) that creates offline purchases. It is inherently measurable, since every transaction happens online. This is distinctively different from the directory model in that the addition of payment helps quantify performance and close the loop.

In retrospect, the fact that this is "big", or that Groupon has been able to grow high-margin revenues faster than almost any other company in the history of the Internet, seems pretty obvious. Your average ecommerce shopper spends about $1000 per day. Let's say your average American earns about $40000 per year. What happens to the other $39000? The delta is higher when you consider that ecommerce shoppers are higher-income Americans than most, but the point is the same.

Most of it (disposable income after taxes) is spent locally. You spend money at coffee shops, bars, gyms, restaurants, gas stations, plumbers, dry-cleaners, and hair salons. Excluding travel, online B2C commerce is largely stuff that you order online and gets shipped to you in a box. It is boring, although the e-commerce industry has figured out an increasing number of items to sell online.

FedEx can't deliver social experiences like restaurants, bars, yoga, sailing, and tennis lessons, but Groupon does. Moreover, for your locally owned and operated yoga studio, there is little marginal cost to add customers to a partially filled class, meaning that the business model of reselling "local" is often more lucrative than the traditional E-Commerce model of buying commodity inventory low, selling it higher, and keeping the difference while managing perishable or depreciating inventory.

The important thing about companies like O2O commerce companies is that performance is readily quantifiable, which is one of the tenets of O2O commerce. Traditional E-Commerce tracks conversion using things like cookies and pixels. Zappos can determine their ROI (Return on Investment) for online marketing because every completed order has "tracking code" on the confirmation page. Offline commerce doesn't have this luxury. The bouncer at the bar isn't examining your iPhone's browsing history. But O2O makes this easy. Because the transaction happens online, the same tools are now available to the offline world, and the whole thing is brokered via intermediaries like Open Table or SpaFinder. This has proven to be a far more profitable and scalable model than selling advertising to local advertising to local establishments. It's entirely due to the collection of payment by the online intermediary.

O2O commerce dwarfs traditional (stuff in a box) E-Commerce—simply because offline commerce itself dwarfs online commerce, and O2O is simply shifting the discovery and payment online. Venture capitalists and entrepreneurs would be wise to think beyond cloning the "deal of the day" concept—and instead think about how the discovery, payment, and performance measurement of offline commerce can move online. This will have ripple effects across the whole Internet industry—advertising, payments, and commerce—as trillions of dollars in local consumer spending increasingly begin online.

Keywords

bouncer	n.	酒保，保镖	measurable	adj.	可量的，可测量的，适当的
clone	n.	复制品			
conversion	n.	变换，转化，改造，变更	online-to-offline commerce		线上—线下电子商务
depreciate	v.	贬值，减价			
dwarf	v.	使……受阻碍，使相形见绌	perishable	adj.	易腐败的，脆弱的
			plumber	n.	管道工
entrepreneur	n.	企业家，创业人，主办人	profitable	adj.	有益的，有用的，合算的
extraordinary	adj.	非常的，异常的，特别的	readily	adv.	容易地，毫不犹豫地
grease	n.	油脂，贿赂，润滑油	retrospect	v.	回顾，对证，参照
gym	n.	健身房，体育馆，体育课	revenue	n.	税收，收益，所得
in the view of		按……的意见	ripple	v. n.	使泛起涟漪，涟漪，脉动
inherently	adv.	内在地，固有地	tenet	n.	教条，信条，原则
lucrative	adj.	有利的，赚钱的	yoga	n.	瑜伽
marginal cost		边际成本			

24.3　Reading Material 2：A Mobile Web

The mobile Web refers to access to the world wide web, i.e. the use of browser-based Internet services, from a handheld mobile device, such as a smart phone or a feature phone, connected to a mobile network or other wireless network.

Traditionally, access to the Web has been via fixed-line services on large-screen laptops and desktop computers. However, the Web is becoming more accessible by portable and wireless devices. An early 2010 ITU (International Telecommunication Union) report said that with the current growth rates, Web access by people on the go—via laptops and smart mobile devices—is likely to exceed Web access from desktop computers within the next five years. The shift to mobile Web access has been accelerating with the rise since 2007 of large multi-touch smart phones, and of multi-touch tablet computers since 2010. Both platforms provide better Internet access, screens, and mobile browsers—or application-based user Web experiences than previous generations of mobile devices have done. Web designers may work separately on such pages, or pages may be automatically converted as in Mobile Wikipedia.

The distinction between mobile Web applications (mobile apps) and native applications is anticipated to become increasingly blurred, as mobile browsers gain direct access to the hardware of mobile devices (including accelerometers and GPS chips), and the speed and abilities of browser-based applications improve. Persistent storage and access to sophisticated user interface graphics functions may further reduce the need for the development of platform-specific native applications.

Mobile Web access today still suffers from interoperability and usability problems. Interoperability issues stem from the platform fragmentation of mobile devices, mobile operating systems, and browsers. Usability problems are centered around the small physical size of the mobile phone form factors (limits on display resolution and user input/operating). Despite these shortcomings, many mobile developers choose to create apps using mobile Web.

Mobile Internet refers to access to the Internet via a cellular telephone service provider. It is wireless access that can hand off to another radio tower while it is moving across the service area. It can refer an immobile device that stays connected to one tower, but this is not the meaning of "mobile" here. Wi-Fi and other better methods are commonly available for users not on the move. Cellular base stations are more expensive to provide than a wireless base station that connects directly to an Internet service provider, rather than through the telephone system.

A mobile phone, such as a smart phone, that connects to data or voice services without going through the cellular base station is not on mobile Internet. A laptop with a broadband modem and a cellular service provider subscription that is traveling on a bus through the city is on mobile Internet.

A mobile broadband modem "tethers" the smart phone to one or more computers or other end user devices to provide access to the Internet via the protocols that those cellular telephone service providers may offer.

Keywords

accelerometer	n.	加速计，加速仪	interoperability	n.	互用性，协同工作能力
anticipate	v.	预期，预料，预测	stem from		起源于，出自，由……造成
blur	n. v.	污点，难以区分	tether	v. n.	拘束，束缚，限度，范围
fragmentation	n.	破碎，破裂	usability	n.	可用，有用，适用

24.4 Terms

1. PDA

PDA is the acronym for "personal digital assistant". It is a lightweight palmtop computer designed to provide specific functions like personal organization as well as communications. Many PDA devices rely on a pen or other pointing device for input instead of a keyboard or mouse, although some offer a keyboard too small for touch typing to use in conjunction with pen or pointing device. For data storage, a PDA relies on flash memory instead of power-hungry disk drives.

2. EDI

EDI is the acronym for "electronic data interchange". It is a standard format for exchanging business data. An EDI message contains a string of data elements, each of which represents a singular fact, such as a price, product model number, and so forth, separated by delimiter. The entire string is called a data segment. One or more data segments framed by a header and trailer form a transaction set, which is the EDI unit of transmission.

3. Protocol

A protocol is an agreed-upon format for transmitting data between two devices. There are a variety of standard protocols from which programmers can choose. From a user's point of view, the only interesting aspect about protocols is that your computer or device must support the right ones if you want to communicate with other computers. The protocol can be implemented either in hardware or in software.

4. Toolbar

A toolbar consists of a set of icons or buttons called toolbar buttons and usually stretches horizontally across the screen. Each toolbar button has a name, which is displayed if you point to the button.

5. Pipelining

In computers, a pipeline is the continuous and somewhat overlapped movement of instruction to the processor or in the arithmetic steps taken by the processor to perform an instruction. With pipelining, the computer architecture allows the next instructions to be fetched while the processor is performing arithmetic operations, holding them in a buffer close to the processor until each instruction operation can be performed.

6. Virtual Machine

A virtual machine is an environment, usually a program or operating system, which does not physically exist but is created within another environment. Virtual machines are often created to execute an instruction set different than that of the host environment.

7. Middleware

Middleware is used to describe separate products that serve as the glue between two applications. It is, therefore, distinct from import and export features that may be built into one of the applications. Middleware is sometimes called plumbing because it connects two sides of an application and passes data between them.

8. Computer Animation

Computer animation is the art of creating moving images via the use of computers. It is a subfield of computer graphics and animation. Increasingly it is created by means of 3D computer graphics. Sometimes the target of the animation is the computer itself, but sometimes the target is another medium, such as film. To create the illusion of movement, an image is displayed on the computer screen then quickly replaced by a new image that is similar to the previous image, but shifted slightly.

24.5 Exercises

1. **Translate the following phrases.**

   mobile inventory

   mobile commerce

   relationship-oriented company

   retail store

   mobile user

   content provider

client-side programs
mobile middleware
cellular network environment
payment provider
payment details
mobile devices

2. **Fill in the blanks with appropriate words or phrases.**

   a. examples   b. sales   c. commodities   d. just-in-time
   e. security   f. vehicles   g. smart phone   h. costs

   (1) With the introduction of the WWW, electronic commerce has revolutionized traditional commerce and boosted _____ and exchanges of merchandise and information.

   (2) Commerce is the exchange or buying and selling of _____ on a large scale involving transportation from place to place.

   (3) There are many _____ showing how mobile commerce helps commerce.

   (4) _____ delivery is critical for the success of today's businesses.

   (5) The passengers in _____ or pedestrians are mobile objects, the ideal clients of mobile commerce.

   (6) Mobile commerce could help reduce operational _____ by providing mobile travel management services to business travelers.

   (7) It is a short number of steps from owning a _____ or tablet, to searching for products and services, browsing, and then purchasing.

   (8) Without trust and _____, there is no mobile commerce period.

3. **Match the following terms to the appropriate definition.**

   a. hertz   b. hub   c. header
   d. high-level language   e. hard disk drive   f. hardware

   (1) The actual parts that make up a computer system, like the monitor, keyboard, printer, and mouse.

   (2) A unit of frequency equivalent to one cycle per second.

   (3) A signal distribution point for part of an overall system.

   (4) Protocol control information located at the beginning of a protocol data unit.

   (5) The mechanism that controls the positioning, reading, and writing of the hard disk, which furnishes the largest amount of data storage for the PC.

   (6) It allows programmers to express algorithms more concisely, take care of much of detail, and often support naturally the use of structured programming or object-oriented design.

Unit 25　　Network Security

25.1　Text

This increase in terrorist attacks on the general population, as well as on special segments of our society's infrastructure, comes at a time when large and small organizations of all types are becoming increasingly dependent on networks to carry on their activities. Networks have become assets like computers, data and information. Communications with customers, suppliers, employees, and other organizations are handled primarily through networks. The loss of the network communications channel could effectively cripple and organization. No longer can a business be operated without having access to information and a reliable communication system. The network asset must be protected like other assets and surrounded with proper controls and the appropriate security. No wonder there is such a strong interest in the subjects of network and computer security.

With the rise of the Internet and its use for conducting business, parts of most organizations' networks are more open and vulnerable to unauthorized access, computer viruses, and attacks that are more aggressive than ever. The potential disruptions of application systems running on computer networks or the corruption of the underlying data are good reasons for organizations to take network security very seriously and to take action.[1] Usually, the value of the data stored on networked computers far exceeds the cost of the networks themselves.

Security Threats

Security threats to a network can be divided into those that involve some sort of unauthorized access and all others. Once someone gains unauthorized access to the network, the range of things they can do is large. Some people are just interested in the challenge of breaking through the security and have no interest in doing anything further. They may, however choose to monitor network traffic, called eavesdropping, for the purpose of learning something specific and perhaps disclosing it to others, or for the purpose of analyzing traffic patterns, traffic analysis could lead to the observation. The types of unauthorized access we usually think of, however, are the active security attacks whereby, after someone gains unauthorized access to the network, they take some overt action. Active attacks include altering message contents, masquerading as someone else, denial of service, and planting viruses.

Network Security Recommendations

- Keep Your Computer up to Date

To help keep the computers on your network safer, turn on automatic updating on each computer. Windows can automatically install important and recommended updates, or important updates only. Important updates provide significant benefits, such as improved security and reliability. Recommended updates can address non-critical problems and help enhance your computing experience. Optional updates are not downloaded or installed automatically.

♦ Use a Firewall

A firewall can help prevent hackers or malicious software from gaining access to your computer through a network. A firewall can also help stop your computer from sending malicious software to other computers.

♦ Run Antivirus Software on Each Computer

Firewalls help keep out worms and hackers, but they are not designed to protect against viruses. So you should install and use antivirus software. Viruses can come from attachments in E-mail messages, files on CDs or DVDs, or files downloaded from the Internet. Make sure that the antivirus software is up to date and set to scan your computer regularly.

Antivirus programs scan E-mail and other files on your computer for viruses, worms, and Trojan horses. If one is found, the antivirus program either quarantines it or deletes it entirely before it damages your computer and files.[2] Windows does not have a built-in antivirus program, but your computer manufacturer might have installed one. Check security center to find out if your computer has antivirus protection. If not, go to the Microsoft Antivirus Partners webpage to find an antivirus program.

Because new viruses are identified every day, it is important to select an antivirus program with an automatic update capability. When the antivirus software is updated, it adds new viruses to its list of viruses to check for, helping to protect your computer from new attacks. If the list of viruses is out of date, your computer is vulnerable to new threats. Updates usually require an annual subscription fee. Keep the subscription current to receive regular updates. If you do not use antivirus software, you expose your computer to damage from malicious software. You also run the risk of spreading viruses to other computers.

♦ Use a Router to Share an Internet Connection

Consider using a router to share an Internet connection. These devices usually have built-in firewalls and other features that can help keep your network better protected against hackers.

♦ Don't Stay Logged on as an Administrator

When you are using programs that require Internet access, such as a Web browser or an E-mail program, we recommend that you log on as a standard user rather than an administrator.[3] That's because many viruses and worms can not be stored and run on your computer unless you are logged on as an administrator.

♦ Use Spyware Protection

Spyware is software that can display advertisements, collect information about you, or change settings on your computer, generally without appropriately obtaining your consent. For example, spyware can install unwanted toolbars, links, or favorites in your web browser, change your default home page, or display pop-up ads frequently.

Some spyware displays no symptoms that you can detect, but it secretly collects sensitive information, such as which websites you visit or text that you type. Most spyware is installed through free software that you download, but in some cases simply visiting a website results in a spyware indection. To help protect your computer from spyware, use an antispyware program.

Management's Responsibility

The network security policy is management's statement of the importance and their commitment to network security. The policy needs to describe in general terms what will be done, but does not deal with the way the protection is to be achieved. Writing the policy is complex, because in reality, it is normally a part of a broader document: the organization's information security policy. The network security policy needs to clearly state management's position about the importance of network security and the items that are to be protected. Management must understand that there is no such thing as a perfectly secure network.

Furthermore, network security is a constantly moving target because of advances in technology and the creativity of people who would like to break into a network or its attached computers.[4] Measures put in place to minimize security risks today will need to be upgraded in the future, and upgrades usually have a price attached, for which management will have to pay.

Simply writing the policy does not put the practices, procedures, or software in place to improve the security situation. That requires follow through and communication with all employees so that they understand the emphasis and importance senior management is placing on security. Management, at all levels, needs to support the policy and periodically reinforce it with employees in various ways.[5] IT and network staff may need to install additional hardware, software, and procedures to perform automated security checking.

Keywords

aggressive	adj.	侵略的，攻势的	overt	adj.	明显的，公然的
annual	adj.	每年的，一年一次的	primarily	adv.	主要地
asset	n.	资产，财产，有用的资源	quarantine	v.	隔离，封锁
attachment	n.	附件	recommendation	n.	推荐，介绍
consent	v.	同意，赞成	regularly	adv.	有规律地，定期地
cripple	v.	削弱	reinforce	v.	增强，加固，补充
disclose	v.	泄露，揭露	subscription	n.	订购，预定
disruption	n.	破裂，分裂	surround	v.	围绕，环绕
eavesdropping	v.	窃听	terrorist	n.	恐怖主义者，恐怖分子
firewall	n.	防火墙			
malicious	adj.	有恶意的，蓄意的	vulnerable	adj.	脆弱的，有弱点的
masquerade	v.	伪装，掩饰	worm	n.	蠕虫

Notes

[1] The potential disruptions of application systems running on computer networks or the corruption of the underlying data are good reasons for organizations to take network security very seriously and to take action.

译文：运行在计算机网络上的应用系统的潜在破坏或基本数据的讹误是组织对网络安全十分重视并采取行动的很好的理由。

说明：本句的主语有两个"The potential disruptions"和"the corruption"，分别有定语修饰；"for organizations…"作定语，修饰"good reasons"。

[2] If one is found, the antivirus program either quarantines it or deletes it entirely before it damages your computer and files.

译文：如果病毒被找到，防病毒程序就会在其破坏计算机和文件之前将其隔离，或者将其完全删除。

说明："If one is found"是条件状语，"before"引导的是时间状语。

[3] When you are using programs that require Internet access, such as a Web browser or an E-mail program, we recommend that you log on as a standard user rather than an administrator.

译文：使用需要 Internet 访问权限的程序时，如 Web 浏览器或电子邮件程序，建议以标准用户身份登录，而不是以管理员身份登录。

说明：本句中的"When"引导的是时间状语，"such as a Web browser or an E-mail program"作"programs"的同位语，"that you log on…"是宾语从句。

[4] Furthermore, network security is a constantly moving target because of advances in technology and the creativity of people who would like to break into a network or its attached computers.

译文：此外，由于科技的进步和那些想入侵网络和计算机的人的创造性，网络安全是一个不断变化的目标。

说明：句中的"because of"引导原因状语，"who would like…"作定语，修饰"people"。

[5] Management, at all levels, needs to support the policy and periodically reinforce it with employees in various ways.

译文：在各个层面，管理层都需要去支持这项政策，并定期与员工用各种方式强化它。

说明：句中的插入语"at all levels"作状语，宾语是不定式结构。

25.2　Reading Material 1：Network Firewall

When you connect your LAN to the Internet, you are enabling your users to reach and communicate with the outside world. At the same time, however, you are enabling the outside world to reach and interact with your LAN.

The purpose of a network firewall is to provide a shell around the network which will protect the systems connected to the network from various threats. The types of threats a firewall can protect against include:

♦ Unauthorized Access to Network Resources

An intruder may break into a host on the network and gain unauthorized access to files.

♦ Denial of Service

An individual outside the network could, for example, send thousands of mail messages to a host on the net in an attempt to fill available disk space or load the network links.

♦ Masquerading

Electronic mail appearing to have originated from one individual could have been forged by another with the intent to embarrass or cause harm.

Basically, a firewall is a standalone process or a set of integrated processes that runs on a router or server to control the flow of networked application traffic passing through it. Typically, firewalls are placed on the entry point to a public network such as the Internet. They could be considered traffic cops. The firewall's role is to ensure that all communication between an organization's network and the Internet conform to the organization's security policies. Primarily these systems are TCP/IP based and, depending on the implementation, can enforce security roadblocks as well as provide administrators with answers to the following questions:

- Who's been using my network?
- Who failed to enter my network?
- When were they using my network?
- Where were they going on my network?
- What were they doing on my network?

A firewall can reduce risks to network systems by filtering out inherently insecure network services. Network file system (NFS) services, for example, could be prevented from being used from outside of a network by blocking all NFS traffic to or from the network. This protects the individual hosts while still allowing the service, which is useful in a LAN environment, on the internal network. One way to avoid the problems associated with network computing would be to completely disconnect an organization's internal network from any other external system. This, of course, is not the preferred method. Instead what is needed is a way to filter access to the network while still allowing users access to the "outside world".

In this configuration, the internal network is separated from external networks by a firewall gateway. A gateway is normally used to perform relay services between two networks. In the case of a firewall gateway, it also provides a filtering service which limits the types of information that can be passed to or from hosts located on the internal network.

In general, there are three types of firewall implementations, some of which can be used together to create a more secure environment. These implementations are: packet filtering, application proxies, and circuit-level or generic-application proxies.

Packet Filtering

Consider your network data a neat little package that you have to driver somewhere. This data could be part of an E-mail, file transfer, etc. With packet filtering, you have access to deliver the package yourself. The packet filter acts like a traffic cop; it analyzes where you are going and what you are bringing with you. However, the packet filter does not open the data package, and you still get to drive it to the destination if allowed.

Most commercial routers have some kind of built-in packet filtering capability. However, some routers that are controlled by ISPs may not offer administrators the ability to control the configuration of the router. In those cases, administrators may opt to use a standalone packet filter behind the router.

Application Proxy

To understand the application proxy, consider this scenario where you needed to deliver your neat little package of network data. With application-level proxies, the scenario is similar, but now you need to rely on someone else to deliver the package for you. Hence the term proxy illustrates this new scenario. The same rules apply as they do for packet filtering, except that you don't get to deliver your package past the gate. Someone will do it for you, but that agent needs to look inside the package first to confirm its contents. If the agent has permission to deliver the contents of the package for you, he will.

Of course, security and encryption also come into play, since the proxy must be able to open the "package" to look at it or decode its contents.

Circuit-level or Generic-application Proxy

As with application-level proxies, you need to rely on someone to deliver your package for you. The difference is that if these circuit-level proxies have access to deliver the package to your requested destination, they will. They don't need to know what is inside. Circuit-level proxies work outside of the application layers of the protocol. These servers allow clients to pass through this centralized service and connect to whatever TCP port the clients specify.

Keywords

conform	v.	使一致，依照	opt	v.	选择，赞成
cop	n.	警察	prefer	v.	更喜欢
denial	n.	拒绝接受，否认	proxy	n.	代理，代表
embarrass	v.	妨碍，阻碍	roadblock	n.	路障
intruder	n.	入侵者	scenario	n.	方案
masquerade	v.	冒充，假装	shell	n.	外壳
neat	adj.	干净的，匀整的			

25.3 Reading Material 2：Digital Signature and Certificates

One potential security problem unique to online commerce is the identification of both customers and commerce sites. The way to identify an individual or a company is to use a digital signature.

A digital signature is the electronic equivalent of a personal signature that cannot be forged. A user creates a digital signature by encrypting any phrase, such as I like green vegetables, with a private key. Next, the encoded phrase is attached to a message before it is sent to another person or Web site, similar to how you would sign a letter you are about to mail. Finally, the sender encrypts the entire message, including the encoded phrase, with the recipient's public key, and then the sender sends the message.

Upon receipt, the recipient decrypts the message with a private key. Finally, the recipient decrypts the signature phrase using the sender's public key. If the recipient successfully decrypts the phrase (I like green vegetables), the sender blows that the message came from the sender and not from an imposter.

The problem with using a digital signature to protect a transmitted message is there is no guarantee that the sender is actually who he or she claims to be. The message could have come from an imposter who obtained the real sender's private key.

Digital certificates take care of this problem. A digital certificate, also known as a digital ID, is an electronic signature that verifies the identity of a user or Web site. A certification authority (CA) issues a digital certificate to an organization or individual. The CA requires entities applying for digital certificates to supply an appropriate proof of identity. Once the CA is satisfied that the entity is valid, the CA issues a certificate to the entity. The certificate contains information about the entity, the certificate's expiration date, and the entity's public key. Then the CA signs the certificate by affixing its stamp of approval and encrypts the certificate with the CA's public encryption key. The CA guarantees that the entity presenting the certificate is authentic.

Certificates are classified as low, medium, or high assurance based largely on the identification requirements they impose on certificate seekers. Certificates exist for individuals (personal certificates), software publishers (software publisher certificates), and Web site servers (site certificates). When you browse to a Web site, you can view the Web site's certificate, if it has one, with your Web browser.

Digital certificates are valid for approximately one year. Near the end of a certificate's validity period, the company or individual must renew the certificate. Creating digital certificates with short life spans decreases the danger that a company's certificate will be valid after the company ceases to exist.

Keywords

affixing	v.	附加上，签署	equivalent	adj.	等价的，相当的
approval	n.	认可，批准	expiration	n.	截止，满期
authentic	adj.	可靠的，可信的	renew	v.	更新，更换
blow	v.	声称	stamp	n.	戳记，图章

25.4　Terms

1. FTP

FTP is the acronym for "file transfer protocol". It is a standard Internet protocol and is the simplest way to exchange files between computers on the Internet. FTP is an application protocol that uses the Internet's TCP/IP protocols. FTP is commonly used to transfer Web page files from their creator to the computer that acts as their server for everyone on the Internet. It's also commonly used to download programs and other files to your computer from other servers.

2. Genetic Algorithm

Genetic algorithms are based on a biological metaphor. They view learning as a competition among a population of evolving candidate problem solutions. A 'fitness' function evaluates each solution to decide whether it will contribute to the next generation of solution. Then, through operations analogous to gene transfer in sexual reproduction, the algorithm creates a new population of candidate solutions.

3. Phishing

Phishing is an E-mail fraud method in which the perpetrator sends out legitimate-looking E-mail in an attempt to gather personal and financial information from recipients. Typically, the messages appear to come from well known and trustworthy Web sites. Phishers use a number of different social engineering and E-mail spoofing ploys to try to trick their victims.

4. OLE

OLE is the acronym for "object linking and embedding". It is Microsoft's framework for a compound document technology. Briefly, a compound document is something like a display desktop that can contain visual and information objects of all kinds: text, calendars, animations, sound, motion video, 3D, continually updated news, controls, and so forth. Each desktop object is an independent program entity that can interact with a user and also communicate with other objects on the desktop.

5. Speech Recognition

Speech recognition is the field of computer science that deals with designing computer systems that can recognize spoken words. Note that voice recognition implies only that the computer can take dictation, not that it understands what is being said. Comprehending human languages falls under a different field of computer science called natural processing.

6. Smart Card

A smart card is a small electronic device about the size of a credit card that contains electronic memory, and possibly an embedded integrated circuit. To use a smart card, either to pull information from it or add data to it, you need a smart card reader, a small device into which you insert the smart card.

7. Blog

A blog is a personal online journal that is frequently updated and intended for general public consumption. Blogs are defined by their format: a series of entries posted to a single page in reverse-chronological order. Blogs generally represent the personality of the author or reflect the purpose of the Web site that hosts the blog.

8. Agent

An agent is given a very small and well-defined task. Although the theory behind agents has been around for some time, agents have become more prominent with the growth of the Internet. Many companies now sell software that enables you to configure an agent to search the Internet for certain types of information.

25.5 Exercises

1. **Translate the following phrases.**
   security threat
   unauthorized access
   traffic analysis
   active security attack
   automatic updating
   malicious software
   antivirus software
   information security policy
   network firewall
   network file system
   packet filtering
   circuit-level proxy

2. **Fill in the blanks with appropriate words or phrases.**

   | a. security situation | b. challenge | c. all others | d. built-in firewalls |
   | e. automatic updating | f. computer | g. information | h. customers |

(1) Some people are just interested in the _____ of breaking through the security and have no interest in doing anything further.
(2) Communications with _____, suppliers, employees, and other organizations are handled primarily through networks.
(3) Simply writing the policy does not put the practices, procedures, or software in place to improve the _____.
(4) No longer can a business be operated without having access to _____ and a reliable communication system.
(5) Security threats to a network can be divided into those that involve some sort of unauthorized access and _____.
(6) Routers usually have _____ and other features that can help keep your network better protected against hackers.
(7) To help protect your _____ from spyware, use an antispyware program.
(8) To help keep the computers on your network safer, turn on _____ on each computer.

3. **Match the following terms to the appropriate definition.**

   a. wizard   b. word processing software   c. webmaster
   d. whiteboard   e. Web address   f. Web community

(1) An area on a display screen that multiple users can write or draw on. It is a principal component of teleconferencing applications because they enable visual as well as audio communication.
(2) Unique address for a Web page. It is also called a Uniform Resource Locator.
(3) It is a web site that joins a specific group of people with similar interests or relationships.
(4) A utility within an application that helps you use the application to perform a particular task.
(5) Using a computer to create, edit and print documents. Of all computer applications, it is the most common.
(6) It is also called the system administrator, the author, or the website administrator. It is the person responsible for designing, developing, marketing or maintaining a website.

参 考 译 文

第1单元 计算机概述

计算机就是一个数字的电子数据处理系统，这是目前公认的一种定义。数字计算机是一个执行各种计算任务的数字系统。"数字的"这个词意味着计算机里的信息是用取值为有限个数的离散值的变量表示的。这些值由可以维持有限个离散状态的元件进行内部处理。数字计算机采用二进制数码系统，包括两个数值：0和1。一个二进制数位称为一个位。在数字计算机中信息是用位的组合来表示的。通过应用各种编码技术，字位的组合不但可以用来表示二进制数，还可以表示其他离散符号，如十进制数字和字母表上的字母等。通过巧妙地运用二进制数码排列和各种编码技术，二进制码组合可用于开发出执行各种类型计算的完整的指令系统。

计算机的发展

- 第一代计算机（1946—1959年）

第一代计算机ENIAC以真空管为显著标志，到1950年又制造了另外几台著名的计算机，而且每一台都取得了很大的进展，如二进制运算、随机存取和存储程序的概念。这些概念在当今的计算机中仍是普遍使用的。

- 第二代计算机（1959—1964年）

对大多数人来说，晶体管的发明是指小型袖珍收音机，但对从事数据处理业务的人则表明第二代计算机的开始。晶体管意味着功能更强、更可靠、更价廉的计算机，它与真空管计算机相比占地面积小，功耗小。

- 第三代计算机（1964—1971年）

第三代计算机中的集成电路完成第二代计算机中的晶体管所完成的任务。第三代计算机基本上解决了第二代计算机中的兼容性问题。然而第三代计算机与第二代有本质上的差别，其变化是完全的、彻底的，而不是在此基础上的改良，这使成千上万的计算机用户感到震惊。很快，信息系统从第二代向第三代硬件的转换就完成了。

- 第四代计算机（1971年至今）

第四代计算机比其他三代更难以定义。这一代计算机的特征是一个芯片上包含越来越多的晶体管。首先，出现了一个芯片上具有数百和数千个晶体管的大规模集成电路（LSI），接着出现了一个芯片上具有数万和数十万个晶体管的超大规模集成电路（VLSI）。这个趋势在今天仍在持续。微处理器对第四代计算机的问世做出了重要贡献。

目前大多数计算机销售商将他们的计算机归为第四代计算机，少数则称是属于第五代。前三代计算机是以电子技术的重大突破为标志，即依次使用电子管、晶体管和集成电路。有些人主张把推出大规模集成电路（单位空间上有更多的电路）的1971年作为第四代计算机的开始，然而有其他的计算机设计者争辩说，如果接受这个观点，则可能在1971年后已经有了第五代、第六代，甚至第七代计算机。

计算机的分类

根据规模和功能可将计算机划分如下。

微机：微机通常是指常用术语"个人计算机"或PC的同义词。个人计算机是微机基础上的小型单用户计算机。除了微型处理器，个人计算机还有输入数据的键盘、显示信息的显示器和存储数据的存储装置。

工作站：工作站是功能强大的单用户计算机。它与个人计算机相似，但其微型处理器功能更强大，并且显示器的质量更好。

大型机：大型机或大型计算机是一种功能强大的多用户计算机。它能支持成千上万用户同时使用。现在通常指"大型服务器"。

小型计算机：小型计算机（这个词已经不太常用了）是一种多用户计算机，尺寸介于微型机和大型机之间。

巨型计算机：巨型计算机是一种速度极快的计算机，每秒钟能完成上亿条指令，但现在指"特大型服务器"，有时也包括采用并行处理的计算机系统。

台式计算机

当一台计算机在它整个工作期间一直都被放在同一个地方时，台式计算机是一个理所当然的选择。一台台式计算机的模块化设计使得根据性能、功能和具体需求进行配置更加容易。

在台式计算机的内部通常有一个或更多个设计标准，因此用一个其他制造商的配件去替换一个损坏的配件通常是可行的。并且当要增加内存、一个大的硬盘或一个显示器时，不必非在一个销售商那里购买。如果机箱的品牌是Compaq或Gateway，还可以去一些大的机箱零售商那里在很多不同品牌中进行选择。这种模块化的设计和市场竞争使得大多数台式计算机配件比那些非标准的配件更低廉。此外，一般的配件说明书中都允许维修车间维持较少的库存量，这是因为在不同的品牌和模式的台式计算机上可以使用相同的配件。

台式计算机采用模块化设计，使其在满足每位用户的特殊需求时能够很容易地增加或替换单个的部件。如果一台计算机要求能够画图或带有计算机辅助设计系统，则要配置一个高质量的显卡及显示设备，而一个采购人要求使用字处理软件和报表系统也并不是什么过分的事情。大多数计算机制造商都能够让用户购买到满足要求的性能及规格合适的产品。

当需要改变配置时，除非计算机使用的是享有专利的部件，否则打开机箱来重新配置系统通常是容易的事情。确信主板上的插座及电源的插孔和新的扩展卡或硬盘之间是匹配的，控制系统其他部分的板和新的部件之间是匹配的。

模块化设计也意味着当要扔掉已经废弃的计算机时，可以将一些旧的配件放到新的计算机上。当然，这种灵活的设计也有一些局限性。不能使用一个新品牌的内存或近期生产的硬盘去匹配十年前生产的主板，因为设计要求它们匹配更新的、更好的处理器和其他设备。

可以通过增加新部件替换旧部件的方式提高计算机的性能，使其具有更快的速度、更大的容量和更多的功能。模块化设计再次使主机的工作变得更加容易。当然，正是由于每台台式计算机都有改进的空间，因而可以以较低的价格买一台更好的机器。大多数普通的主板都有一个或更多的插槽用于扩展内存，因此可以通过买一个或更多的内存条来增加内存的容量。也可以用每个容量都更大的内存条来替代现有的内存条。在台式计算机中，增加内存容量是非常容易的，因为在机箱中有很多的空间。

CPU 是台式计算机系统的中央处理器，用来控制系统中的所有功能。移走它或用一个相似结构的、能够匹配同样插槽的快速 CPU 来替换，相对来讲也是容易的。一个新的 CPU 能比原始计算机配置的 CPU 提供更快的处理速度和更好的性能。与其他主板上的集成电路不同，CPU 安装在一个特殊的插槽上，这个特殊的插槽用一个封闭的机制固定在一个地方。

阅读材料 1：主板

主板，也称主机板或系统板。它被安装在机箱内，是计算机最重要的部件之一。它一般为矩形电路板。如果把 CPU 看成是计算机的大脑，那么主板就是计算机的身躯。当计算机拥有了一个优异的大脑（CPU）后，同样也需要一个健康强壮的身体（主板）来运作。主板影响着整个计算机系统的性能。

主板结构

主板通常由一个 CPU 插槽、一个 AGP 插槽、一个 CNR 插槽、5 个 PCI 插槽、3 个 DIMM 插槽、两个 IDE 接口、一个软驱接口、两个串行口、一个并行口、一个 PS/2 键盘接口、一个 PS/2 鼠标接口、两个 USB 接口，以及可擦写 BIOS、控制芯片组等组成。

CPU 插槽主要分为 Socket 系列和 Slot 系列。Socket 系列 CPU 插槽采用 ZIF（零插拔力插座）标准。在插座旁边有一个杠杆，拉起杠杆 CPU 的每一根引脚就可以轻松地插进插座的每一个孔位里，然后把杠杆压回原来的位置，就可将 CPU 固定住。Slot 系列采用插槽的形式，看上去像主板上常见的扩展槽一样。任何外界的接口卡，如显卡、声卡、网卡等，都要插在主板上的扩展槽上才能够与主板连接，输出图像和声音。

AGP 插槽专门用于高速图像处理。AGP 插槽为褐色，在形状上与 PCI 扩展槽相似。AGP 只能插显卡，因此在主板上 AGP 接口只有一个。现在主板大都采用 AGP 8×接口，配合 AGP 8×的显示卡，大大提高了计算机的 3D 处理能力。

内存插槽的作用是安装内存条。内存插槽有 30 线、72 线、168 线和 184 线等，现在采用的是 168 线。

BIOS（基本输入输出系统）是被安装在主板上的一个 ROM 芯片，其中保存有计算机系统最重要的基本输入输出程序、CMOS 设置、自检程序等。

主板上的软驱接口一般为一个 34 针双排针插座，标注为 Floppy、FDC 或 FDD。IDE 接口为 40 针双排针插座，用于和 IDE 硬盘驱动器或光盘驱动器相连接。现在为提高数据传输的可靠性，改用 80 针的排线。

主板上为连接键盘提供了一个标准的 AT 键盘 DIN 连接器。可以直接将键盘电缆插到该连接器上。

主板上为 PS/2 鼠标电缆（可选）提供了一个 5 针连接器。可以直接把 PS/2 鼠标电缆插到它上面。

电源插座共有两种规格，AT 规格和 ATX 规格。ATX 是目前广泛使用的电源插座规格，配合 ATX 电源使用。ATX 电源插座是 20 针双列插座，具有防插错结构，如果插头拿反了就插不进去，所以不必担心会烧毁主板。

计算机系统与时钟是紧密相关的，当计算机关机以后，由电池来提供系统时钟所需的电源。主板上使用的电池通常是纽扣电池，被插在主板的电池插槽里。

声卡和显卡

声卡即音频卡，其基本功能是产生和处理声音信号，然后将信号送给扬声器。这些信号在计算机里是以数字的形式表示的。数字信号经声卡芯片处理后，再经由声卡上的数模转换器转换成人耳能识别的模拟信号。如果要处理外部输入的声音信号，首先声卡要通过模数转换器先将这些信号转换为计算机能识别的数字信号，然后，再交给声卡芯片处理。

显卡是显示卡的简称。早期的显卡并没有受到太多的重视，它的作用只是如实地将处理器输出的数据转换后显示在显示器上。随着图形软件的不断增多，尤其是图形系统的逐渐普及，单单依靠 CPU 去处理所有的数据会极大地影响系统的整体性能，于是出现了显卡。

显卡的作用更像是一个协处理器，正是用它来对图形处理操作进行加速，CPU 才可以处理其他更重要的任务。实际上，现在的显卡都已经是图形加速卡了，都可以执行一些图形处理功能。通常所说的加速卡的性能，是指加速卡上的芯片集能提供图形函数计算能力，这个芯片集通常也被称为加速器或图形处理器。一般来说，在芯片集的内部会有一个时钟发生器、VGA 核心和硬件加速器。

第 2 单元　中央处理器和存储器

中央处理器

CPU 即中央处理器，是计算机系统的心脏。处理器是解释并执行指令的功能部件。每个处理器都有一个独特的诸如 ADD、STORE 或 LOAD 这样的操作集，这个操作集就是该处理器的指令系统。计算机系统设计者习惯将计算机称为机器，所以该指令系统有时也称为机器指令系统，而书写它们的二进制语言称为机器语言。

CPU 可以是一个单独的微处理器芯片、一组芯片，或者是一个带有晶体管、芯片、导线和接点的插件板。依据 CPU 方面的差别可以区分大型、小型和微型计算机。处理器由两个功能部件（控制部件和算术逻辑部件）和一组称为寄存器的特殊工作单元组成。

CPU 的两个主要性能指标如下。

时钟频率：计算机有一个发出脉冲以控制所有系统操作同步的系统时钟。系统时钟与保存每天时间的"实时时钟"不同。系统时钟设置数据传输和指令执行的速度或频率。系统时钟的频率决定了计算机执行指令的速度，因此限制了计算机在一定时间内所能执行的指令数。完成一个指令周期的时间用兆赫兹（MHz）或每秒百万个周期表示。

最初 IBM PC 的微处理器时钟频率是 4.77MHz。现在的处理器的执行速度可以超过 2GHz。如果其他条件一样，时钟频率越高，就意味着处理速度越快。

字长：字长是指中央处理器（CPU）可以同时处理的位数。字长由 CPU 的寄存器大小和总线的数据线根数所决定。例如，字长为 8 位的 CPU 被称为 8 位处理器，它的寄存器是 8 位宽，可以同时处理 8 位数据。

字长较长的计算机在一个指令周期中要比字长短的计算机处理更多的数据。每个周期内处理的数据越多，处理器的性能就越高。例如，最初微型计算机的微处理器是 8 位，而现在较快的计算机都是 32 位或 64 位的微处理器。

64 位微处理器

处理器设计中的最新情况是 64 位 ALU，今后十年内，这种处理器可望在家用计算机中出现。还有一种趋势是用特殊指令使某个运算特别有效，以及在处理器芯片上增加硬件虚拟内存支持和 L1 缓存。所有这些趋势都要增加晶体管个数，从而出现了当今几百万个晶体管的微处理器。这样的处理器每秒可以执行十亿条指令。

64 位处理器从 1992 年就已经出现，21 世纪已开始成为主流。Intel 与 AMD 公司都引入了 64 位芯片，Mac G5 支持 64 位处理器。64 位处理器具有 64 位 ALU、64 位寄存器、64 位总线等。

之所以需要 64 位处理器，是因为其扩展的地址空间。32 位芯片通常限于最多 2GB 或 4GB 的 RAM 存取。但是 4GB 的局限对服务器和运行大型数据库的机器可能造成严重问题。按照目前的趋势发展下去，就连家用计算机也很快会遇上 2GB 或 4GB 的极限。64 位芯片没有这种限制，因为 64 位 RAM 地址空间在可以预见的将来实际上是无限的，因为 2^{64} 字节内存是千万亿 GB 内存数量级的。

利用主板上的 64 位地址总线与宽的高速数据总线，64 位机器也为硬盘驱动器和显示卡等提供了较快的 I/O 速度。这些特性可以大大提高系统性能。

服务器明显可以受益于 64 位结构，但普通用户呢？除了 RAM 方案，64 位芯片目前还看不出对普通用户有任何真实的具体的好处。它们可以更快地处理数据（由许多实数构成的复杂数据）。这类计算功能对进行视频编码和进行超大图像照片编辑的人有好处。一旦将它们利用 64 位特性的优势进行重新编码，则高端游戏也将受益。

多处理器

多处理器是指采用结构化技术，将许多处理器组成单一的计算机系统，从而提高该系统在其应用环境中的性能，使其超过单个处理器的性能。

多处理器系统可以分成四类：单指令流单数据流（SISD）、单指令流多数据流（SIMD）、多指令流单数据流（MISD）及多指令流多数据流（MIMD）。其中，多指令流单数据流系统比较少见，而其他三类结构可根据其指令周期的不同而加以区别。

在单指令流单数据流结构中，有单一的指令周期，在执行前，单个处理器按序取操作数。

单指令流多数据流结构也只有单一的指令周期。但是，其多个处理器可以取多个操作数，且可以在单一指令周期内同时执行。多功能部件处理器、阵列处理器、向量处理器及流水处理器均属此类。

在多指令流多数据流结构中，在任何给定时间内可以有多个指令周期，且每个指令周期均独立地取指令和操作数，送到多个处理器中，且以并行方式执行。此类结构包括的多处理器系统中的每个处理器均有自己的程序控制，而不是共享单一控制器。

存储器

任何一台数字计算机都需要存储 CPU 所执行的程序，因此，存储器是计算机最重要的部件之一。RAM 和 ROM 在存储部件中扮演了重要的角色。RAM 是随机访问存储器 3 个词首字母的缩写。RAM 主要用来作内存条。ROM 的意思是只读存储器。控制和专用程序存放在 ROM 芯片中，用户不能修改。基本操作指令集存储在 ROM 中，关闭计算机也不会被删除。计算机中有可以升级的模块，称为 EEPROM（电可擦除可编程只读存储器）。BIOS 指令和配置应用程序都存储在计算机的快速 EEPROM 中。

计算机工作期间，所得到的指令及所处理的信息都保存在 RAM 中。RAM 不是一个长久存储信息的地方。关闭计算机后，内存中就不再保存工作期间所输入的信息了。由于 RAM 只在开机时有效，因此计算机用磁盘驱动器来存储信息，这些信息在关机后依然存在。

计算机内存以信息的千字节或兆字节来度量（一字节等于一个字符、一个字母或数字的存储量）。1KB 等于 1024 字节，1MB 约等于 1 000 000 字节。软件需要一定数量的内存来正常工作。如要给计算机增加新的软件，在软件包装上通常可以找到该软件所需要的确切内存容量。

高速缓冲存储器是影响 CPU 性能的另一个因素。高速缓冲存储器是一个特别的高速存储器，可以使 CPU 非常迅速地访问数据。高速 CPU 处理指令的速度非常快，以至于大部分时间不得不等待从处理速度很慢的 RAM 传送数据，这影响了处理速度。高速缓冲存储器可以保证一旦 CPU 请求就可以迅速获得数据。

阅读材料 1：硬盘

现在几乎人们使用的每台台式计算机和服务器都有一个或多个硬盘驱动器。每个大型机和超级计算机通常都连接着上百个硬盘驱动器。这些成千上亿的硬盘都做了一件很出色的事——把变化的数据存储为相对永久的形式。它们使计算机在断电时可以保存资料。

硬盘是在 20 世纪 50 年代发明的。起初直径 20 英寸大的硬盘只能存储几兆字节。新式的硬盘可以在很小空间存储大量的信息。硬盘可以在很短的时间内访问任何信息。一个典型的台式计算机拥有 10～40GB 的硬盘容量。数据是以文件的形式存放在硬盘上的。文件是字节的简单集合。字节也可以是文本文件中的 ASCII 码，或者是计算机执行的软件的指令集，或者是数据库的记录，或者是 GIF 图像的像素颜色。不管它包含什么，文件都是简单的字符串。

有以下两种方式来测量硬盘的性能。

◆ 数据速率

数据速率是指驱动器每秒可以发送给 CPU 的字节数，通常为 5～40MB/s。

◆ 寻道时间

寻道时间是 CPU 请求读取文件到文件第一个字节发送到 CPU 的时间间隔。通常时间间隔为 10～20ms。

另一个重要的参数是驱动器的容量，也就是它可以存储的字节数。

一个典型的硬盘驱动器是一个带有电子控制器的密封铝盒。电子设备控制读写机制和薄盘片里旋转的马达。它同时能将驱动器磁性区域转换为数据（读），并且把数据转换成磁性区域（写）。电子设备被包含在小盒里，与驱动器的其余部分分开。

在电子板下面是旋转薄盘片马达的线路，还有一个使内外气压相等的高度过滤排气孔。移开驱动器外壳后，可见一个简单但精密的内部结构。

◆ 盘片

当系统运行时盘片一般以 3600r/min 或 7200r/min 的速度旋转。这些人造的盘片有惊人的承受力和镜子般的平滑。

◆ 臂状物

臂状物支撑着读写磁头，并受左上角的机械装置控制。臂状物可以把磁头从驱动器中心移到边缘。臂状物和它的运动装置非常小巧迅速。它在典型的硬盘驱动器中每秒可以从中心移到边缘并且返回 50 次。

为了增加磁盘的存储信息容量，大部分的硬盘有多个薄盘片。在硬盘上，移动臂状物的机械装置必须非常快速和准确。它可以用一个高速的线性马达来构建。很多驱动器使用一种叫作音圈的方法（这和带动立体声的锥形扬声器所用技术相同）来移动臂状物。

数据存储在薄盘片表面的扇区和磁道上。磁道是同心圆，扇区是磁道上一个个楔形区。一个扇区包含固定数目的字节，如 256 或 512。在驱动器或操作系统的层面上，扇区常常被组成簇。低级格式化的过程就是驱动器在盘片上建立磁道和扇区的过程。每个扇区的开始点和结束点都被写入盘片。这个过程为驱动器存储字节块做准备，接着高级格式化把文件存储结构写入扇区，如文件分配表。这个过程为驱动器存储文件做准备。

第 3 单元　输入输出系统

计算机是功能强大的机器，它可以完成人们交给它的任何事情。然而计算机并不能够直接与人交流。借助于输入、输出设备，计算机和人便可以相互"了解"了。使用输入设备，人们告诉计算机该做些什么，而计算机通过输出设备反馈结果。输入输出设备是人机接口。它通常包括键盘、鼠标、显示器、打印机、磁盘（硬盘或软盘）、输入笔、扫描仪和麦克风等。让我们来看看一些常用的输入和输出设备。

键盘

键盘用来向计算机中键入或输入信息。键盘的布局和尺寸的大小有多种不同的设计方法，最常用的是以拉丁语为基础的 QWERTY 布局（以键盘上前 6 个键命名的）。标准键盘有 101 个键，笔记本电脑具有内置的键，可以通过专门的键或按键组合来访问。标准键盘上的某些键具有特殊的用途，有些键称为命令键。最常用的 3 个命令键是控制键（Ctrl）、替换键（Alt）及轮换键（Shift）。标准键盘上的每个键上面都有 1~2 个字符，按下一个键将得到下排的字符，按住 Shift 的同时将得到上排的字符。

鼠标

鼠标是一种小型的设备，用户可以在桌面上移动，以指向屏幕上的某个位置并从该位置选择一项或多项操作。最常见的鼠标顶部有两个按键，左键用得最多。鼠标由以下几个部分组成：一个金属或塑性的盒体、一个凸出于盒体底部并可以在平面上滚动的球体、位于盒体上部的一个或多个按键及一条连接到计算机的电缆线。球体在平面上沿任意方向滚动时，鼠标内部的传感器将相应的脉冲信号传输到计算机，这时支持鼠标操作的程序随即做出响应，将可视化的指示器（光标）在屏幕上重新定位。光标的定位相对于其开始位置。看到光标的当前位置后，用户可以移动鼠标再次进行调整。

阴极射线管显示器

CRT 显示器就是阴极射线管显示器。阴极射线管主要由五部分组成：电子枪、偏转线圈、荫罩、磷光粉层及玻璃外壳。它是目前应用最广泛的显示器之一。CRT 显示器的核心部件是阴极射线管，其工作原理和电视机的显像管的工作原理相似，可以把它看作是一个图像更加精细的电视机。经典的阴极射线管使用电子枪发射高速电子，经过垂直和水平的偏转线圈控制高速电子的偏转角度，最后高速电子击打屏幕上的磷光物质使其发光。通过电压来调节电子束的功率，就会在屏幕上形成明暗不同的光点以形成各种图案和文字。

CRT 纯平显示器具有如下优点：可视角度大、无坏点、色彩还原度高、色度均匀、可调节的多分辨率模式、响应时间短。另一个重要的优点是 CRT 显示器要比许多 LCD 显示器便宜。

液晶显示器

LCD（液晶显示器）采用液晶控制透光度技术来实现色彩。从液晶显示器的结构来看，无论是笔记本电脑还是桌面系统，采用的 LCD 显示屏都是由不同部分组成的分层结构。LCD 由两块玻璃板构成，厚约 1mm，中间由包含有液晶材料的 5μm 均匀间隔隔开。因为液晶材料本身并不发光，所以在显示屏两边都没有作为光源的灯管，而在液晶显示屏背面有一块背光板（或称匀光板）和反光膜。背光板是由荧光物质组成的，可以发射光线，它主要用于提供均匀的背景光源。

背光板发出的光线在穿过第一层偏振过滤层之后进入包含成千上万液晶液滴的液晶层。在玻璃板与液晶材料之间是透明的电极。电极分为行和列，在行与列的交叉点上，通过改变电压而改变液晶的旋光状态。液晶材料的作用类似于一个个小的光阀。在液晶材料周边是控制电路部分和驱动电路部分。当 LCD 中的电极产生电场时，液晶分子就会产生扭曲，从而将穿越其中的光线进行有规则的折射，然后经过第二层过滤层的过滤在屏幕上显示出来。

与 CRT 显示器相比，LCD 显示器的优点是很明显的。由于通过控制是否透光来控制亮和暗，当色彩不变时，液晶也保持不变，这样就无须考虑刷新率的问题。对于画面稳定、无闪烁感的液晶显示器，即使刷新率不高图像也很稳定。LCD 显示器还采用液晶控制透光度技术的原理让底板整体发光，所以它真正是完全平面。一些高档的数字 LCD 显示器采用了数字技术传输数据、显示图像，这样就不会产生由于显卡造成的色彩偏差或损失。它还有完全没有辐射的优点，即使长时间观看 LCD 显示器屏幕也不会对眼睛造成很大伤害。体积小、能耗低也是 CRT 显示器所没有的优点。

与目前的 CRT 显示器相比，LCD 显示器的图像质量仍不够完善。在色彩表现和饱和度方面，LCD 显示器都在不同程度上输给了 CRT 显示器，而且液晶显示器的响应时间也比 CRT 显示器长。

I/O 子系统的组成和接口

每一个 I/O 设备与计算机系统的地址总线、数据总线和控制总线相连接，它们都包括 I/O 接口电路。与总线交互的实际上正是这一电路，同时，它与实际的 I/O 设备交互来传输数据。

对于一个输入设备的一般接口电路，从输入设备来的数据传送到三态缓冲器。当地址总线和控制总线上的值正确时，缓冲器设为有效，数据传到数据总线上，然后，CPU 可以读取数据。当条件不正确时，逻辑块不会使缓冲器有效，它们保持高阻态，而且不把数据传到总线上。这一设计的关键在于使能逻辑。正如每一个存储单元都有一个唯一的地址一样，每一个 I/O 设备也有一个唯一的地址，除非从地址总线得到了正确的地址，否则使能逻辑不置缓冲器有效。同时，它还必须从控制总线上得到正确的控制信号。

输出设备接口电路的设计与输入设备的设计有所不同，寄存器代替了三态缓冲器。输入设备中使用了三态缓冲器是为了确保在任何时刻都只有一个设备向总线写数据。而输出设备是从总线读取数据，不是写数据，因此不需要缓冲器。数据对于所有的输出设备都可获得，但只有具有正确地址的设备才会读取它。

有些设备既用于输入又用于输出，个人计算机中的硬盘驱动器就属于这一类。这样的设备需要一个组合接口，本质上是两个接口，一个用于输入，另一个用于输出。一些逻辑元件（如检查地址总线上的地址是否正确的门电路）既可以用来产生缓冲器的使能信号，又可以用来产生寄存器的装载信号。

I/O 设备比 CPU 和存储器慢得多。基于这个原因，当它们与 CPU 交互时，就可能存在时序上的问题。为了说明这一点，考虑当 CPU 想要从硬盘中读取数据时会发生的情况，这可能要消耗磁盘驱动器几个毫秒来正确地定位磁头，以便读取想要的数值，而在这段时间里，CPU 可能已经读入了不正确的数据，并且读取、译码和执行了成千上万条指令。

大多数 CPU 都有一个控制输入信号，称为就绪信号 READY（或其他意思相近的名称），通常它为高电平。当 CPU 输出某 I/O 设备的地址和正确的控制信号，促使 I/O 设备接口的三态缓冲器有效时，该 I/O 设备置 READY 信号为低电平。CPU 读取这一信号，并且继续输出同样的地址信号和控制信号，使缓冲器保持有效。在硬盘驱动器的例子中，此时驱动器旋转磁头，并且定位读写头，直到读到想要的数据为止。这时 CPU 才从总线上读入数据，之后继续它的正确操作。设置 READY 为低电平而生成的附加时钟周期称为等待状态。

阅读材料 1：鼠标

1984 年，在苹果公司的引导下，鼠标首次进入公众舞台。从此，它完全重新定义了人们使用计算机的方式。鼠标的简单构造和巨大作用令人惊叹不已，而它成为日常生活一部分所经历的时间之长，更是不可思议。

鼠标单击屏幕的响应是即时实现的，简单地说是自然而然的。与图形输入面板相比，鼠标价格低廉且外形小巧。在个人计算机领域，鼠标占据一席之地较晚，主要是因为缺乏操作系统的支持。一旦 Windows 3.1 使图形用户界面作为标准，鼠标很快就成为人机交互的首选部件。鼠标的主要目的是把手部运动转化为计算机可以识别的信号。现今大多数的鼠标用 5 个部件完成这个转换。

（1）一个接触桌面的内置小球，能随着鼠标的移动而转动。

（2）内部有两个滚轴与小球接触。一个滚轴用于控制 x 方向的运动；另一个转过 90 度，控制 y 方向的运动。当小球转动时，它们中的一个或全部也会跟着转动。

（3）每个滚筒都与一个轴连接，轴带动一个有孔的圆盘旋转。当滚筒滚动时，轴和圆盘就随之一起转动。

（4）在圆盘的两侧，分别是一个红外线的二极管和传感器。发光二极管发出的光束从圆盘上的孔穿过，传感器便接收到了光脉冲。脉冲的速度与鼠标的移动速度及移动距离有关。

（5）处理器芯片读出红外传感器的脉冲并把它们翻译成机器可识别的二进制数据，再通过鼠标线把数据传入计算机。

随着鼠标定位技术的发展，早期的滚轮鼠标将被淘汰。现在首选的点击工具是光电鼠标。这种鼠标有一个小型红色发光二极管，把光线经过反射传到 CMOS 传感器表面上来，这使它几乎能在任何表面上工作。

CMOS 传感器把各个图像发送给数字信号处理器（DSP）分析。DSP 以每秒 1800 万条指令的速度操作，可以探测图像的模式及看清那些模式怎样从之前的图像上转变过来。基于一连串的图像模式的变化，DSP 确定鼠标已经移动多远，并且把相应数据送到计算机。计算机根据从鼠标收到的信息相应地移动光标。这个过程每秒可进行数百次，使得光标的移动看起来非常平滑。

光电鼠标较滚轮鼠标有以下几个优点。

（1）不需要滚动，所以减少了磨损和操作失败。

（2）递增地跟踪感应，使得感应更加平滑。

（3）不要求特别的表面环境。

（4）脏东西没有办法进入鼠标内部，也不会有东西妨碍传感器跟踪。

如今大多数的鼠标都使用 PS/2 标准接口。当移动或点击鼠标时，它给计算机发送 3 字节的数据。接下来的两个字节分别包含了 x 及 y 方向的移动数据。它包含这两个方向上探测到的从上一次数据传送完毕后的脉冲数。数据从鼠标通过数据线串行传输到计算机，时钟脉冲告诉计算机每位的开始点和结束点。

第 4 单元　C++语言

程序设计语言是指导计算机实现某些具体任务的一套特殊词汇和一组语法规则。从广义的角度说，它包含一组既能被人所理解又能被计算机所识别的声明和表达式。人们能理解这些指令是因为它们使用的是人类的表达方式。另一方面，计算机通过使用专门的程序来处理这些指令，这些专门的程序就是我们所熟知的翻译程序，它能解码我们发出的指令并生成机器语言代码。这些指令用 1 和 0 的长串表示。而这种用 0 和 1 来表示程序的语言就是机器语言。

简单地与人类语言相比，高级程序设计语言比计算机实际识别的语言，也就是机器语言，复杂得多。每种型号的 CPU 都有它独自的一套机器语言。处于机器语言和高级语言之间的是汇编语言，它直接与机器语言相关。也就是说，它可将一条汇编指令生成一条机器语言指令。人们几乎不可能去读和写那些只包含数字的机器语言。虽然汇编语言具有和机器语言相同的结构和命令集，但是编程人员可以使用助记符来代替数字。

位于高级语言之上的是第四代程序设计语言（简称 4GL）。它与机器语言差异更大，代表了与人类语言更为接近的那类计算机程序设计语言。大多数 4GL 被用于进行访问数据库的操作。

C++是一种通用的计算机编程语言，具有高级和低级处理的能力。它是静态输入、自由格式、多范例、编译程序支持的过程编程、数据抽象、面向对象编程及通用编程的语言。

C++通常被认为是 C 的扩展集，事实上并非如此。大部分 C 代码能在 C++中正确编译，但是有一些区别，使得一些合法的 C 代码在 C++中变为不合法的，或者在 C++中变得不同。

C++是 C 语言的加强版。C++包括 C 的一部分并且增加了对面向对象编程的支持。C 是一种面向过程的编程语言。C++包括了很多的改进和特性。

C 和 C++的基本语法和语义是相同的。如果你对 C 较熟悉，那么可以很快地学会 C++编程。C++拥有在 C 中定义的类似的数据类型、操作符及其他工具，能直接适用于计算机体系结构。

面向对象的程序设计是一种程序设计技术，使得你能把一些概念看作各种各样的对象。通过使用对象，你能表示要被执行的任务、它们之间的相互作用和必须观察的某些给定的条件。一种数据结构经常形成某个对象的基础；因此，在 C 或 C++中，结构类型能形成某种基本对象。与对象的通信能通过使用消息来完成。消息的使用类似于在面向过程的程序中对函数的调用。

当某对象收到一个消息时，包含在该对象内的一些方法做出响应。方法类似于面向过程程序设计的函数。然而，方法是对象的一部分。

C++允许程序员创建类，有些类似于C中的结构。在C++中，有各种原型的相关方法、函数，能够在类中访问和操作，就类似于C函数操作支持的处理指针。C++类是C语言结构的扩展。由于结构与类的唯一区别在于结构成员的默认访问权限是公共的，而类成员的默认访问权限是私有的，因此可以使用关键字类或结构来定义相同的类。

C++的类是对C和C++结构类型的扩充，并且形成了面向对象程序设计所需要的抽象数据类型。类能包含紧密相关的一些项，它们共享一些属性。更正式地说，对象只不过是类的实例。

最终，应该出现包含很多对象类型的类库，你能使用这些对象类型的实例去组织程序代码。

典型地，一个对象的描述是一个C++类的一部分，且包括该对象内部结构的描述、该对象如何与其他对象相关，以及把该对象的功能细节和该类的外部相隔离的某种形式的保护。C++类结构做到了所有这些。

在一个C++类中，你使用私有的、公共的或受保护的描述符来控制对象的功能细节。在面向对象的程序设计中，公共部分一般用于接口信息（方法），使得该类可在各应用中重用。如果数据或方法被包含在公共部分，它们在该类外部也可用。类的私有部分把数据或方法的可用性局限于该类本身。包含数据或方法的受保护部分被局限于该类和任何派生子类。

C++类实际上用作创建对象的模板或模式。从类描述形成的对象都是该类的实例。开发类层次结构是可能的，其中有一个主类和几个子类。在C++中，做这事的基础是派生类。父类表示更一般化的任务，而派生子类执行一些特定的任务。

面向对象程序设计中的继承使得一个类能继承某对象类的一些性质。父类用作派生类的模式，且能以几种方式被改变。如果某个对象从多个父类继承其属性，便称为多继承。继承是一个重要概念，因为它使得无须对代码做大的改变就能重用类定义。继承鼓励重用代码，因为子类是对父类的扩充。

与类层次结构相关的另一个重要的面向对象概念是公共消息能被发送到各个父类对象和所有派生子类对象。按正式的术语，这称为多态性。

多态性使每个子类对象能以一种对其定义来说适当的方式对消息格式做出响应。试设想收集数据的一个类层次结构。父类可能负责收集某个个体的姓名、社会安全号、职业和雇用年数，那么你能使用子类来决定根据职业将添加什么附加信息。一种情况是一个管理职位会包括年薪，而另一种情况是销售员职位会包括小时工资和回扣信息。因此，父类收集一切子类公共的通用信息，而子类收集与特定工作描述相关的附加信息。多态性使得公共的数据收集消息能被发送到每个类。父类和子类两者都对该消息以恰当的方式做出响应。

多态性赋予对象这种能力，当对象的精确类型还未知时响应来自例行程序的消息。在C++中这种能力是迟绑定的结果。使用迟绑定，地址在运行时刻动态地确定，而不是如同传统的编译型语言在编译时刻静态地确定。这种静态的方法往往称为早绑定。函数名被替换为存储地址。使用虚函数来完成迟绑定。一个父类，当随后的各派生类通过重定义一个函数的实现而重载该函数时，便在其中定义了虚函数。当你使用虚函数时，消息不是直接传给对象，而是作为指向对象的指针传送。

虚函数利用了地址信息表，该表在运行时刻通过使用构造符初始化。每当创建它的类的一个对象时调用一个构造符。这里构造符的工作是把虚函数与地址信息表链接，在编译进行期间虚函数的地址是未知的；相反，给出的是地址表中将包含该函数地址的位置。

阅读材料1：计算机软件的种类

今天，有七大类计算机软件给软件工程师不断提出挑战。

◆ 系统软件

系统软件是指为服务于其他程序而编写的一批程序。有些系统软件（如编译程序、编辑程序及文件管理实用程序）处理复杂的但确定的信息结构。其他的系统应用程序（如操作系统组件、驱动程序、联网软件、电信处理器）主要处理不确定数据。无论这两种情况的哪一种，系统软件领域都具有如下特征：与计算机硬件的大量交互；多个用户的大量使用；需要调度、资源共享和复杂进程管理的并行操作；复杂的数据结构及多个外部界面。

◆ 应用软件

应用软件是指解决特定业务需要的独立程序。该领域的应用程序以促进业务运作或管理/技术决策的方式处理业务或技术数据。除了常规数据处理应用程序外，应用软件还用于实时控制业务功能（如销售点交易处理、实时制造过程控制）。

◆ 工程/科学软件

工程/科学软件应用的范围包括从天文学到火山学、从汽车应力分析到航天飞机轨道动力学、从分子生物学到自动化制造等诸多领域。然而，工程/科学领域里的现代应用程序正逐渐放弃常规数值算法。计算机辅助设计、系统仿真及其他的交互式应用程序已经开始呈现实时的甚至系统软件的特点。

◆ 嵌入式软件

嵌入式软件常驻在产品或系统之内，并用于为终端用户及系统本身实现和控制功能特征。嵌入式软件可执行有限的和难懂的功能（如微波炉的小键盘控制）或提供重要的功能与控制能力（如汽车的燃料控制、仪表板显示、刹车系统等数字功能）。

◆ 产品线软件

产品线软件可聚焦于一个有限的和难懂的市场（如库存控制产品）或针对大众消费市场（如文字处理、电子表格、计算机图形、多媒体、娱乐、数据库管理及个人与企业财务应用程序）。

◆ 万维网应用程序

这个称为"WebApps"的以网络为中心的软件种类包括一系列广泛的应用程序。就其最简单的形式而言，WebApps可能只是一套被链接起来、使用文本和有限图形呈现信息的超文本文件而已。然而，随着Web 2.0的出现，WebApps不断演变成复杂的计算环境，不仅向终端用户提供独立特征、计算功能及内容，而且与公司数据库及商业应用程序集成在一起。

◆ 人工智能软件

人工智能软件利用非数值算法，来解决不易进行计算或直接分析的复杂问题。该领域里的应用程序涉及机器人技术、专家系统、模式识别（图像与话音）、人工神经网络、定理证明及博弈等。

世界各地的不计其数的软件工程师在努力从事以上种类中的一种或多种软件项目。在有些情况下，他们在构建新的系统，但是在许多其他情况下，他们在校正、改造和增强现有的应用程序。

♦ 开放源码

开放源码是一种日益增强的趋势。这种趋势带来的结果是开放系统应用程序（如操作系统、数据库和开发环境）的源代码，以便许许多多的人可为其开发做出贡献。这对软件工程师提出的挑战是，编写自我描述的源代码，但更为重要的是开发使客户和开发者均能知道已经做出哪些改变，以及这些改变在软件中是如何显现的相关技术。

♦ 开放世界计算

无线联网的迅速发展可能不久就会带来真正的普遍性分布式计算。这对软件工程师提出的挑战是，开发出允许移动设备、个人计算机和企业系统横跨巨大网络进行通信的系统和应用软件。

万维网正在迅速成为一个计算引擎，而不仅仅是一个内容提供者。这对软件工程师提出的挑战是，设计出可使世界各地的目标终端用户市场都收益的简单（如个人财务计划）和复杂的应用程序。

这些新挑战每一个都无疑将服从非预期后果法则，并产生今天无法预测的影响（如对商家、软件工程师和终端用户）。然而，软件工程师可通过实例演示一个进程来做出准备，这个进程具有足够的灵活性和适应性，可适应技术上发生的戏剧性变化，以及未来十年必定出现的商务规则。

第 5 单元　操作系统

　　任何一台计算机中最重要的程序就是操作系统或 OS。操作系统是一个由很多小程序组成的大程序。它控制 CPU 与硬件间的通信，也方便不懂得编程语言的用户操作计算机。换句话说，操作系统使用户更友好地操作计算机。

　　操作系统对于其生产厂家及其运行的硬件环境通常是特有的。一般来说，安装一台新计算机系统的同时也购买了与该硬件相对应的操作系统。用户需要有效地支持其处理工作的可靠的操作系统软件。尽管各厂家的操作系统软件各不相同，但特性都是相似的。对于现代硬件系统，由于其复杂性，因此要求其操作系统满足某些特定的标准。例如，考虑到该领域的现状，操作系统必须支持某种形式的联机处理。

　　操作系统是管理、控制计算机活动的系统软件。它监控 CPU 的操作，控制输入、输出及数据存储活动，并且提供各种支持服务。它是计算机系统的主要管理者。操作系统决定调用解决某个问题所需要的计算机资源及资源的使用顺序。

　　操作系统有以下主要功能。

♦ 分配系统资源

系统资源主控程序，也称为监控程序、执行程序或监督程序，它负责计算机的运行，协调计算机的所有活动。因监控程序常驻内存，必要时监控程序可将其他程序从辅助存储器调入主存储器。每当一个程序被激活，主控程序便将控制权转交给这个具体的程序。一旦程序运行结束，控制权又返回到监控程序。命令语言解释程序负责分配系统资源。该解释程序将特殊指令读入操作系统，这些特殊指令不仅包括文件的检索、保存、删除、复制、移动等，还包括选择 I/O 设备、选择编程语言和应用程序，以及为执行某特殊应用程序而执行其他处理需求。这些指令称为命令语言。

◆ 作业调度及资源的利用

操作系统软件的一个非常重要的职责是对计算机系统要处理的作业进行调度。这是作业管理功能的主要任务之一。计算机可以同时执行多个任务。操作系统选定处理这些任务的时机，因为计算机处理任务的顺序不一定是按任务下达的顺序进行的。例如，工资发放任务或网上订货任务的优先级就高于其他任务。诸如软件程序测试之类的工作，就得等到上述任务完成或有足够的资源后才能进行。操作系统协调计算机方方面面的工作，这样，不同任务中的不同部分方能同时运行。例如，当某些程序执行时，操作系统也可以调度输入和输出设备的使用。

◆ 监控计算机系统的活动

在计算机系统中，操作系统也要负责跟踪计算机的各种活动。它保存着作业操作的日志，将任何非正常终止或错误情况通知给终端用户或操作员。它也要负责终止那些已经超过所允许的最大运行时间的程序。操作系统软件也可以包括安全监控功能，如记录谁正在使用或曾经使用过系统，他们所运行的程序及非法访问系统的企图。

◆ I/O 操作控制

系统资源的分配与控制 I/O 操作的软件密切相关。由于 I/O 操作开始之前往往需要对特定设备进行访问，因此操作系统必须协调 I/O 操作和所使用设备间的关系，实际上操作系统建立了一个执行程序和完成 I/O 操作所必需的设备的目录。

为便于 I/O 的执行，大多数操作系统都有一套标准的控制指令集来处理所有输入输出指令。这些标准指令称为输入输出控制系统（IOCS），是大多数操作系统不可分割的部分，它们简化了由处理的程序所承担的 I/O 操作。控制软件调用 IOCS 软件以实际完成 I/O 操作。由于大多数程序都考虑 I/O 操作的级别，因此 IOCS 指令至关重要。

在操作系统的庞大家族里，根据所控制的计算机类型和所支持的应用程序的种类将操作系统分为以下四类。

◆ 实时操作系统

实时操作系统用来控制机器、科学仪器和工业系统。实时操作系统的一个很重要的部分是管理计算机资源，这样它能根据资源的每一次发生数量精确产生一个相应的动作。

◆ 单用户、单任务操作系统

正如名称所揭示的，这一操作系统是为单用户在某一时刻有效地完成一个任务而设计的。用于 Palm 手持电脑的 Palm OS 就是现代化的单一用户、单一任务操作系统的很好样例。

◆ 单用户、多任务操作系统

当今，大多数人在他们的台式机和笔记本电脑上用的操作系统就是这一种。微软公司的 Windows 和苹果公司的 Mac OS 平台都是单用户、多任务操作系统的例子。例如，Windows 操作系统用户就完全可能做到一边从互联网上下载文件，一边打印电子邮件的信息，一边还在用文字处理器编辑文本。

◆ 多用户操作系统

该操作系统允许不同的用户同时使用一台计算机的资源。多用户操作系统必须保证不同用户的需求的平衡，每一个用户所使用的程序有独立的足够的资源，这样一个用户的问题不会影响到整个用户群体。UNIX、VMS 和大型机的操作系统，如 MVS 都是多用户操作系统的例子。

随着技术水平的提高，操作系统经常被改进和升级。升级一个操作系统会有很多好处，如简化任务，便于导航；但是也有缺点，以前的很多程序在新的操作系统下都不能运行了。到目前为止，最受欢迎的微机操作系统是微软公司的 Windows 操作系统。Windows 操作系统是专为运行 Intel 微处理器及 Intel 兼容处理器（如 Pentium Ⅳ）而设计制造的。

阅读材料 1：Windows Vista

　　Windows Vista 是微软的最新一代操作系统。和它的前辈 Windows XP、Service Pack 2 相比，它提供了许多增强的功能。一启动计算机进入 Windows 界面，就会发现有很大的不同。新的"开始"菜单、新的桌面背景，还有在屏幕右侧的新的工具条，所有这些都会告诉用户，正在进行一场新的体验。现在移动鼠标单击"开始"按钮，并且启动一个新程序，屏幕上不会再有直达屏幕边缘的多层数的菜单。相反，每次打开一个菜单，它都覆盖在前一个已打开菜单的上面，这样很容易找到想要的程序。

　　其他新特点包括一些新程序(如 Windows 日历、Windows 相集、Windows 媒体中心、Windows 传真和扫描)、得心应手的网络工具（如网络和共享中心）、一张被重新设计的控制面板和许多额外的改进。

　　作为微软的最新的操作系统，Vista 第一次引入了"Life Immersion"概念，即在系统中集成许多人性的因素，一切以人为本，使得操作系统尽最大可能贴近用户，了解用户的感受，从而方便用户。

　　Windows Vista 的最低系统要求较高，但还是比较合适的。它的最低系统要求包括 800MHz 的处理器、512MB 的 RAM 和一个 20GB 的硬盘。符合这些标准的计算机系统能运行 Windows Vista，并能达到 Windows Vista 所能达到的运行性能。

　　微软保证 Windows Vista 将给你的世界带来透明度，因此你能更加确实地和安心地依靠你的个人计算机。在个人计算机上装载了 Windows Vista 后，你完全能利用动态音频—视频输出、音乐和电视来得到更加生动的多媒体体验。

　　就安全问题而言，Windows 系列的老版本有很多安全问题。黑客离你的系统几千英里却可以用他们的计算机侵略你的系统，得到他们所需要的数据，这些数据可以是存储在你计算机里的个人信息、银行账户等信息。Windows Vista 的默认浏览器 Internet Explorer 被升级到了 7.0 版本。它包含了许多安全特征，如保护模式浏览、反钓鱼式欺骗软件、向外和入站防火墙、标准用户账号功能、用户账号控制、窗口防御和父母控制等。

　　Windows Vista 提供了新的令人惊异的图像功能。凭借 Windows Vista，微软在图像方面已超过了 Mac OS。这也是第一次微软提供了大量的高质量的包含 3D 效果的图像。这将有助于一些 3D 游戏软件的推广。

　　在无线网络功能方面，我们可以看到在 Windows Vista 的最终版本里有非常多的变化。在 Windows Vista 中，我们可以保存无线网络连接的设置并可为它命名，也可以调用这些设置进行网络再连接。

　　Windows Vista 家庭高级版是家用台式机和笔记本电脑的首选 Windows 版本。它是一个令你在家或外出时让你的计算机易于使用、更加安全和轻松娱乐的操作系统。它包括 Windows 媒体中心，它可以使你更加轻松地享受你的数字照片、电视、电影及音乐。

　　Windows Aero 用户界面让用户体验到动态视觉效果、透明玻璃样条菜单和三维窗口切换功能。即时搜索和组织用户信息的新方法让用户可以轻松、快速地找到并且使用电子邮件、文档、照片、音乐文件和其他用户想要的信息。Windows Defender 可以自动保护电脑免受间谍软件和其他有害软件所带来的阻止窗口弹出、性能降低和安全威胁。

  Windows Vista 商用版是第一款专门设计用于满足中小型企业需要的 Windows 操作系统。你可以在技术相关支持问题上花费更少的时间，这样就可以有更多的时间取得业务上的成功。使用经过改进的、简单易用的界面，会使你的业务有更高的工作效率，也会帮助你更快更容易地在计算机或网上找到你需要的信息。它强大的、新的安全特性，让你能够控制和保护业务上的重要信息并且建立你与客户之间的信任。

第 6 单元　数据结构

  数据结构是用来组织和存储数据的一种特殊形式。一般的数据结构类型包括数组、文件、记录、表、树等。任何数据结构都是用来组织数据以适应某种特定的目标，从而以适当的方法达到存取和操作的目的。在计算机程序设计中，可能会选择或指定一种数据结构来存储数据，以便在这种结构上实现各种算法。

数据结构与算法

  数据结构使用令算法更有效率的方法组织数据。例如，考虑一些我们用来查找数据的组织方式。一种过分简单的方式是将数据放置到数组中，并且采用对元素遍历的方法找到需要的元素。然而，这种方法是低效率的，因为在许多情况下，我们需要遍历所有元素才能完成。使用其他类型的数据结构，如哈希表和二叉树，我们能够相当快速地搜寻数据。

  过程抽象或算法抽象是隐藏算法细节的，允许算法在各个细节层次上可见或被描述。建立子程序是抽象的一个实例，子程序名描述了子程序的功能，子程序内部的代码表示了处理过程是如何完成的。类似地，数据抽象隐藏了描述的细节。一个明显的例子是通过把几种数据类型组合起来来构建新的数据类型，每种新类型描述了一些更复杂的对象类型的属性或组成。数据结构中面向对象的方法通过把对象类的表示整合而将数据抽象和过程抽象组合在一起。

  一旦问题被抽象化以后，看起来似乎是不同的，但在深层意义下本质上是类似的甚至是相同的问题就变得明显化了。例如，维护一个学生的听课名单与组织一个编译程序词典结构有很多相同的地方。二者都要对确定的事物进行存储和操作，而这些事物又都有特定的属性或性质。抽象概念允许对看似不同的问题具有相同的解法。利用抽象算法和数据结构求解问题具有最大的实用性和被重用的机会。

数据类型

  数据结构是一种数据类型，其值是由与某些结构有关的组成元素所构成的。它有一组在其值上的操作。此外，可能有一些操作是定义在其组成元素上的。由此我们可知：结构数据类型可以有定义在构成它的值之上的操作，也可以有定义在这些值的组成元素之上的操作。

  整型：整数是关于一组特定的必须遵循的原理或规则的组合。整数的表达形式并不重要，只要所有读者明白这些符号——二进制、八进制、十进制、十六进制、2 的补码或符号和量值等意思就行了。

  表：表是一个灵活的抽象数据类型。在一个程序运行之前，当要存储的元素个数不知道时，或者运行过程中元素个数会改变时，这种结构是非常有用的。它非常适合元素的顺序处理，因为表中当前元素的下一个元素极易得到。

数组和记录：数组和记录在大多数编程语言中都作为固有数据类型。通过使用指针数据类型和动态存储分配，大多数编程语言也能为用户提供建立链接结构的机制。数组、记录和链接结构是更高一级抽象数据类型的基本构造单元。

栈和队列

栈是一种数据类型，它的主要性质是由对其元素的插入与删除的控制规则来确定的，被删除或移去的元素只能是刚刚插入的，就是所谓具有后进先出（LIFO）性质或协议的结构。当进行一个新函数调用时，所有局部于调用程序的变量都需要由系统存储起来，否则新函数将要重写调用程序的变量。而且调用程序的当前位置也必须保存，以便新函数知道它执行完后返回何处。变量通常由编译器分配给机器寄存器，而且尤其是涉及递归时，肯定会有冲突。

堆栈尾部可以进行插入和删除操作的记录称为堆栈的栈顶，另一端称为栈底。为了表示如何限制堆栈只能从栈顶访问，我们用特殊的术语来表示插入和删除操作。把一个对象插入堆栈的操作称为进栈操作，而从堆栈中删除一个对象的操作称为出栈操作。

为了在计算机存储器中实现栈结构，一般采取的方法是保留一块足够容纳栈大小变化的连续的内存空间。通常来说，确定块的大小是一个很重要的决定。如果保留的空间过小，那么栈最后可能从所分配的存储空间中溢出；而如果保留的空间过大，将造成存储空间的浪费。块的一端作为栈底，栈的第一条数据会被存储在这里，以后的条目被依次放置在它之后的存储单元中，也就是堆栈向另外一端增长。

有几种采用队列的算法能给出有效的运行时间。现在举几个使用队列的简单例子。当作业提交给打印机时，它们是按到达顺序排列的。所以基本上送往打印机排队的作业都是按队列放置的。另一个例子与计算机网络有关，很多网络与个人计算机连接，其中硬盘与一台称为文件服务器的机器相连。可以在先来先受服务的基础上给予其他机器上的用户访问文件，所以这个数据结构是队列。

用来在计算机中控制队列的最一般的方法是为队列分配一块存储器，从存储块的一端开始存储队列，并且让队列向另一端增长。当队尾到达块的末端时，我们开始将新的条目反向于末端的方向插入，此时它是空闲的。同样，当队列的最后一条成为队头并被移出时，调整头指针回到块的开端，同时在此等待。在此方法下，队列在一块区域内循环而不会出现内存溢出情况。采用此技术的实现方法称为循环队列，因为分配给队列的一块存储单元组成了一个环。就一个队列而言，存储块的最后一个单元与它的第一个单元相邻。

多级反馈队列

在计算机科学中，一个多级反馈队列是一种调度算法，其目的在于符合一个多模式系统的设计要求。

- ♦ 对简短的工作给予优先。
- ♦ 对专门进行输入/输出的工作给予优先。
- ♦ 快速确定处理过程的本质，从而确定处理的时间表。

采用了多级先进先出（FIFO）队列，其操作如下：

- ♦ 把新的处理项目置于先进先出队列的顶层。
- ♦ 在某个阶段，某个处理过程到达了队列的头部，则被分配给CPU进行处理。
- ♦ 如果某过程完成就离开系统。
- ♦ 如果某过程自动地放弃控制，它就离开队列网络；如果过程又重新准备好，它就再进入队列中的同一级别。
- 如果过程用尽了额定时间，它就预先清空并被置于下一级别的末端。
- 这种做法继续进行，直到过程完成或到达队列的底层。

在底层队列，过程按循环赛的方式进行循环，直到这些过程都已经完成并离开系统。

在多级反馈队列中，一个过程在被强制降到了下一级队列之前只被给予一次在给定队列级别上完成任务的机会。

阅读材料1：.NET 框架

.NET 是一个几乎可以解决任何开发问题并对网络编程提供大量支持的开发平台。事实上，.NET 比微软开发的其他平台在本质上更支持网络应用。一个网络程序是使用计算机网络与其他应用程序传输信息的所有应用程序。例如，普遍存在的浏览器 Internet Explorer，或者是你使用的接收电子邮件的程序。所有这些软件模块均具有与其他计算机通信的能力，并且这样做对终端用户更加有用。

在浏览器模式下，你浏览的每个网站实际上是存放在互联网其他地方的计算机上的文件。利用电子邮件，你可以同保留有你的电子邮件的网络服务提供商或公司的计算机进行收发电子邮件。用户通常信任网络应用程序。因此，这些程序对于运行它们的计算机来说，拥有比网站更大的控制权。这使得一个网络应用程序可以管理本地计算机的文件，而拥有各种实用功能的网站是不能做这个工作的。更重要的是，从网络的角度来说，一个应用程序拥有更大的权限去控制它怎样与互联网上的其他计算机通信。

.NET 不是编程语言，它是一个融合 4 种官方编程语言的开发框架：C#、VB.NET、Managed C++、J#.NET。在 4 种语言的对象类型上是有交叉的，.NET 框架定义了框架类库（FCL）。在框架中的 4 种语言均共享框架类库和一个提供.NET 应用运行环境的面向对象的开发平台，即公共语言运行时系统（CLR），这个 CLR 与 Java 语言的虚拟机类似，但它是为 Windows 设计的，而不像虚拟机可跨平台那样使用。

.NET 语言不是面向过程的，而是面向对象的语言。这就在同样的逻辑结构中提供了一个自然的机制去封装相互联系的数据及修改这些数据的方法。一个对象是一个程序化的具有某些特征或能够执行操作的结构。面向对象的核心概念是一个类继承另一个类的特征和方法的能力。

程序集通常是一个包含.NET 类集的预编译代码的.DLL 文件，不像标准 Win32 动态链接库那样，开发者必须依赖于文档，如头文件等。利用给定的 DLL，.NET 汇集了一定的元数据，它利用在程序集中恰当地包含一些方法来提供任何.NET 应用足够的信息。元数据也用来描述程序集的其他特征，如它的版本号、代码的创始人，以及被加到类中的任何自定义属性。

.NET 在多个应用程序共享程序集的问题上能够提供一个唯一的解决方案。通常，当一个程序集被指定仅为一个应用程序使用时，它将被包含在和应用程序相同的文件夹（或子文件夹）中。这就是所谓的私有程序集。一个共享程序集被复制到一个本地系统所有的.NET 应用都能访问的地方。而且，这个共享程序集被设计成为版本化的、唯一的且防作弊的。这是由于有一个优秀的安全模型。共享程序集存放的地方被称为全局程序集缓冲器。

在.NET 中编写的应用程序的代码是指托管的、类型安全的代码。这意味着代码被编译成被严格控制的中间语言，这样它就不能包含那些可能潜在地导致计算机崩溃的代码了。但以原生代码编写的应用程序能够任意修改计算机内存的地址，其中一些有可能导致系统崩溃，或者常规的保护错误。

在.NET 出现之前组件的设计是通过原生代码编写的，因此是非托管的、被认为是不安全的。在.NET 应用程序中结合不安全的代码没有技术上的困难。然而，如果一个基本组件能够使一台计算机瘫痪，那么整个应用程序也被认为是不安全的。不安全的应用程序可能要受到限制，如当它们执行网络共享操作时，可能会被阻止操作。

第 7 单元　数据库原理

一个数据库由一个文件或文件集合组成。这些文件中的信息可分解成一个个记录，每个记录有一个或多个字段。字段是数据存储的基本单位，每个字段一般含有由数据库描述的属于实体的一个方面或一个特性的信息。用户使用关键字和各种排序命令，能够在许多记录中快速查找、重排、分组和选择相应的字段，以检索或建立特定数据集上的报表。数据库记录和文件必须被组织起来才能确保能对信息进行检索。大型数据库的许多用户必须能够在任何时间内快速使用数据库中的信息。

数据抽象

显然，在处理二进制位的计算机和处理像航班或把乘务组人员分派到飞机这样一些抽象的最终用户之间将存在很多层抽象。

最底层，也称物理层，把数据存储在硬件设备上。用户程序不能直接访问它们。它们必须通过逻辑层存取数据。外部层按外部或用户程序的需要定义数据库的不同视图。一个用户程序不可能需要数据库中的所有数据，因此用户/应用程序仅观察所需的数据库信息。这意味着取决于对数据的需求，不同的程序对同一个数据库将有不同的视图。这样一些视图是数据库外部的，因而在外部层规定。不同的视图也不必要包含完全不同的数据。在不同的视图中会存在公共的信息。

概念层描述整个数据库，它是由数据库管理员使用的，他们必须决定什么信息要被保存在数据库中。

数据模型

数据模型是一组概念工具，用来描述数据、数据关系、数据语义和数据约束。数据模型分成三类，即基于对象的逻辑模型、基于记录的逻辑模型和物理数据模型。

基于对象的逻辑模型用于描述概念层与视图层的数据。它们非常接近于人的逻辑。很多不同的模型可用来描述基于对象的逻辑模型，其中最重要的是语义数据模型和实体关系模型。语义数据模型提供了表达数据库中数据意义的工具。实体关系模型（E-R 模型）是基于这样的认识：现实世界是由一组称为实体的对象和这些对象之间的关系组成的。实体是对象，它能与其他对象唯一地区别开。例如，名称、物理尺寸和每单位长度的重量唯一地描述一段特定的钢材。相同类型的所有实体之集合和相同类型的各种关系分别称为实体集合和关系集合。

实体和关系要被区别开，并且数据库模型应指明这如何能实现。这用主关键字概念来实现。实体关系模型可以定义数据库的内容必须遵守的某些约束。一个重要的约束是某个实体通过关系所关联的其他实体的个数。对于包含两个实体集合的关系，可能存在像一对一、一对多、多对一和多对多那样的关系。

基于记录的逻辑模型定义数据库的整个逻辑结构及其实现的更高级描述。3 种不同的基于记录的逻辑模型被广泛使用。它们是层次模型、网络模型和关系模型。

物理数据模型用于描述最底层的数据。只有很少几种物理数据模型在使用。广泛知道的是统一模型和框架存储。

关系型数据库的术语

表：表是包含数据类型和原始数据的对象。

字段或列：这是表中用来保存数据的部分。每一个列必须指定一个数据类型和唯一的名称。

数据类型：有许多种数据类型可供选择，如字符型、数值型、日期型和其他类型。

存储过程：它有点像宏，其中的 SQL 代码能按名称来存储。运行这个名称，实际就是运行它的代码。使用 SQL 代码生成周报表就是这样一个例子，SQL 代码作为一个存储过程存储起来；以后只要你运行存储过程就可以产生周报表。

主键：尽管主键本身不是一个对象，主键对于关系型数据库来说是重要的。主键保证各行之间的数据的唯一性。

外键：同样，外键本身也不是一个对象，它是一个文件或列，用来引用另一个表的主键。在查询时，SQL 服务器用主键和外键来关联不同表中的数据。

约束：约束是一种基于服务器的、系统实现的、数据完整性的强制机制。

视图：视图基本上就是一个存储在数据库里的查询，它可以用来引用一个或多个表。可以创建及把它保存起来，以便于将来使用。一个视图一般可以仅包含一个表中的几列，或者将两个或多个表连接在一起。

数据的安全性、完整性和独立性

数据的安全性是指保护数据库以防止不合法的使用所造成的数据泄露、更改或破坏。数据库的安全性和计算机系统的安全性是紧密联系、相互支持的。

数据完整性是指数据的精确性和可靠性。它检查数据库是否与数据和信息的语义规定不一致，以防止因无效的输入输出操作或错误信息而造成的错误。数据完整性分为四类：实体完整性、域完整性、参照完整性和用户自定义的完整性。

◆ 实体完整性

实体完整性规定表的每一行在表中是唯一的实体。使用 Primary Key、Unique 和 Identity 约束就是实体完整性的体现。

◆ 域完整性

域完整性是指表中的每一列必须满足某种特定的数据类型或约束。表中的 Check、Foreign Key 约束和 Default、Not Null 定义都属于域完整性的范畴。

◆ 参照完整性

参照完整性是指两个表中一个表的主关键字应与另一个表的外关键字一致。在 SQL Server 中，参照完整性的作用如下。

（1）禁止向一个表中插入与另一个表中的数据没有任何关系的新数据行。

（2）禁止从一个表中删除与另一个表中的记录有关系的记录。

◆ 用户自定义的完整性

用户自定义的完整性即是特定的关系数据库的约束，它反映了数据的某一具体应用必须满足语义的要求。

可以说数据处理的发展就是数据独立性不断进化的历史。在手工管理阶段，数据和程序完全交织在一起，没有独立性可言。数据独立性包括物理独立性和逻辑独立性。

物理独立性是指用户的应用程序与数据库中数据是相互独立的。数据在磁盘上怎样存储及怎样被访问，用户不需要了解。逻辑独立性是指用户的应用程序与数据库的逻辑结构是相互独立的，即当数据库的逻辑结构改变时，用户程序可以保持不变。

阅读材料1：数据仓库和数据挖掘

1. 数据仓库

数据仓库是一种数据库，它将从不同产品和操作系统中调出的数据组合起来放入这种大型数据库，对管理状况做出报告和进行分析。这种数据库对源于组织核心事务处理系统的数据进行重新组织并与其他信息（包括历史数据）进行合并。这样，这些数据可以用来做出管理方面的决策和分析。

在大多数情况下，数据仓库中的数据只可用来进行报告。数据不可进行更新，所以公司的隐性操作系统的表现就没有受到影响。数据仓库这种侧重解决问题的特性，使众多的公司由于运用了数据仓库而获益匪浅。

数据仓库一般有重新塑造数据的能力。关系数据库的数据视图让用户从两维观察数据。多维数据视图允许用户以多于两维的方式观察数据，如按地区按季度销售。为了提供这种信息，组织可以用一种特殊化的多维数据库，或者用可以在关系数据库中生成数据的多维视图的工具。多维分析能够使用户使用多维的不同方式看到相同的数据。

构造数据仓库面临的问题如下。

- 何时及如何收集数据。在数据收集的源驱动体系结构中，数据源要么连续地在事务处理发生时传送新信息；要么阶段性地，如每天晚上传送新信息。数据仓库常会保留稍微有点儿过时的数据。但这对于决策支持系统来说通常不是问题。
- 采用什么模式。各自独立构造的数据源可能具有不同的模式。事实上，它们甚至可能使用不同的数据模型。数据仓库的部分任务就是做模式集成，并且在数据存储前将数据按集成的模式转化。因此，存储在数据仓库中的数据不仅仅是源端数据的复制，同时它们也可被认为是源端数据的存储视图（或实体化的视图）。
- 如何传播更新。数据源中关系的更新必须被传至数据仓库。
- 汇总什么数据。由事务处理系统产生的原始数据可能太大而不能在线存储。但是，我们可以通过对关系做总计而得到的保留汇总数据回答很多查询，而不必保留整个关系。

2. 数据挖掘

数据挖掘是关于使用自动或半自动化方法去分析数据和寻找隐蔽模式的技术。数据挖掘对企业有重要的商业价值。

- 增加竞争力
- 顾客分割
- 周转分析
- 交叉销售
- 销售预报
- 欺诈检测

♦ 风险管理

在过去的十年中,大量的数据积累并存储在数据库中。其中很多数据来自商业软件,如财务应用、企业资源计划(ERP)、客户关系管理(CRM)和万维网运行记录。这些数据收集的结果,使得各个组织变成数据丰富而知识贫穷者了。数据收集得非常多并且增长得很快,以至于这些存储数据的实际使用很受限制。数据挖掘的主要目的是从手头上的数据中提取模式,增加数据的自身价值,并将这些数据转变为知识。

数据挖掘对数据集提供算法,如判定树、聚类、关联、时序等算法,并分析它们的内容。这一分析产生了模式,它能给出有价值的信息,如图 7-1 所示。与基本算法有关,这些模式可以呈现为树形、规则、聚类或是简单的一组数学公式。在这种模式中发现的信息可用于做出报告,去指导市场策略,而更重要的是预测。

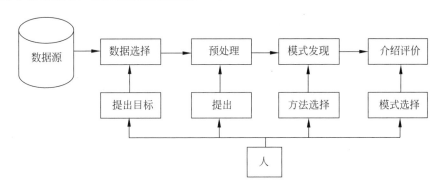

图 7-1 数据挖掘

第 8 单元 软件工程

软件工程第一次作为一个通俗的术语出现是 1968 年在德国 Garmisch 市举行的北约会议的标题中。数字计算机问世不到 25 年,而我们却已面临着"软件危机"问题。首先我们发明了计算机程序设计,然后教人们如何编写程序,下一个任务便是对大型系统的开发,这种系统要可靠、能按时交付使用,并在预算范围内。随着各种技术的发展,我们的目标放在能够成功地做些什么这个界限上。实践证明,我们在准时实现大型系统方面表现得不太好。

软件工程的早期方法坚持要严格地遵守分析、设计、实现及测试的顺序。因此,软件工程师坚持应当在设计之前进行完整的系统分析,同样,设计应该在实现之前完成。这就形成了一个现在称为瀑布模型的开发过程,这是对开发过程只允许以一个方向进行的事实的模拟。

第一阶段是需求定义阶段,是指系统期望的需求分析,即描述功能特征和操作细节的阶段。如果系统是一个在竞争的市场上销售的通用产品,这个分析将会包括一个广泛的调查来发现潜在用户的需求。但是,如果系统是为特殊用户设计的,那么这个过程就是一个更专业的调查。当潜在用户的要求被确定之后,要将这些要求汇编成新系统必须满足的需求。这些需求是从应用的角度来表述的,而不是用技术术语来表达。

一种需求可能是对数据的存取必须限制为有权限的人员。另一种可能是当最后一个工作日结束时,数据必须反映目前的清单状态,或者可能是在计算机屏幕上的数据组织必须按照目前使用的纸质形式的格式来显示。系统的需求被确定以后，它们就转化为更多的系统技术说明书。在需求阶段，一个错误的功能说明会导致不满足需要的设计和实现。如果没有查明让错误发展下去，那么到了测试阶段就会花大量的财力去修正这个错误（包括重新设计和重新实现）。

第二阶段是设计。设计关注这个系统应该怎样来实现目标。正是通过设计建立了软件系统的结构。这个阶段的输入是一份（经过调试和确认过的）需求文档，输出是以某种适当形式表示出的设计，如"伪代码"。确认设计阶段的正确性是非常重要的。需求文档里每一个需求都必须有相应的设计片段与之相符。正规的验证虽然可以达到一定的程度，但却是极其困难的。更多的是整个的设计团队、管理者，甚至是客户的非正式的校阅。

大型软件系统最好的结构是模块化系统，这是一条被充分证实的原则。确实，正是借助模块化的分解方法，大型系统的实现才成为可能。没有这样的分解，在大型系统实现过程中所需要的技术细节可能会超过一个人的理解能力。然而，有了这种模块化设计，仅仅需要熟悉与在考虑中的模块相关的细节。同样，模块化设计对未来的维护是有益的，因为它允许对基本的模块进行修改。

第三阶段是软件实现，它是对第二阶段设计开发的实际编码阶段。设计必须被翻译成机器可识别的形式。编码阶段的任务就是做这样的工作。如果设计做得详细，编码就能由机器完成。这个阶段具有很大的诱惑力，很多鲁莽的程序员没有经过前两个阶段的充足准备就跳到了软件实现阶段。结果是，需求关系没有完全弄清楚，设计也有缺陷。软件实现进行得很盲目，结果是越来越多的问题涌现。

第四阶段是软件测试。测试与实现紧密联系，因为系统中的每一个模块都要在实现的过程中进行正常的测试。在一个大系统的开发中，测试包括若干个阶段。首先，每个程序模块作为一个单独的程序进行测试，通常与系统中的其他程序分离开。这种测试被定义为模块测试或单元测试。在任何可能的时候，单元测试是在控制环境下进行的，这是为了让测试小组能提供被测试模块先前确定的数据并观察产生的输出数据。另外，测试小组要检查内部数据结构、逻辑，以及输入输出数据的边界条件。

当对模块集进行过单元测试后，下一步是确保各模块间接口的定义和处理要适当。综合测试是确定系统里的部件像程序设计和系统设计说明书中所描述的那样一起工作的过程。一旦我们确定信息按照设计要求在模块间传递，我们测试系统来确保它具有预期的功能。功能测试是通过对系统进行评估来确定需求说明书中描述的功能是否在整个系统中都能实现。

不幸的是，成功地进行系统的测试和调试是极其困难的。经验表明，大型的软件系统可能包含众多的错误，甚至是经过关键测试之后。许多这样的错误在软件的生命期中一直潜伏着，但是也有一些会导致关键性的错误。减少这样的错误是软件工程的一个目标。

术语"软件维护"用来描述在软件产品交付给用户以后所进行的软件工程活动。维护活动包含增强软件产品、调整软件产品以适应新的环境和纠正错误。软件产品增强可以包括提供新的功能、改进用户显示和交互模式、升级外部文档和内部文件说明或升级系统的性能指标。软件对新环境的适应可以包括把软件移植到不同的机器。或者，例如修改软件以适合于新的远程通信协议或添加的磁盘驱动器。问题的纠正包括修改和重新确认软件以纠正错误。对有些错误需要立即采取措施，有些则可按计划定期纠正，而其他的错误虽然已知但却永远不能纠正。

软件维护是软件开发周期的缩影。软件的功能增强和适应使开发重新回到分析阶段，而软件纠错可使开发周期回到分析阶段、设计阶段或实现阶段。因此，所有用于开发软件的工具和技术对软件维护都有潜在的用途。

软件质量保证活动与软件生命周期各阶段的验证和有效性确认活动紧密关联。事实上，在许多组织中这些活动没有明显的区别。尽管如此，质量保证实际上与其他验证和有效性确认活动相差甚远，质量保证是一个管理功能，而验证和有效性确认是软件开发过程的一部分。

软件工程项目标准的开发是一个非常困难的过程。一个标准是一个产品的某种抽象表示，它定义了被开发产品必须达到的最低性能、健壮性、结构等。软件的国家标准问题在于这些标准过于笼统，这是必然的，因为不像硬件，人们至今尚不能将大多数软件特性进行量化。因此，有效的质量保证就要求开发更多特定的组织标准。当然，为质量保证而开发软件标准所带来的问题及软件产品的优良程度难于评估是软件质量难以把握的本质。

软件质量保证正成为软件工程的一个新兴分支。质量保证不同于系统测试。系统有效性确认是开发或测试小组的责任，而质量保证小组要报告确认和确认工作是否充分。这自然意味着质量保证与系统最后阶段的综合测试有着密切的联系。

阅读材料 1：企业资源计划

企业资源计划（ERP）软件尝试把一个企业交叉的所有部门和功能整合成一个统一的能满足所有不同部门的特殊需求的计算机系统。

ERP 系统可以成为客户订单有效操作的场所，从代表接受的客户服务开始直到装运货物及财务部门送出发票。在一个系统中获取信息胜过在分散的不同的无法互相通信的系统中获取，企业可以更容易地跟踪订单，并可同时在不同地点安排生产、库存和装运。

ERP 可以使生产过程的流程更加平滑，提高了企业内部订单实施过程的可见度。它还能减少用于制造产品的原材料库存，并能帮助用户更好地计划去给客户发货，尽量减少成品在仓库和码头的库存。

ERP 是接受客户订单，并提供一种软件路线图，来使订单沿着规定的实施路线自动地完成不同的步骤。当一个客户服务代表输入客户订单信息进入 ERP 系统后，他就有了所有用来完成订单所需要的信息。例如，他可以从财务模块中了解客户的信用和订单历史，从仓库模块中了解公司库存水平，从后勤模块中了解货船码头运输时刻表。

不同部门的员工可以看到相同的信息，并可以更新它，当一个部门完成订单的一个步骤后自动地沿着 ERP 系统发送到下一个部门。要想寻找订单在系统的哪一个点，只需要登录 ERP 系统并沿着踪迹找下去。幸运的话，一个订单处理移动的速度就像一道闪电通过你的组织，客户能比以前更快的同时更好地收到订单。ERP 还可以在企业其他商业处理中展现它的魔力，如员工福利或财务报告。

ERP 对很多大公司来说是一个吸引人的解决方案，因为它提供了那么多潜在的用处。例如，同一个系统可以用来预报产品的需求、订购所需的原材料、建立生产进度表、跟踪库存、制定价格和计划主要的金融措施。ERP 为一个公司的核心业务处理起到一个计划中枢的作用。除了指导很多这方面的工作外，该系统还把那些跨公司获取的、不同处理过程的应用数据联结在一起。例如，一个常见的 ERP 系统管理各种职能和活动，包括材料账单、订单条目、采购账、人力资源和库存控制，这里只列出了可用的 60 个模块中的小部分的名称。按照需要，ERP 也可以把从这些过程得到的数据与其他公司的软件系统共享。ERP 另一项重要收益是，用一个单一的、集成的系统来取代复杂的计算机应用程序。

然而，尽管有这些潜在收益，传统的 ERP 系统仍然存在几个缺点。例如，早期的系统往往比较大型、复杂、昂贵。其实现需要从公司的信息技术部门或外界专业人员那里得到大量的时间许诺。此外，由于 ERP 影响到公司大多数主要部门，它们往往在很多业务处理过程中产生变化。要把 ERP 放到适当的位置，就需要有新的处理章程、雇员培训，以及管理和技术上的支持。结果，很多公司发现，完全改变到 ERP 是一个缓慢且痛苦的过程。一旦实现阶段结束，某些业务在量化从 ERP 得到的收益时又会碰到麻烦。

最后，随着技术加速向互联网连接、基于网络的商家对商家关系、电子商务转移，某些公司担心他们的基于大型机的 ERP 系统会太慢。尽管如此，但通过实践，很多公司发现他们的 ERP 系统通过业务过程的标准化、促进信息跨公司共享及创建变革的机构，为公司未来的发展提供了坚实的技术基础。对于公司来说，安装 ERP 系统感觉上既昂贵又费力。然而，这为大范围地使用新的增值应用程序建立了一个很有价值的基础。

第 9 单元　多媒体

多媒体是用计算机和其他电子手段带来的文字、图像、声音、动画和视频的综合。它是感情丰富的呈现。当你将亮丽的图片和动画、动人的音乐和引人入胜的视频片断、原始的文字信息等多媒体情感元素编排在一起时，就可以使人们的思维和行为中枢触电。多媒体刺激着人们的视觉、听觉和触觉，而最重要的则是激励了人们的思维。

多媒体元素

多媒体是一组多于一个媒体元素的集合，用于产生集中和更结构化的通信方式。多媒体同时使用不同来源的数据。多媒体的这些来源被称为媒体元素。随着信息技术的快速增长和日新月异的变化，多媒体已成为计算机世界的一个重要组成部分。

◆ 文本

Windows 操作环境为用户表示文本提供了无限的能力。通过以多种形式显示文本，使人们更易理解多媒体应用程序要表达的信息。

◆ 声音

把声音融入多媒体程序，用户可以得到使用其他通信方式无法得到的信息。某些类型的信息不用声音很难有效表达，如用文字准确描述心脏的跳动声及大海的声音几乎是不可能的。

◆ 静态图像

当你想象图像时，你可能想到静态的图像——也就是像照片或画一样，这种类型的图像是不动的。静态图像是多媒体的重要部分，因为人类是视觉定位的。Windows 也是可视化环境，它比基于 DOS 的环境更容易显示图像。

◆ 动画

动画就是运动的图像。动画和静态图像一样，都是强有力的通信形式，动画在解释涉及运动的概念中特别有用。讲解如何弹吉他或打高尔夫球，只用一幅图是不行的，甚至一系列的图也不行，用文字解释就更困难了，而多媒体应用程序中用动画解释它则轻而易举。

全运动影像，如电视里的图像，可使多媒体的应用更为广泛。虽然全运动影像听起来像是一个往多媒体程序中加入强有力信息的理想方法，但它无法达到人们看电视一样的效果。

多媒体的应用

多媒体的商业应用包括展示、培训、市场、广告、产品演示、数据库、产品目录及网络通信。在很多局域网和广域网上很快将提供语音邮件和视频会议。

◆ 基于计算机的培训

多媒体被广泛地用于各种培训计划中。许多公司使用多媒体应用程序培训其雇员。一家大电话公司使用多媒体应用程序模拟一些重要的紧急情况来培训雇员如何处理这些情况。通过使用多媒体,公司发现它既节省开支又比其他培训员工的方法有效。

◆ 教学

多媒体的教育用途正在迅速增长。其表达信息和其资源集成的能力容许人们建立丰富的学习环境。过去,教师或学生为访问所需信息也许不得不查阅大量的资源并使用多种媒体。通过集成媒体和利用超媒体,我们能够建立用户控制的即时信息学习环境。

多媒体在未来几十年将彻底改变教育过程,特别是对于那些优秀的学生,他们会发现自己可以远远超出传统教育方法的限制。在某些情况下,老师在教学过程中的确更像导师或顾问,而不是基本的信息和理解提供者。教学过程的核心不再是教师,而是学生。在教育家眼中,这是一个敏感和高度政治化的课题,因此,目前教育软件通常只是"丰富"教学过程,而不是潜在地取代传统的以教师为主导的方法。

◆ 娱乐

有一些娱乐是绝对没错的。许多情况下,当今最好的游戏软件中的图形技术将被用于明天的商业软件。如果有一条途径把技术扩展到其相邻层次,你绝对应该把眼光放到多媒体开发者身上。另外,写娱乐软件本身也很有趣。

现在有一种寓教于乐的新型软件,想法就是通过提供一些娱乐使学习过程更有趣。

◆ 其他

从园艺、烹饪到室内设计、改造和修理,直到家谱软件,多媒体已经进入家庭。最终,大多数多媒体项目将通过电视机或具有内置交互用户输入的监视器而走入家庭。在这些装置上看多媒体很可能采取在数据高速公路上付费点播的方式。在宾馆、火车站、购物中心、博物馆和杂货店,多媒体将在单独的终端或服务亭里为人们提供信息和帮助。这些设施将减少对传统信息服务亭和工作人员的需求,并且能全天候,甚至在工作人员下班之后的午夜继续工作。

当公司和商业机构了解了多媒体的功能,并且多媒体设施的成本降低时,更多的能使企业更平稳、更有效地运作的多媒体应用程序将由企业自己或第三方开发出来。

多媒体网络应用

许多应用如视频邮件、视频会议和其他协作工作系统都需要网络化的多媒体。在这些应用中,多媒体对象被存储在服务器上并在客户端上进行播放。这些应用需要向各种远程站点广播多媒体数据或访问大容量多媒体资源。多媒体网络即使是在数据压缩的情况下,也需要一个非常高的传输率或带宽。传统的网络用于提供无误差传输,而大多数多媒体应用能在传输中允许由于损坏或包遗失而产生误差,且无须重传或进行修正。在某些情况下,为满足实时传递需求或实现同步甚至允许丢弃某些包。其结果是可以在多媒体网络上使用轻量级传输协议。由于可能会产生不可接受的延迟,这些协议不允许重传。

多媒体网络必须为交互操作提供低的延时。因为当多媒体数据到达目的站点时必须同步,网络必须用低抖晃提供同步。

在多媒体网络中,与传统网络点对点不同的是大多数的多媒体通信是多点式的。例如,有两个以上参与者的多媒体会议，需要以不同媒体对每一个人发布信息。会议网络使用多播和桥接发布方法。多播技术复制一个单一输入信号并将它传送给多个目标站点，桥接技术将多个输入信号组合成一个或多个输出信号，然后传送给参与者。

过去，由于受很多技术手段的限制，多媒体不能广泛应用。如今，高速计算机的使用和互联网的联通，使得多媒体的实施更加灵活，而其应用也呈指数趋势上升。

阅读材料 1：计算机辅助设计

在广义上，计算机辅助设计（CAD）指的是计算机在解决设计问题中的应用。工程技术人员可以借助于视频显示器、键盘、绘图仪和更多的人机接口与计算机通信。工程技术人员可以提出问题并能很快从计算机得到解答。更确切地说，CAD 是使工程技术人员和计算机协同工作，彼此发挥长处的技术。

CAD 允许设计者以图形方式与计算机交互作用，工程技术人员能够测试一个设计思想，并很快地查看到设计效果，然后对其进行修改和重新评价。如此循环往复，直至形成一个合格的设计。每重复一次，设计方案都会得到进一步的改善。因此，在时间、材料和资金允许的条件下所执行的循环次数越多，设计效果就越好。

CAD 之所以如此有效，是因为计算机和设计师不是用数字而是用图像进行交流，而图像更容易为人类思维所接受。该方法也允许对正在设计的元件进行模拟测试和评价，测试的结果以图形或数字的形式叠加到显示屏幕上。例如，如果该元件正在被加热，其各部分将会用不同的颜色着色，每一种颜色代表某一特定的温度范围。这种有效的图像表示方式有助于设计师精确地发现正在发生的情况。

CAD 涉及中央设计描述的发展，它决定了所有的设计和制造。这意味着用于分析和模拟设计及产生制造指令的以计算机为基础的技术，应该与设计的形式和结构建模技术密切结合。另外，中央设计描述为同时存在的同步工程活动设计的全部方面的开发打下了良好的基础。原则上，CAD 能够应用于设计的全过程，但实际上，它在开始阶段广泛使用诸如草图等非常不准确的描述，所造成的影响使它具有一定的局限性。还必须强调的是，目前 CAD 还不能帮助设计者设计更富有创造性的产品，如产生可能的设计方案，或那些涉及设计复杂推理的方面。

应用计算机辅助设计传统模型技术的驱动力是希望通过自动完成设计上更繁复冗长的程序，从而提高设计者的生产力和提高设计模型的精确性。已经开发的新技术正尝试克服传统实践上的局限性，特别是处理复杂问题，如汽车车体某些设计的复杂形式或集成电路一类产品结构的复杂性。CAD 使设计者能更快、更准确地完成任务，并且，不能用其他方法实现时，它提供了一种途径。

在研究通用 CAD 系统时要考虑该系统应具有尽可能广的应用范围，仔细考虑以下几个方面。

- ◆ 机械工程设计
- ◆ 建筑设计
- ◆ 结构工程设计
- ◆ 电子电路设计
- ◆ 动画和图形设计

对大多数实际应用来讲，应考虑把通用作图系统结合到大型专用系统中，因此作图系统应尽可能简单，而且运行效率高，因而这种结合不需要花很多精力。对分析作图和绘图分析这两个过程而言，在由作图系统所产生的数据和分析程序之间必须有一个简单的、有效的连接。另外，图形数据可用能被分析程序识别但不对作图系统产生影响的方法进行注释。

第10单元　计算机图形和图像

计算机图像的3个主要领域是计算机图形学、图像处理和计算机视觉。这3项技术已开始融合到很多应用中。图像合成或计算机图形学通常是以三维模型输入，以二维图像输出；图像处理通常是将三维实物以二维图像（或二维现实）输入，再以二维图像输出；计算机视觉通常是三维实物以二维投影图输入，以数值、分类或动作输出。这样的分类在某种程度上是暂时的，并且取决于当前的技术。情况的确如此，当前最常用的图形显示器都是二维的。在计算机视觉中，使用最多的输入设备是把从三维景物中采集的信息作为二维投影图的TV摄像机。

图形学

图像合成或称计算机图形学，是研究用计算机创建图像的方法学。在三维计算机图形学中，图像是由数学描述或模型经编制的程序而生成的。计算机图形学获取的是三维模型，经计算成为一个可显示的二维投影图。计算机图形学中的算法大部分是在三维空间中实现的，此空间中的作品在整个处理过程的最后阶段才被映射成二维显示或平面图像。传统的计算机图形学在创建图像时，先要进行非常详细的几何描述，使图像在三维空间中经过一系列转换而面向观察者和对象，然后通过使对象看上去有实体感和真实感来模拟实物——这一处理过程就是渲染。

通常，图形编程软件包提供了一些功能来描述场景，这些功能使用了称为输出图元的基本几何结构并将输出图元组合成更复杂的结构。每个输出图元是由输入的坐标数据和有关物体显示方式的一些信息来指定的。点和直线段是图的最简单的几何成分，其他可以用来构造图形的输出图元有圆及其他圆锥曲线、二次曲面、样条曲线和曲面、多边形填色区域及字符串等。

通过将应用程序提供的单个坐标位置转换成输出设备的相应操作，可以进行点的绘制。例如，对于CRT显示器，则是发射电子束，从而在选中的位置上照亮屏幕的荧光层。电子束的定位方法取决于显示技术。

通过计算沿线路径上两指定端点间的中心距离，可以绘制一条线段，输出设备则按指令在端点间的这些位置直接填充。对于矢量笔绘图仪或随机扫描显示器这类模拟设备，可以从一个端点到另一个端点绘制光滑线段。这是根据x方向和y方向需要修改的实际量，线性地改变水平和垂直偏转电压而实现的。数字设备通过绘制两端点间的离散点来显示线段。线段路径上离散的坐标位置是通过直线方程计算出来的。

图像处理

图像处理是对一幅图像的处理以产生另一幅图像，该图像不同于输入图像。原始图像可以是一个图像文件或是可直接对由TV类型照相机输出的图像进行处理的程序。图像分析也是图像处理的一部分。如今图像处理更具特色的是它移植到了新的应用领域而不是新算法的出现。我们已经看到，在虚拟现实的应用中，基于图像的渲染技术的出现，比起三维计算机图形学的算法来，与图像处理的二维世界有着更多的共同之处。图像处理的另一更普遍的应用是在图形设计方面，可用来处理广告展示中的照片图像等。

为便于显示，图像被存储在快速存取的缓冲存储器内，该存储器以30帧/秒的速率刷新显示器，以形成视觉上连续的显示。小型或微型计算机可通过联网方式（如以太网）通信，控制所有的数字化、存储、处理和显示操作。利用终端可以把程序输入到计算机内，在终端、电视显示器、打印机或绘图仪上输出信息也是很方便的。

通过卫星获取的图像可用来跟踪地球资源，地理测绘，农作物生长、城市发展和气象的预测；防洪、防火及其他环境下的应用。空间图像应用包括对在外层空间探测飞行中所获取图像中的物体进行识别与分析。图像传送和存储的应用体现于广播电视、电话会议、办公室自动化的图像传真（打印的文件和图形）、计算机网络通信、基于安全性监视系统的闭路电视及军事通信之中。雷达和声纳图像被用来侦测和识别各类目标，或者用于飞行器或导弹系统的制导与操纵。

计算机视觉

计算机视觉是从图像中抽取信息。它与图像分析所不同的是，它的目的更热衷于模仿人类的视觉系统。计算机视觉中的原始图像通常是一个真实景物的二维投影图。处理的目的是由二维投影恢复三维信息，如立体投影的深度。除此之外，处理后的输出结果可能是一个数值，或一个标记（如文字识别时），也可以是机器人技术中计算机视觉的一个动作。计算机视觉是一个非常苛求的领域，是因为我们试图从一系列的二维投影图中提取三维信息。尽管最初很乐观，但要为大多数潜在的应用领域找到解决办法，事实证明是很困难的。

24 位彩色

计算机通常是以 8 位（称为字节）组合的结构工作的。这 8 位数字可以计数到 256，因此能够描述从黑色到白色的 256 种灰度色调，这足以令人在视觉上感到满意。使用 8 位描述每个像素的计算机就能在其显示器上产生完美的黑白照片。能显示出所有这些色调的显示器就称为灰色标度显示器。

人眼可以辨别出由红色、绿色和蓝色这 3 种颜色产生的 150～200 种色调。如果使用 8 位来描述彩色，就只有 256 种色彩，这不够——只达到柔和的张贴画效果，尽管抖动处理能以牺牲质量为代价模拟出更多色彩。

为了在计算机显示器上获得完全的彩色照片效果，必须能够为每种颜色产生出 256 种色调。每种颜色占有 8 信息位，总共需要 24 位，这就是在计算机杂志上见到的 24 位彩色。如果让这三种颜色都产生全部 256 种可能的变化，就能获得 1670 万种彩色色调。

如果只做彩色照片的屏幕润色或类似的屏幕"绘画"，实际上只需要 24 位彩色。对于画线和图片布局，8 位彩色显示器就已足够完美了，因为即使不能在屏幕上显示也还能在打印时定义彩色。

阅读材料 1：数字摄影

数字摄影采用了照相机的摄影技术，运用一个电子感应器将图像作为电子数据记录下来，而不是在相片上产生化学变化。这个感应器或是一个高度敏感的电荷耦合装置（CCD），或者是一个 CMOS 半导体装置。数字存储器装置通常被用来存储图像，以后这些图像还可以被转移到计算机里。

这种方法（数码相机）与传统相机相比，其优势有大大减少了每一张图像的成本，可以立即获取所拍摄的影像来欣赏，可以方便地传输更多的图片，不再需要在暗室冲洗照片。除此之外，数码相机比胶片相机更小、更轻。

最近，数码相机的主要生产商，如尼康、佳能，提倡摄影记者采用数码单镜头反射照相机（SLRs）。以2兆像素获取的图像质量被认为足以保证做报纸或杂志复印品的小图片。把6兆像素的先进数码 SLRs 相机与高端镜头相结合照出的照片，就能够匹配甚至超过基于 SLRs 的35mm 胶片冲洗出来的各个细节，并且令人惊讶的是，最近出现的12兆像素模式能够提供细致的图像，其效果几乎好于所有的35mm 光学相机的图像。

因为数字摄影能产生满意的图像质量及其他的一些优点，越来越多的专业新闻摄影师使用这类设备。

数字摄影也被许多业余的快照摄影爱好者所采用。在他们需要通过电子邮件形式传送图像和上传图像到万维网时，这种相机就提供了极大的便利。

数字摄影应用于天文学领域是在它投入普通的公众生活之前很早就开始的，并且早在1980年就已经完全替代了照相金属板，不仅 CCD 对光的敏感度高于金属板，而且所得的信息还能够下载到计算机上做数据分析。另外一些商业摄影师和许多业余爱好者狂热地搜罗数码相机，因为他们相信数字摄影的灵活机动性，以及更加低廉的长期成本已经足以弥补它最初的价格劣势。

可交替的图像文件格式（EXIF）是一系列专用于数码照相机的文件格式。它指定 TIFF 为高质量的图像格式，JPEG 为节约空间但质量较低的图像格式。很多低端相机仅能传送 JPEG 文件。另一种可能遇到的格式是 CCD-RAW，这是一种没有标准化的格式。

数码相机通常专门配有数字图像处理芯片来把从图像传感器获取的原始数据转换为具有标准图形文件格式的、颜色经过校正的图像。从数码相机摄得的照片通常都优于胶片照片。数字图像处理通常是由能以多种方法处理图像的特定的软件来完成的。很多数码相机都可以检视图像的柱状图，这可以作为一种辅助手段供摄影师更好地了解照片中各点的亮度范围。

照片编辑器是一个专供数码相机使用的图像编辑应用程序。它用来修剪和润色相机拍摄出的照片，并把它们组合起来放入相册或作为幻灯片放映。

数字图像处理使用户可以采用很复杂的算法来进行图像处理，因而就可以在简单的任务中提供更精确的性能，同时又可以用上模拟手段所不可能做到的处理方法。数字图像处理最常见的类型就是数字图像编辑。

常见的图像编辑器只处理位图式的图形，如 GIF、JPEG 和 BMP 的图像。但是，有些编辑器支持位图和插图。其常见的功能有对图像进行手工剪裁、更改尺寸，并用过滤器调整亮度、对比度和颜色。过滤器的种类繁多，可用来取得不同的特殊效果。图像过滤器是一个可以通过以某种方式更改像素的阴影和颜色来改变整个图像或部分图像外观的子程序。过滤器可以用来增强亮度和对比度，以及在图像中增加种类繁多的纹理、色调和特殊效果。

第11单元　动画

计算机动画通常指场景中任何随时间而发生的视觉变化。除了通过平移、旋转来改变对象的位置外，计算机生成的动画还可以随时间进展而改变对象大小、颜色、透明性和表面纹理等。

许多计算机动画的应用要求有真实感的显示。利用数值模型来描述的雷暴雨的形态或其他自然现象的精确表示对评价该模型的可靠性是很重要的。

娱乐和广告应用有时较为关心视觉效果。因此可能使用夸张的形体和非真实感的运动和变换来显示场景。但确实有许多娱乐和广告应用要求计算机生成场景的精确表示。在有些科学和工程研究中，真实感并不是一个目标。例如，物理量经常使用随时间而变化的伪彩色或抽象形体来显示，以帮助研究人员理解物理过程的本质。

剧本是动作的轮廓。它将动画序列定义为一组要发生的基本事件。依赖于要生成的动画类型，剧本可能包含一组粗略的草图或运动的一系列基本思路。

为动作的每一个参加者给出对象定义。对象可能使用基本形状，如多边形或样条曲线进行定义。另外，每一对象的相关运动则根据形体而指定。

一个关键帧是动画序列中特定时刻的一个场景的详细图示。在每一个关键帧中，每一个对象的位置依赖于该帧的时刻。选择某些关键帧作为行为的极端位置。另一些则以不太大的时间间隔进行安排。对于复杂的运动，要比简单的缓慢变化运动安排更多的关键帧。

插值帧是关键帧之间过渡的帧。插值帧的数量取决于用来显示动画的介质。电影胶片要求每秒 24 帧，而图形终端按每秒 30~60 帧刷新。一般情况下，运动的时间间隔设定为每一对关键帧之间有 3~5 个插值帧。依赖于为运动指定的速度，有些关键帧可重复使用。一分钟没有重复的电影胶片需要 1440 帧。如果每两个关键帧之间有 5 个插值帧，则需要 288 对关键帧。如果运动并不是很复杂，我们可以将关键帧安排得少一些。

可能还要求其他一些依赖于应用的任务，包括运动的验证、编辑和声音的生成与同步。生成一般动画的许多功能现在都由计算机来完成。

开发动画序列中的某几步工作很适合由计算机进行处理。其中包括对象管理和绘制、照相机运动和生成插值帧。动画软件包，如 Wave-front，提供了设计动画和处理单个对象的专门功能。

动画软件包中有存储和管理对象数据库的功能。对象形状及其参数存于数据库中并可更新。其他的对象功能包括运动的生成和对象绘制。运动可依赖指定的约束，使用二维或三维变换而生成。然后可使用标准函数来识别可见曲面并应用渲染算法。

另一种典型功能是模拟照相机的运动，标准的运动有拉镜头、摇镜头和倾斜镜头。最后，给出对关键帧的描述，然后自动生成插值帧。

Flash

Micromedia 公司的 Flash 一经推出便在很短时间内就成为网页动画的一种主要制作工具和格式。Flash 成功的部分原因在于它的两重性：既是一种编辑工具又是一种文件格式。Flash 不但比动态网页制作语言 DHTML 更容易学习，而且还集成了许多重要的动画特性，如关键帧的插入、运动路径、动态遮罩、图像变形及洋葱皮绘画效果等。利用如此丰富的功能，不仅可以制作 Flash 电影，而且还可以导入 QuickTime 格式和许多不同格式的动画文件。对每一种格式的文件，Flash 都能处理得更好。以前在网上即使是观看最简单的动画也必须等待很长的时间下载，而 Flash 使用流技术与矢量图改变了这一切。

Flash 电影是专为网页服务的图像和动画。它主要由矢量图形构成，但是也可以包含导入的位图和音效。Flash 能结合交互性以允许输入信息，可以创建能与其他 Web 程序交互的非线性电影。网页设计师可以利用 Flash 来创建导航控制器、动态 logos、含有同步音效的长篇动画，甚至可以产生完整的、富于感性的网站。

Flash 的标准工作环境包括菜单栏、工具栏、舞台、时间轴窗口和工具面板。除了这几个主要的部分，打开 Window 菜单，还可以调出素材窗口等小窗口。

工作区域就是 Flash 的工作平台，它是一个比较大的区域，实际上涵盖了下面要说的舞台及画图和编辑电影片段的工作对象。可以把它看作是后台和舞台的结合。舞台是 Flash 电影片段中各个元素的表演平台，它将显示当前选择的帧的内容。与工作区域不同的是，电影片段发布后只有在舞台上的内容才能被看到，而舞台之外的工作区域中的内容就如同在后台的演员和工作人员一样不会被观众看见。就像戏剧可以有几幕一样，舞台上也可以放下几个场景。

3ds Max

3ds Max 是由 Autodesk Media and Entertainment 开发的，配备周全的 3D 制图应用软件。它运行于 Win32 和 Win64 平台。3ds Max 是一个应用最广泛的供内容创作的专业人员使用的 3D 动画程序。它具有很强的建模能力、一个灵活的"即插即用"结构和悠长的在微软 Windows 平台上工作的传统。它主要被游戏开发人员、电视商业广告设计人员和建筑结构可视化工作室使用。它也用于电影效果制作和电影的可视预览。

除了作为建模和动画工具外，最新版本的 3ds Max 也配置了高级梯度功能（如周围环境的禁锢和渲染）、动态模拟、粒子系统、辐射密度、常规地图的创作与渲染、球形照明、一个直观的和完全定制的用户界面、自带脚本语言及很多其他功能。

多边形建模在游戏设计方面比其他的建模技术更为常见，因为它对个体多边形的非常特殊的控制可以进行极度的优化。而且，在实时计算中也相对快得多。通常，建模者都从 3ds Max 的一个图元开始，使用像斜角、突出、多边形切口那样的工具以增加细节并精细加工模型。版本 4 及后续版本都配备有"可编辑的"多边形对象，它简化了大多数的网眼编辑操作，并提供了达到专用化水平的细分和平滑。

阅读材料 1：视频图像压缩

视频图像压缩是一组用来缩小视频图像文件大小的技术。它包含在称为 codecs（压缩/解压）的产品中，这些方法分为两大类：帧间压缩和帧内压缩。

帧间压缩利用关键帧和 δ 帧系统。δ 帧或"差异"帧，仅记录帧间变化。解压时，CPU 从关键帧和累加的 δ 帧构造各个帧。

帧内压缩是完全在单个的帧内进行的。在帧内压缩时，codecs 使用各种技术把像素转换成更紧凑的数学公式。最简单的技术称为行程长度编码（RLE）。按这种编码法，各行中相邻的相同像素串归在一起。

帧内压缩技术从简单的 RLE 到文档标准，如 JPEG，到一些特殊数学方法，如小波和分形变换。不是所有的 codecs 既用帧间压缩技术，又用帧内压缩技术——一些只用帧内压缩。那些使用两种压缩技术的 codecs 在消除了帧间冗余后，对关键帧和 δ 帧内保留的信息进行帧内压缩。

标准的视频图像源，如摄像机、录像机或光盘播放机把模拟视频信号传送给视频捕获卡，与此同时模拟音频信号被发送给 PC 内的声卡。

捕获卡利用模数转换器（ADC）把模拟视频信号转换成二进制代码。这些视频图像信号能被捕获作为连续的原始视频帧，发送到并保持在系统的随机存取存储器中，在那里用软件对它们进行压缩。

同时，音频信号经历了由声卡的转换器对它进行的模/数转换。此信息也发送到 PC 的主系统随机存取存储器。

在已捕获视频和音频信道后,这些被捕获的信号可以直接存储到硬盘上,也可用软件对它们进行压缩。通常这些数字视频和音频信号被存储为同步的或交错的.AVI文件,放在硬盘上。

MPEG(运动图像专家组)标准是用于压缩视频的主要算法,已于1993年成为国际标准。它是用于数字压缩格式编码视听信息(如电影、视频和音乐)的标准的名称。MPEG与其他视频、音频编码格式相比,MPEG的主要优势在于其文件更小。主要的MPEG标准包括以下类型。

MPEG-1:它的目标是以1.2Mbps的比特率产生视频录制质量的输出。它能在适当的距离上通过双绞线传输线进行数据传输,并且可以以CD-I及CD-Video格式将电影存储在光盘存储器上。

MPEG-2:它最初是为了将广播性质的视频压缩到4~6Mbps而设计的,因此它能装配于NTSC或PAL广播频道。它是用于移动图像及相关音频信息的普通编码的标准。MPEG-2是MPEG-1的扩展集,具有附加的特性、帧格式及编码选项。

MPEG-3:它支持更高的分辨率,包括HDTV。

MPEG-4:它是由MPEG开发的ISO/IEC标准。MPEG-4具有低的帧速率及低带宽,用于媒体解决方案的视频会议。MPEG-4文件比JPEG或QuickTime文件要小,因此它们被设计成能通过更窄的带宽传输视频和图像,并且能够将视频和文本、图形、2D和3D动画层混合在一起。

MPEG-7:它是一种用于固定和移动网络的多媒体标准,允许综合多种范例。MPEG-7通常被称为多媒体内容描述接口。它提供了一种完整地描述多媒体内容的工具设置,被设计成用于普通而非特殊的应用。

MPEG-21:MPEG-21描述了一种标准,而不像其他的MPEG标准那样是用来描述压缩编码方式的,它定义了内容的描述,以及访问、查询、存储和保护内容版权的处理。

第12单元　大数据

"大数据"通常指的是所有来自不同来源的10^{15}~10^{18}内的数据,即数十亿到数万亿的记录。"大数据"能够被更大量地产生,并且产生的速度远远快于传统数据。这些数据也许是无结构的或半结构化的,因此使得它们不适用于以行和列的形式组织数据的关系数据库产品。流行术语"大数据"指的就是这种蜂拥而至的数字化数据。这些数据很大一部分是来自于网站和互联网点击数据流向世界各地的公司。由于这些数据的量过于庞大,以至于传统的数据库管理系统无法在适当的时间内获得、存储和分析它们。尽管每条微博被限制为140个字符,但Twitter公司每天仍然产生了超过8太字节的数据。根据IDC技术研究公司的调研,数据每两年都会增长一倍以上,因此对组织而言可用的数据量不断激增。理解数据的高速增长以便获取市场优势至关重要。

在大约五年以前,各种企业收集的大多数数据都是结构化的事务处理数据。这些数据可以简单地填充到关系数据库管理系统的行和列当中。从那时以后,来自Web流量、电子邮件及社交媒体内容(微博、状态信息),甚至音乐列表及来自于传感器的机器产生的数据带来了数据爆炸。

根据IBM最近的一项研究,在不断创建的所有信息中,90%都是在过去的两年中产生的。

那就是大数据，其中许多如此之大，以至于要高功效地处理已超出常规数据库管理系统的能力。它包括由社交媒体、电子邮件、视频、音频、文本和网页产生的数据。数据来自智能手机、数码相机、全球定位系统（GPS）、工业传感器，社交网络，甚至公共监控和交通监控系统，以及其他一些来源。每次你使用手机、进行谷歌或百度搜索、使用你的信用评级，或者从亚马逊买东西，都扩展了你的数字足迹，创建更多的数据。

随着谷歌、百度和亚马逊这些互联网公司的探索改进他们的网络规划、广告部署和顾客照管的方式，他们创造了创新的大数据分析新应用。今天，最大的数据生产者是 Twitter、Facebook、LinkedIn 及其他一些新兴的社交网络服务。

企业对"大数据"感兴趣，是因为它们与更小的数据集相比包含了更多的模式与有意思的特殊情况，"大数据"具有能够提供新的针对顾客行为、天气模式、金融市场活动或其他现象的洞察力的潜力。然而，要从这些数据中获取商业价值，组织需要能够管理和分析非传统数据及其传统企业数据的新的技术和工具。

大数据分析需要基于云的解决方案——一种能够跟上数据传输量和速度加速增长的联网方法。数据的巨大增长与"云"的开发相一致，"云"实际上只是全球互联网的一个新的别名。随着数据的激增，特别是使用 Apple 和 Android 智能手机及操作系统的移动数据应用程序的增长，对数据传输和存储的需求也增长了。例如，microsoft.com 或 fedex.com 这些网站，它们为全球的客户通信服务，有数以百万计的事务，对于从它们得到的巨大量的数据，没有单个的网站服务器配置能够提供足够的能力。用户想要快速的响应时间和即时的结果，特别是在最大和最复杂的基于网络的商业环境中，以互联网速度操作。

一些制造复杂工业设备的大公司，现在能够从仅仅检测设备故障到预测故障。这使得设备能在一个严重问题形成之前就被更换。

一些大的互联网商务公司，如亚马逊和淘宝，使用大数据分析来预测买家活动，以及了解仓储需求和地理定位。

政府医疗机构和医学科学家们使用大数据用于早期发现和跟踪潜在的流行病。急诊室就诊突然增加，甚至某些非处方药的销售增加，可能是某种流行病的早期警告，使医生和应急部门能启动一些控制和遏制程序。

许多大数据应用程序被创建来帮助使用网络的公司与意想不到的数据量做斗争。只有最大的一些公司有大数据的开发能力和预算。但是数据存储、宽带和计算的成本持续下降，意味着大数据正迅速成为中型，甚至一些小公司的有用和负担得起的工具。大数据分析已经成为许多新的、高度专业化的商业模式的基础。

正如通用电气、沃尔玛和谷歌等大企业使用数据分析，一些小公司也将在较小规模上使用同样一些处理，以提高自己的竞争地位和更有效地利用稀缺资源。

云服务允许个人和企业使用由第三方远程管理的软件和硬件。云服务的例子包括在线文件存储、社交网络、网络邮件和在线商务应用程序。云计算模式允许从任何网络可达的地方获取信息和计算机资源。云计算提供共享的资源池，包括数据存储空间、网络、计算机处理能力，以及专业化的企业和用户应用程序。

云计算的特征包括按需自助服务、宽带网络、资源缓冲池、快速的伸缩性及可测量的服务。按需自助服务是指客户（通常是机构）可以请求并管理自己的计算资源。宽带网络接入提供经互联网或专用网络访问的服务。缓冲资源是指客户从通常位于远程的数据中心等资源池中取出资源。服务规模可以被放大或缩小，而且服务的使用可以被计量并对客户进行相应计费。

　　云计算的最大优势主要体现在，由于数据是在线存储而不是本地存储，用户可以在任何互联网覆盖的地区获取数据，并且当设备丢失、被盗或损坏时，数据依然安全。此外，基于Web的应用程序通常比需要安装的软件更加便宜。云计算的缺陷包括，当应用程序在云端的运行速度大大低于本地的运行速度时，可能会降低应用的性能，以及包括对于使用高带宽公司和个人来说，与数据传输相关的潜在的高费用。同时，还存在在线存储的数据被非授权访问或数据丢失等情况的安全隐患。尽管存在这些潜在的风险，很多人还是相信云计算是未来的潮流。

　　云计算允许企业将其信息技术基础设施的元素外包，或者承包给第三方的供应商。它们只需要支付它们实际使用的计算能力、存储、带宽，以及访问应用程序的相关费用。因此，企业不再需要在设备方面进行大规模的投资并雇佣专员来维护它们。

阅读材料1：使用即时通信进行商务沟通

　　即时通信是一种在互联网上沟通的形式，它提供了从发送端到接收端的基于文本消息的瞬时传输。在两个或更多使用个人计算机或其他设备的用户模式下，即时通信基本上可为共享用户提供实时直接的以语言（书写）为主的在线聊天模式。使用者的文本转达到网络，如互联网，它便可能会解决点对点通信，以及从一个发送者到多个接收者的组播通信。更先进的即时消息可以增强沟通模式，如真人语音或视频通话、视频聊天，并加入媒体的超链接。

　　即时通信已经成为一种与客户和同事沟通的流行方式，而且理由充足。在互联网上实时通信可以简化通信并且可以节省时间和金钱。另外，大多数的即时通信软件是免费的，因此在技术上你并不需要投入更多的资金。

　　但是在使用即时通信之前，要先思考是否真的会提高你的商务沟通效率，或者只是作为你的另一种娱乐方式。所以，首先要考虑使用即时通信的原因。

- ◆ 节省长途话费。使用即时消息与世界其他地区的员工和客户沟通，可以减少你的长途电话费。
- ◆ 进行实时互动。当你有繁忙日程的同事或在不同的办公室的合作者共同进行一个项目的工作时，即时通信方便快捷、沟通简单。
- ◆ 减少垃圾邮件。与电子邮件相比，即时通信不会接收很多的无用信息。
- ◆ 主持聊天和会议。公共聊天室往往是混乱的，很难进行有重点的对话，特别是用于商业目的的对话。虽然即时通信是最常见的双向对话，但是大多数程序可提供会议或聊天设置从而满足你的工作组要求。

　　现在，你已经考虑了即时通信的优势，再来看一下它的缺点。与大多数技术一样，即时通信确实有一些缺点。

- ◆ 专有软件。虽然现在可以使用软件，让你连接到多个即时信息服务，但只有使用相同的即时信息软件，你才能发送消息给其他用户，或者将他们添加到你的联系人列表中。
- ◆ 字数限制。大多数通信程序都会限制消息的字符数。如果你需要大量的信息沟通，电子邮件或电话是更好的选择。
- ◆ 安全。通过聊天或即时通信分享的信息一定要谨慎使用。看看在你的个人资料中什么样的信息可以公开。如果该软件创建了用户的公共目录，一定要找到它。
- 滥用。在工作时间内，员工可能会与他们的朋友尝试去交流，因为即时通信很方便也很有趣。

第13单元 云计算

云计算基础

术语云计算指的是一种由"云"服务器进行的计算，这种计算通常通过互联网来实现。这种类型的网络已经投入使用多年，曾经主要用于建立技术研究和大功耗应用所需的超级计算机。在当时的情况下，它通常指的是网格计算。现在，云计算指的是通过个人计算机、手机，以及其他可以登录互联网的设备来访问基于Web的应用程序和数据。云计算的概念为在任何地方、任何设备上都可以使用的应用和数据。例如，当在线存储或访问文档、照片、视频和其他媒体时，当在线使用程序或应用（如电子邮件、办公软件、游戏等）时，当在线与他人分享观点、想法及内容（如社交网站）时，你都会用到云计算的功能。

采用云计算这个名词的灵感来自于在流程图和示意图中经常用云符号表示Internet。云可分类为公有云、私有云和混合云。企业通常会使用公有云上可用的应用，它们也经常会创建私有云管理自己公司的数据和应用。

云计算是一种技术的营销用语，它提供计算、软件、数据访问、存储的服务而不需要为终端用户提供服务系统的物理位置和配置信息。这个概念与电力网格是平行的关系，在电力网格中终端用户可以节约电能，而无须了解组件提供服务所需的设备或基础设施。

云计算提供商通过互联网提供应用程序服务。这些程序可以从网络浏览器、桌面和移动应用程序进行访问，而商业软件和数据存储在服务器上的远程位置。在某些情况下，遗留应用程序（业务应用程序到现在为止已普遍在瘦客户机窗口计算业务应用）通过屏幕共享技术提供，而计算机资源合并在一个远程数据中心的位置；在其他情况下，整个业务应用已使用网络技术，如AJAX进行编码。

云计算依赖于资源共享以达到聚合性和规模经济。这类似于网络上的公用设施，如电网，云计算的基础是汇聚的基础设施和共享服务这个更广义的概念。

云计算也专注于使共享资源的效用最大化。云资源通常不仅仅为多个用户共享，而且也按需被动态地重分配。它能够把资源分配给用户。例如，一个在欧洲工作时间用特定应用程序（如email）服务欧洲用户的云计算机设备，可以在北美工作时间用不同的应用程序（如Web服务器）重新分配该相同的资源服务北美用户。这种方法将使计算能力的使用最大化，从而减少对环境的破坏，因为对于各种各样的功能只需要较少的电力、空调、架子的空间等。用云计算，多个用户能够访问单个服务器来检索和更新他们的数据，而不需要为不同应用购买多个许可证。

那么云计算是如何运作的呢？

比如说，你是一家大公司的高管。你的特定的职责包括确保所有员工有合适的、可供他们做好自己工作所需的硬件和软件。为大家购买计算机是不够的——你还必须购买软件或软件许可给你的员工，为他们提供所需要的工具。只要你有一个新员工，你就必须购买更多的软件，或者确保你当前的软件许可其他用户使用。如果你面对这样一大笔开销，你会有非常大的压力，每天晚上都难以入睡。

很快，可能又有一个像你这样的高管。这时，你只需要加载一个应用程序而不必为每台计算机安装一套软件。该应用程序将允许员工登录到一个基于 Web 的服务系统，它承载着用户工作所需的所有程序。由另一家公司所拥有的远程机器将会运行电子邮件、文字处理、复杂的数据分析等程序。这就是所谓的"云计算"，它可以改变整个计算机行业。

在云计算系统中有一个重要的工作量的转变。本地计算机在运行应用程序的时候不再需要去完成所有繁重的任务。计算机的网络汇集成的云将会处理这些工作。用户端的硬件和软件的需求会减少。用户的计算机需要运行的唯一的任务就是云计算系统的接口软件，它可以作为简单的 Web 浏览器，云的网络将会完成其他的工作。

如果你已经使用过某种形式的云计算，那么有一个很好的机会。如果你有一个基于 Web 的电子邮件服务的电邮账户，如 Hotmail 电邮、Yahoo 电邮、Mail 或 Gmail 电邮，那么你已经有一些云计算的经验了。你不是在你的计算机上运行了一个电子邮件程序而是远程登录到一个 Web 电邮账户。你的账户使用的软件和存储的信息在你的计算机上是不存在的——它存在于服务系统的计算机云当中。

云计算的服务

云计算提供者按 3 种基本模式提供他们的服务：基础设施即服务（IaaS），平台即服务（PaaS）和软件即服务（SaaS）。

基础设施作为服务，如亚马逊 Web 服务用唯一的 IP 地址和按需分配的一些存储块提供虚拟服务器实例。客户使用提供者的应用程序接口（API）来开始、停止、访问和配置他们的虚拟服务器和存储。在企业里，云计算允许一个公司仅仅支付所需容量的费用，并且一旦需要可在线获得更多的容量。因为这种"支付你所用的"的模式很像消费电、燃料和水的方式，它有时称为公共设施计算。

在云中，平台作为服务被定义为放在提供者的基础设施上的一套软件和产品开发工具。开发人员通过 Internet 在提供者的平台上创建应用程序。PaaS 提供者可以使用一些 API，Web 门户网站或安装在客户计算机上的网关软件。一些 GoogleApp 是 PaaS 的例子。某些提供者不允许他们的客户创建的软件搬离提供者的平台。

在软件作为服务的云模式中，供应商提供硬件基础设施、软件产品，并且通过前端门户与用户交互。SaaS 是一个很广阔的市场。服务可以是任何一种，从基于 Web 的 E-mail 到库存控制和数据库处理。因为服务提供者既拥有应用程序，又拥有数据，因此终端用户可以自由地从任何地方使用该服务。

阅读材料 1：云计算和云存储

1. 云计算

云计算是一种计算，在这种计算中几大组远程服务器由网络连接，以允许集中式的数据存储和在线访问计算机服务。云计算是通过互联网交付的计算服务。

云计算服务模型是软件即服务（SaaS）、平台即服务（PaaS）和基础架构即服务（IaaS）。在软件即服务模式中，可提供预定的应用，连同所需的软件、操作系统、硬件和网络。在平台即服务模型中，可提供操作系统、硬件和网络，并且客户可以安装或自行开发自己的软件和应用。基础架构即服务模式只提供硬件和网络，客户需安装或开发自己的操作系统、软件和各种应用。

云服务有3个显著的特性区别于传统的托管。它是按需销售，通常按分钟或小时；它是有弹性的——一个用户可以在任何给定时间内拥有他们所需的或多或少的服务；而服务完全由提供者管理（客户只需要一台个人计算机和对 Internet 的访问）。虚拟化和分布式计算方面的一些重要革新，以及改善的高速 Internet 访问和疲软的经济已导致了云计算的增长。云供应商正经历每年 50%的增长率。

云计算的目标是对计算资源和 IT 服务提供容易、可伸缩的访问。

通常，云服务是通过私有云、社区云、公有云或混合云提供的。

一般来说，通过公有云提供的服务是经互联网提供的，并由云提供商拥有和管理。针对一般公共服务的例子包括在线照片存储服务、电子邮件服务、社交网站服务等。而且，也可以在公有云中提供针对企业的服务。

在私有云中，云基础设施只为某个特定的组织运营，并由该组织或第三方进行管理。

在社区云中，云服务由多个机构共享，并只向那些机构提供服务。基础设施可以由这些机构或云服务提供商所拥有和运营。

混合云是不同资源缓存方法的组合，如综合公有云和社区云。

2．云存储

云存储是一种数据存储模式，在那里数字化的数据存储在一些逻辑池中，物理存储分布在多个服务器（经常是多个位置），并且物理环境通常由托管公司所有和管理。这些云存储提供者负责保持数据可用和可访问，并且保护其物理环境和运行。个人和机构从提供者那里购买或租用存储容量用以存放用户、机构或应用数据，如图 13-1 所示。

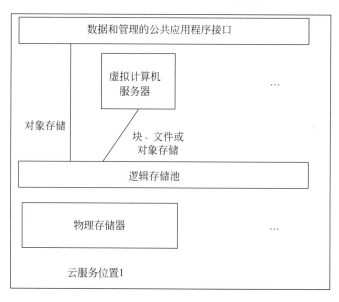

图 13-1　云存储结构

云存储服务可以通过本地协作的云计算机服务、Web 服务应用程序接口来访问，或者通过利用 API 的应用程序，如云台式机存储、云存储网关或基于 Web 的内容管理系统来访问。

云存储是基于高度虚拟化的基础设施，并且从可访问接口、近乎即时的灵活性和可缩放性、多租户和计量资源这几方面来说，它像更广义的云计算。可以从一个远程服务或所部署的本地服务来利用云存储服务。

第 14 单元　计算机网络基础

　　如果用户能把他们的计算机与其他计算机互联在一起的话，他们就可以共享其他计算机上的数据，包括高性能的打印机。一组计算机和其他的设备连接在一起构成的系统称为网络，互联的计算机共享资源这样一种技术概念称为网络技术。联网的计算机可以共享数据、消息、图形、打印机、传真机、调制解调器、光盘、硬盘及其他数据存储设备。

　　计算机网络可用于多种服务，既为公司也为个人。对公司而言，由使用共享服务器的个人计算机组成的网络提供了灵活的方式和很好的性价比。对个人而言，网络提供访问各种信息和娱乐资源的接口。

　　网络可以分为局域网、城域网、广域网和互联网，每一种网络都有自己的特点、技术、速度和位置。局域网覆盖一栋建筑物，城域网覆盖一座城市，广域网覆盖一个国家或大陆。局域网和城域网是非交换网络（也就是没有路由器），广域网是交换网。

　　网络软件由协议或过程通信的规则组成。协议既可以是无连接的也可以是面向连接的。大部分的网络支持分层的协议，每一层为它的下层提供服务。典型的协议栈不是基于 OSI 模式，就是基于 TCP/IP 模式。这两种模式都有网络、传输和应用层，但其他层有所不同。

◆ 局域网（LAN）

　　局域网是专有的通信网络，它可以覆盖一个有限的地域，如一个办公室、一幢建筑或一群建筑等。局域网是通过一个通信信道把一系列计算机终端连接到一个小型机上，或者更普遍的是把若干台个人计算机连接到一起而形成的。复杂的局域网可以连接各种办公设备，如字处理设备、计算机终端、视频设备及个人计算机等。

　　局域网的两个基本应用是硬件资源共享和信息资源共享。硬件资源共享可使网上的每一台计算机访问并使用由于太昂贵而无法为每人配备的设备。信息资源共享允许局域网上每一个计算机用户访问存储于网上其他计算机中的数据。在实际应用中，硬件资源共享和信息资源共享是常常结合在一起的。

◆ 广域网（WAN）

　　广域网是位于不同场所的局域网的组成，这些局域网由互联网上的虚拟个人网络或租用了通信方法的虚拟个人网络把它们相互连接起来。广域网相对于局域网，在覆盖的地理范围上要更大一些，它使用电话线、微波、卫星或这些通信信道的组合来传递信息。广域网用于将局域网互联起来，使位于同一个地点的用户和计算机能够和位于其他地点的用户和计算机通信。许多广域网是为一个特定组织建造，并且是私有的。其余的广域网由 Internet 服务提供商组建，用于将某一组织的局域网连接到 Internet。

　　广域网大多使用租借线路组建，在每条租借线路的末端，路由器连接到局域网的一端，位于广域网内部的集线器连接另一端。租借线路非常昂贵，也可以用价格稍微便宜一点的电路交换或分组交换方式代替租借线路。网络协议包括 TCP（传输控制协议）交付传输和寻址功能。协议包括 SONET（同步光纤网）、MPLS（多协议标记交换）、ATM（异步传输模式）和帧中继的报文，这些协议通常被服务提供商用来提供用于广域网的链接。X.25 是一个重要的广域网早期协议，常常被认为是帧中继的鼻祖，X.25 的许多基本协议和功能（升级版本）现在仍然通过帧中继在使用。

- 城域网（MAN）

城域网的定义不可能是精确的。它介于局域网和广域网之间，具有局域网和广域网的某些特征。一个局域网运行在一个有限的地理区域内，而且几乎始终运转在一个机构中。许多局域网技术可以在 10km 以内的范围低成本地提供非常高的性能。另一方面，广域网使用不受地理范围限制的技术，并且希望连接位于不同管理区域中的各节点。一个 MAN 的目标是为一个超出局域网技术范围的地理区域提供服务，然而它受到某些明确的团体利益的制约，通常为一座城市。一个 MAN 将提供一个机构内各个节点间的互联，同时还提供各机构间的互联。这就意味着 MAN 必须提供与局域网相类似的性能，但却要控制多个管理域之间的相互作用。

区分城域网与局域网或广域网有以下 3 个重要特征。

（1）网络规模介于局域网和广域网之间。一个 MAN 通常覆盖 5~50km 范围的一个区域。许多 MAN 覆盖城市大小的一个区域。

（2）一个 MAN 通常不属于一个单独的机构。MAN 及它的通信链路和设备通常属于一个用户协会或一个向用户出售服务的网络供应商。因此，向每个用户提供的服务级别必须与 MAN 操作员进行协商。

（3）一个 MAN 经常可以作为一个允许共享区域资源的高速网（类似于一个大型局域网）。利用一个到广域网的链接，它也经常被用于提供与其他网络的一个共享链接。

在过去几年中，一些通信公司已经开始城域网业务了，它们先在城市范围内安装光缆，然后将其售给使用它的其他公司，这些公司将他们在市内的各个地点互联起来以建设城域网。可行的方法有很多种，如用电线杆、用地铁隧道或埋在城市的街道下面。这些公司主要是卖高速数字带宽，而且一般来说，不提供增值或其他服务。

这项服务的主要市场是那些需要大量高速通信服务的客户，其范围往往是城市这样的相对小的地理区域。这些客户通常是在一个城市中有多个办事处和其他工厂的大公司。城域网供给者通常提供比其他通信运营商和紧急情况提供备份的另外的路由更优惠的价格。他们还声称能比其他运营商提供更快的安装和更好的服务。

城域网技术已经得到电子电气工程师协会（IEEE）的标准化，并由其开发了我们熟知的城域网标准 IEEE 802.6。这项工作是定义局域网的协会做的，并解决了关于局域网怎样连接横跨几千米或更大距离的悬而未决的问题。

阅读材料 1：通信介质

通信介质承载一个或多个通信信道，并且在发送和接收设备之间提供一个链接。当今通信系统最常用的介质包括双绞线、同轴电缆、光纤等。

- 双绞线

全世界的电话系统和局域网都在使用数百万英里的双绞线。双绞线由成对的铜线扭在一起。这些双绞线通常终端使用塑料的 RJ-11 或 RJ-45 连接器。主要有两种类型的双绞线。在屏蔽双绞线中，电缆线外面包了一层箔屏蔽罩，以减少数据传输时信号噪声的干扰。非屏蔽双绞线没有屏蔽层，它比屏蔽双绞线更便宜，但也容易受信号噪声的影响。

♦ 同轴电缆

同轴电缆是由铜导体、绝缘体、保护箔、金属编织的外部防御物和塑料外壳组成的高容量通信电缆。同轴电缆通常用于承载有线电视信号，因为它的高容量允许同时传送超过 100 个电视频道的信号。它也为数据通信提供了很大的容量，可用于使用双绞线不够传输大容量数据的情况。同轴电缆的带宽超过了 100Mbps，尽管它的带宽较好，但和双绞线比起来，它更不耐用、更昂贵而且工作起来也更困难。

♦ 光缆

光缆是一束极细的玻璃管。每一根比人的头发还细的管子被称为光纤。光缆通常由坚固的内部支撑金属线、多股覆盖塑料绝缘体的光纤和外部坚硬的覆盖物组成。与双绞线和同轴电缆不同，光缆不能传导或传输电信号。相反，激光可通过光纤发送表示数据的光脉冲。光纤接收端的电子设备把光脉冲转换回电信号。每根光纤只有一路通信信道，也就是说至少需要两根光纤才能提供一条双向通信链路。

♦ 无线电和红外线链路

无线电波为移动通信（如蜂窝电话），或者为很难或不可能安装电缆的地方的固定通信（像地理位置偏远且地形崎岖的地区）提供无线传输。无线电波网络可使用的频率为 1M～3GHz。

红外线传输仅使用可见光谱以下的频率范围进行数据传输。红外线传输是光路可达通信的一个例子，即当它工作时，发送信号的发送者和接收者之间必须有一条无障碍的路径。

♦ 微波和卫星链路

微波是可以提供声音和数据的高速传输的无线电波。数据通过空气以类似无线电信号的传送方式从一个微波站传送到另一个微波站。微波传输的缺点是它们只限于视距传输。这意味着微波必须以直线的方式传输，在微波站之间不能有建筑物或高山等障碍物。正是由于这一原因，微波站通常都是把天线置于建筑物、塔或高山的顶上。

通信卫星接收来自地球的信号，把接收到的信号放大后再传回地球。地球上的卫星接收站是使用很大的蝶形天线向卫星发射信号，并接收由卫星传回信号的通信设施。

第 15 单元　无线网络

在无线网络中，计算机通过无线电信号而不是电线或电缆进行连接。无线网络的优点包括移动灵活和没有电线。缺点包括连接速度比有线网络慢，并且会受到其他无线设备的干扰。计算机需要内部或外部无线网络适配器。若要查看计算机是否具有无线网络适配器，请执行下列操作。

通过单击"开始"按钮，再执行"控制面板"→"网络和 Internet"→"网络和共享中心"命令，然后单击"管理网络连接"命令，打开"网络连接"图标，列出计算机中安装的适配器。

在可用的无线网络列表中，可以看到一个显示每个网络无线信号强度的符号。强信号通常意味着无线网络很近或没有干扰。为了获得最佳性能，请连接到信号最强的无线网络。但是，当不安全网络中的信号比启用了安全保护的网络中的信号更强时，为了使数据更安全，最好还是连接到启用了安全保护的网络。若要提高信号强度，可以将计算机移动到距离无线路由器或访问点更近的位置。或者将路由器或访问点移动到远离干扰源（如砖墙或包含金属支撑梁的墙体）的位置。

公共无线网络非常方便，但是如果没有适当的安全保护，连接到公共无线网络可能就很危险。尽可能只连接到要求网络安全密钥或具有某些其他安全形式（如证书）的无线网络。通过此类网络发送的信息是经过加密的信息，这有助于保护计算机免受未经授权的访问。在可用的无线网络列表中，每个网络会标记为已启用安全保护或标记为不安全。如果确实要连接到不安全的网络，请注意拥有适当工具的用户可能会看到你执行的所有操作，其中包括访问的网站、处理的文档，以及使用的用户名和密码。连接到此类网络时，不应处理文档或访问包含个人信息（如银行记录）的网站。

无线设备供应商面临着应用的挑战，这些应用要求支持的带宽越来越大，如 IP 电话、流式视频和视频会议等。为了显著提高总处理能力，802.11a 的支持者必须解决室内射频这一重大挑战。他们必须开发出一种方法，解决目前 2.4GHz 单载波延迟扩展系统的延迟扩展问题。

延迟扩展是由被发射的射频的反射引起的。在这些信号到达某一点（如无线天线）的过程中，它们常常因物体、墙壁、家具和地板等而产生反射，由于路径长度不同，信号到达天线的时间也不同。故需要基带处理器或均衡器"拆分"有歧义的射频信号。延迟扩展必须小于符号速率，或者小于为传输而进行的数据编码的速率。如果不是这样，有些延迟的信号会扩散到下一个符号的传输中。这就给能被持续获得的最大位速率加了个上限。

在采用目前位速率技术的情况下，这个上限为 10～20Mbps。802.11a 标准通过一种称为正交频分多址（COFDM）的创新的调制技术，巧妙地解决了这个挑战，COFDM 早已用于欧洲的数字电视和音频的传输。COFDM 通过两种方法突破了数据位速率的上限：①以大规模并行的方式发送数据；②放慢符号速率，致使每个符号的传输比典型的延迟扩展长得多。为了让所有的延迟信号在基带处理器对数据解调之前"确定下来"，在符号传输的起始处插入保护间隔（有时也称周期前缀）。

无线设备供应商现在的目标是将无线总处理能力提高到 100Mbps 以上。802.11a 标准目前最高能在带宽为 20MHz 的信道上达到 54Mbps，但已有几家公司正在开发和建议 802.11a 标准的高速扩展。这些建议一般都是设想把总处理能力至少提高一倍，达到 108～155Mbps 的某个值。

对于局域网有一个新开发的网络：无线局域网。无线局域网使用电磁波通过空气传送和接收数据，最低限度地减少了有线连接。这样，无线局域网把数据连接和用户移动性结合起来，通过简化的配置，形成了移动的局域网。随着近几年的发展，无线局域网在一些市场领域已经获得了广泛的普及，其中包括健康保健、零售、制造业、仓储业和学术界。这些工业利用手提终端和笔记本电脑将实时信息传送到中央主机进行处理，从而从生产力增加中获益匪浅。

如今，对于广泛的商业客户来说，无线局域网正成为公认的通用连接的替代品。下面描述了通过无线局域网的灵活性和功能可能实现的许多应用中的一部分。

- 医院的医生和护士工作起来更有效率，因为无线局域网连接性能的笔记本电脑及时传递了病人的信息。
- 公司的培训点和大学的学生使用无线连接便于访问信息、进行信息交换和学习。
- 在会议室里的高级行政官因为手头有实时信息可供使用，所以可以做出快速的决定。

日益增长的移动用户也成为无线局域网坚实的后备力量。使用笔记本电脑和无线网络接口卡就可以实现移动访问无线局域网，这就使得用户可以穿梭在会议室、门厅、休息室、自助食堂、教室等不同地方，但仍然可以访问其网络数据。假如没有无线局域网，用户就不得不寻找网络插头了。

无线以太网允许以太网标准用于无线网连接。它也称为 Wi-Fi，虽然从技术上讲，标签 Wi-Fi 只能用于由无线以太网兼容性联盟认可的无线以太网产品。Wi-Fi 认可产品的用户获得保证，他们的硬件将与所有其他认可产品兼容。IEEE 802.11 标准扩充了以太网技术所用的载波侦听多路存取原理以适于无线通信特性。802.11 打算支持相应位于大约 150 米以内的计算机之间以最高速度 54Mbps 通信。无线以太网对那些希望扩充它们的有线以太网的组织是一种正在增长的选择。

WiMAX 是无线数字通信系统，也称为 IEEE 802.16，打算成为无线城域网。WiMAX 对固定站点能提供宽带无线访问达 30 英里（5 万米），对移动站点达 3~10 英里（5000 米~1.5 万米）。相对照，Wi-Fi/802.11 无线局域网标准在大多数情况被限于仅仅 100~300 英尺（30~100 米）。

用 WiMAX 很容易支持类似 Wi-Fi 的数据传输率，但是干扰的问题减少了。倘若无线载波有一个受控制的环境和可行的经济模式，则 WiMAX 既工作在许可的频率，也工作在不要许可的频率上。WiMAX 能够以更常规的 Wi-Fi 协议相同的方式用于无线联网。WiMAX 是第二代协议，允许在更长的距离有更高的数据传输率。

无线网有许多用途。例如，在旅途中可使用计算机发送邮件、接收电话和传真及阅读远程文件。另外，对卡车、出租车、汽车和维修人员与基地保持联系极其有用。尽管无线网容易安装，但也有一些缺点。一般传输效率较低，它比有线局域网慢得多，差错率通常也比较高，并且不同计算机的传输可能会互相影响。

阅读材料 1：互联网上的加密技术

现代网络浏览器所使用的是安全套接层（SSL）协议，来保证像电子商务和电子银行的安全传输。SSL 通过使用一个公钥来加密和一个完全不同的私钥来解密。由于 SSL 加密技术完全依靠密钥，因此，通常衡量 SSL 加密技术的效用或力度就是根据密钥长度，即密钥中包含的二进制位的位数。

早期在网页浏览器上实施 SSL 协议，从最初的 Netscape 3 到随后的 Microsoft Internet Explorer 3，使用的都是一个 40 位的 SSL 加密标准。不幸的是，40 位的编码证明了它在实际操作中太容易被解密或破译。为了破译一个 SSL 通信，只需简单地算出正确的解密密钥即可。

在密码系统中，一种最普通的破译密码的方法都具备强大的解密功能，实际上，可以使用计算机一个挨一个地穷尽计算和尝试每一种可能的密钥值。举例来说，一个两位的密码包括了 4 种可能的密钥值：00、01、10 和 11；3 位密码就包括了 8 种可能的密钥值等。从算术的角度讲，2^n 个组合密钥值就存在于一个 n 位密码中。

与 40 位加密密钥相比，128 位加密密钥在密钥长度增加了 88 个附加位。这样的标准之下就能破译出 $2^{88} \times 2^{40}$ 种可能的密钥值，来应对强大的破译技术。基于过去一段时期内计算机在性能方面的改进完善，安全专家们期望在今后至少 10 年的时间里，128 位加密标准将在互联网上正常运作。

认证是日常生活中的重要部分。缺少强有力的认证制约了电子商务的发展。写在纸上的合同、法律文件和官方信函仍是必要的。如果互联网用于电子商务，强有力的认证是一个关键要求。强有力的认证通常是建立在现代版的一次性密码本技术上的。例如，令牌用来代替昔日的一次性密码本，而且储存在智能卡或磁盘上。数字证书相当于电脑世界的身份证。当一个人想获得数字证书时，他生成自己的一对密钥，把公钥和其他的鉴定证据送达证书授权机构（CA），证书授权机构将核实这个人的证明，来确定申请人的身份。如果申请人确如自己所声称的，证书授权机构将授予带有申请人姓名、电子邮件地址和申请人公钥的数字证书，并且该数字证书由证书授权机构用其私有密钥做了数字签名。

在计算机文件上有不可伪造的签名就好了，但是这种想法存在一些问题。首先，比特流易于复制；只出现这样的一个签名没有意义。即使一个人的签名被制作得难以伪造，例如，如果它是通过指纹的图形图像来实现的，利用当今的剪贴软件将一个有效的签名从一个文件移植到另一个文件非常容易。其次，签名之后，在不留下任何改动迹象的情况下文件很容易被改动。

一个数字签名就是一种电子签名，它被用来鉴别信息发送者或文件签名者的身份，用于确保已发送的信息或文档的原始内容没有被改动。一个数字签名是被附在一份电子文档上的二进制位串，该文档可能是一份文字处理文件或是一封电子邮件。这个二进制位串由签名者生成，它是基于文件的数据和个人的保密口令。数字签名易于传输，不能被其他人所假冒，而且可以自动地被打上时间戳。接收文件的人能证明签名人确实在文件上签名。如果文件被改动，签名者还可以证明他没有在被改动的文件上签名。

公钥加密可用于数字签名，公钥加密使用特殊的加密算法和两个不同的密钥：一个每个人都知道的公有密钥和一个只有一个人知道的私有密钥。公钥算法使用两个密钥对电子文件的内容进行加密。所产生的文件是公有密钥、私有密钥及原始文件内容的混合物。

第16单元　物联网

物联网是未来互联网的一个集成部分，可以定义为具有自动配置能力的动态全球网络结构。它基于标准的共同遵守的通信协议。物联网中的物理和虚拟"物"有身份、物理属性和虚拟人格，并使用智能接口无缝整合到信息网络中。

物联网建立在现有技术的基础上，如射频识别技术（RFID），并且随着低成本传感器的可用性，数据存储费用的降低，可以处理万亿数据的"大数据"分析软件的发展，以及允许为所有这些新设备分配互联网地址的IPv6技术的实现，使物联网正在成为可能。欧盟和中国（被称为遥感地球），以及美国的公司（如IBM的智慧地球计划）正在带头资助和研究物联网。尽管在物联网完全实现之前，挑战还是存在的，但已经越来越接近成功了。

物联网将由成千上万的无线标识物组成动态网络，这些物可以与其他物通信，可以进一步与像普适计算、泛在计算及情景智能等概念整合。物联网的远景是泛在计算与环境智能，这通过完全的通信和完备的计算能力，以及整合各种通信、识别和交互要素来增强。物联网通过不同的概念和技术成分融合了数字世界和现实世界：普适网络、设备小型化、移动通信和业务进程的新模式。

在物联网中，"物"可以成为业务、信息和社会进程的主动参与者。它们可以互相通信与影响，自治地响应"真实／物理世界"的事件并通过触发行为和创建需要人直接干涉或无须干涉的服务，同时也能够通过交换数据和"感知到的"环境信息与环境交互。

以服务形式出现的接口促使与这些"智能物"在互联网上交互、查询并改变它们的状态和与之相关的任何信息，并考虑其安全与隐私问题。

应用、服务、中间件、网络和端点将以全新的方法结为一体。起初，建立全球泛在网络会有商业上和物理上的挑战，而且许多连接物和设备进行双向网络连通的能力有限。因此物联网结构设计应该支持双向缓存和数据同步技术，以及用于虚拟表示连接的物和设备状况的网络端点，这十分重要。这些网络端点既可以用来监控物和设备的位置、条件和状态，也可以向它们发送请求和指令。

来自于个体、群体、团体、物体、产品、数据、服务、过程的任何东西都将通过物联网相互连接。连通性将成为物联网中的一种日用品，所有人都可以用非常低的成本来使用而不仅仅属于任何一个私有实体。在这种情况下，需要建立恰当的有情景意识的开发环境来激发创建服务和适当智能中间件、了解和解释信息、确保防止欺骗和恶意攻击并保护隐私。

从技术角度来看，物联网不是单个新技术的结果，相反，它是几个互补技术的发展共同提供的功能，这些功能组合在一起有助于弥合虚拟世界和物理世界之间的差距。这些功能包括以下几方面。

- 沟通与合作：物体拥有将互联网资源连成网络，甚至实现物体间相互连接，以及使用数据和服务，并且更新物体状态的能力。在此，无线技术如 GSM、UMTS、无线网络、蓝牙、ZigBee 及其他各种无线网络标准都与此紧密关联。
- 寻址能力：在物联网内，物体可以被定位，并通过发现、查找或命名服务进行寻址，因此可以远程查询或配置。
- 识别：物体是唯一可识别的。RFID、NFC（近场通信）和光可读条形码都是这类技术的应用实例。采用这些技术后即使没有内置能源的被动物体都可以被识别（借助于如 RFID 阅读器或移动手机这样的"中介"）。如果"中介"被接入网络，则识别功能就可使物体被链接到与该特定物体相关联的信息，该信息还可以由服务器检索到。
- 感知：物体利用传感器收集其周围的环境信息，记录、转发该信息，或者直接对该信息做出反应。
- 执行：物体包含执行机构（如将电信号转换成机械运动）来操纵其周围环境。这种执行机构可以通过互联网远程控制真实的过程。
- 嵌入式信息处理：智能物体带有处理器或微控制器，以及存储空间。例如，这些资源可被用于处理和解释传感器信息，或者向物体中存入有关物体是如何被使用的信息。
- 本地化：智能物体能知道其物理位置，或者被定位。全球定位系统或移动电话网络是适于实现该功能的技术，另外还有超声时间测量、UWB（超宽带）、无线电信标（例如，相邻的 WLAN 基站或具有已知坐标的 RFID 阅读器）和光学技术。
- 用户界面：智能物体能以适当方式与人通信（无论直接或间接，如通过智能手机）。与此有关的创新互动方式有诸如可触摸的用户界面、基于聚合特性的柔性显示，以及语音、图像或手势的识别方法。

最为特定的应用程序只需这些功能的一个子集，特别是由于实现所有上述功能通常很昂贵，并且需要复杂的技术。例如，在后勤应用中，目前集中在近似定位功能（即上一次读取点的位置）并对使用 RFID 或条形码的物体进行相对低成本的识别，传感器数据（如用于监视低温运输）或嵌入式处理器只被用于不可或缺这些信息的物流应用中，如有温度控制的疫苗运输。

阅读材料 1：物联网的未来

讨论互联网的未来却不提及物联网（IoT）是不完整的。物联网有时也被称为工业互联网。互联网技术广泛地应用在台式计算机、笔记本电脑、平板电脑、智能手机、消费类电子产品、电气设备、汽车、医疗设备、公共事业系统、各种类型的机器，甚至衣服上——几乎所有可以装备传感器的东西。传感器可收集数据并连接互联网，能够用数据分析软件分析数据。

未来的网中网将制定为公有/私有结构并由相互连接的"物"所建立的边界点来动态延伸和改进。实际上，物联网的通信不只在人们之间进行，也在人与环境之间进行。

通信将更多地出现在终端和数据中心之间而不是当前网络的节点之间。存储的成本越来越低，存储能力却不断加强，将使人或物所需的大部分信息都可以本地实现。有了增强的处理能力和永远在线的连通性，终端将成为通信的主角。

终端能够建立本地通信网络也可以作为通信网络之间的桥梁，因而扩展了整体基础设施的能力，尤其是在城市环境中。这或许可以确定网络结构的不同视角。未来的互联网将呈现高度的异质性，因为在功能、技术和应用领域完全不同的东西有望属于同一通信环境。

物联网也会带来切实的商业利益，如进行资产和产品的高精准管理、改进生命周期管理及在企业间实现良好的协作。其中许多益处是通过使用各个物的唯一标识和搜索与发现服务来实现的，使每个物体都可以互动，建立每个物的行为与交互记录。

物联网允许人和物随时随地与任何东西和任何人连接，理想情况下可以使用任何路径/网络和任何服务。这意味着要应对的问题有：在人与物、物与物之间无缝互连的环境中处理收集、内容、集合、计算、通信及连通性等要素。

在物联网景象中，可以预见的是任何"物"将来至少有鉴别身份的唯一方法，可以建立可设定地址的"物"的集合，如计算机、传感器、人、激发器、电冰箱、电视、车辆、移动电话、衣服、食品、药品、书籍、护照、行李等。由于具有寻址和相互通信能力并校验其身份，所有的"物"都能够交换信息，如果有必要也可以确定身份。也可以期待某些"物"在不同的场合中、不同的"人格"下有多个虚拟地址和身份。

许多"物"将具有内嵌的通信能力并能建立与周围其他"物"相结合的本地通信网络。这个特殊网络可以与其他本地和全球的通信网络连接，这些物的"功能"会受到通信能力和环境的影响。"物"可以检索参考信息并根据环境启动新的通信方法。

第 17 单元　计算机病毒

病毒就是一个能够通过把自我附加到可执行文件之后来复制其自身代码的程序，当被感染的可执行文件执行时，病毒代码也能够被执行。这里的"程序"（COM 或 EXE 文件）指的是一个可执行文件。附加可能指的是物理地址添加到一个文件的尾部，插入到一个文件的中间，或者只是简单地将指针放置到病毒可以在磁盘中找到的某处的不同位置。

许多病毒将可自我复制的代码放置到其他程序。这样当被感染的程序被执行时，更多的程序就被这些可自我复制的代码感染了。当这些自我复制代码被一些事件触发时，或许会做出一些对计算机有潜在危害的行为。

计算机病毒种类

宏病毒会感染称为"宏"的指令集。宏实质上是一个微型的程序，所包含的合法指令通常用来自动完成文档和工作表的操作。专家们认为目前存在 2000 多种病毒，但是，病毒所造成的损失是由不到 10 种病毒引起的。在这些病毒中，宏病毒占 90%以上。

现代的特洛伊木马是一种计算机程序，它表面上完成一种功能，而实际上还同时做其他一些事情。特洛伊木马不是病毒，但是可能携带病毒。如果不携带病毒的话，特洛伊木马也可能设计用来破坏数据，或者盗窃口令。

时间炸弹和逻辑炸弹是两种计算机病毒。它们在没有发觉的情况下潜伏在计算机系统中。时间炸弹是一种计算机程序，它在不被发现的情况下，滞留在计算机中，直到某个特定的时刻被引发，如计算机的系统时钟到达了某一天。时间炸弹通常是由病毒或特洛伊木马携带的。逻辑炸弹是一种计算机程序，它是通过某些特定数据的出现或消失触发的。

蠕虫是一种通过安全漏洞来进入计算机系统通常是网络的程序。与病毒类似，蠕虫也可以自我复制，但与病毒不同，蠕虫不需要附着在文档或可执行文件上来进行复制。

计算机病毒的传播

病毒以有趣图片、贺卡或视频音频文件的附件伪装起来。病毒能通过网上下载的形式进行传播。它们可能隐藏在违法软件及可能下载的其他文件或程序中。当运行这些文件或程序时，它就会传播出去。一些病毒能在互联网上在特定时间传播，也许是在某一天的某一时刻。当然，当在互联网上使用电子邮件时，对于不确定的邮件必须要小心谨慎。如果不确定是谁发给你的邮件，就不应该打开它，因为它很可能存在着病毒。

当使用 U 盘、MP3 等时，病毒能够通过 USB 接口进行传播。现今，有很多病毒专门设计用于 USB 接口。当插入 USB 设备时，反病毒软件不能自动扫描 USB 设备，该 USB 设备就会成为计算机的病毒源了。因此，当我们使用 MP3 及其他可移动存储设备时，必须要十分小心。

病毒感染的症状

病毒只要在未被发现的状态下，就可以自由繁殖。因此，常见的计算机病毒的感染并不显现任何症状。有必要用防病毒工具来识别感染。然而，很多病毒有缺陷，确实常常露出马脚。以下是一些值得留意的迹象。

- 程序长度的变化
- 文件日期或时间记录的变化
- 程序装入时间加长
- 系统运行速度降低
- 内存或磁盘空间减少
- 软盘上出现坏扇区
- 异常出错信息
- 异常屏幕显示
- 程序运行失败

- 异常磁盘写入

病毒的危害

病毒造成的主要危害如下。

- 格式化整个磁盘或磁盘的某个扇区。
- 删除硬盘上的重要信息。
- 删除软盘、硬盘和网上的数据文件及执行文件。
- 占用磁盘空间。
- 修改和破坏文件中的数据。
- 降低 CPU 的运行速度。
- 干扰用户的操作及破坏计算机的正常显示。

预防策略

有两种通用的策略来检测病毒。首先，也是迄今为止最常见的病毒检测方法，是使用反病毒软件。这种检测方法的缺点是用户仅仅能够防治那些最新更新定义的病毒。第二种方法是使用一种启发式算法来查找基于一般行为的病毒。这种方法具有检测病毒的能力，反病毒安全企业仍未能为其创建一个签名。

很多用户都安装反病毒软件，在计算机下载或运行可执行文件后可检测和清除已知病毒。它们通过审查计算机内存的内容及存放在固定或可移动驱动器上的文件，将这些文件与病毒库进行比较的方式进行工作。用户必须定期更新软件来修补安全漏洞。反病毒软件也需要经常更新，以获取最新的威胁信息。

恢复策略

一旦计算机已经受病毒危害，在未完全重装操作系统的情况下使用同一台计算机通常是不安全的。然而，在计算机有病毒后，有一些恢复选项。

- 数据恢复

我们可以通过定期在不同媒介上备份数据（和操作系统）的方法来避免病毒所造成的危害，它们与系统不连接，或者由于其他缘由而不可访问，如使用不同的文件系统。这样，如果由于病毒而使数据丢失，可以使用备份重新开始。

- 病毒清除

一种可能性是一个被称为系统恢复的工具，它将注册和关键系统文件重存到预先的检查点。通常，病毒会导致系统挂起，而后来的硬启动将提供当天的一个系统恢复点。

- 操作系统重装

作为最后的努力，如果你的系统上有病毒而反病毒软件又不能清除它，那么就可能需要重装操作系统了。如果这样做，硬盘驱动器要彻底删除，并且从媒介（如只读光盘）来安装操作系统。不要忘记，在重装操作系统前首先要把重要的文件备份。

阅读材料 1：计算机病毒年鉴

计算机病毒是 20 世纪 80 年代计算机飞速发展带来的结果。计算机病毒这一名词起因于在计算机上传染的有害程序和生物学上病毒的相似性。计算机病毒是自己复制和传播的一个程序，通常它的存在会造成明显的损害。

1981 年：苹果病毒 1、2 和 3 号是首批被发现四处流传的病毒中的几个。它们在 Apple Ⅱ 操作系统中被发现，并借助盗版计算机游戏通过 Texas A&M 传播。

1983 年：弗雷德·科恩在他撰写的论文中，正式定义计算机病毒为"可以通过包括进行自我复制（进化）在内的手段，感染其他计算机程序的计算机程序"。

1986 年：两位名为巴斯特和阿姆捷特的程序员把软盘启动扇区的可执行代码修改为他们自己设计的代码，从而感染任何一张被访问的 360KB 软盘。

1987 年：Lehigh 病毒，早期的文件病毒之一，感染 command.com 文件。

1988 年：最常见的病毒之一"耶路撒冷"被释放。每逢星期五的日子发作，病毒会感染带有.exe 和.com 后缀的文件，并把当天所运行的程序都删除掉。

1990 年：Symantec 公司发布了诺顿防病毒软件 Norton Antivirus，它是由大型公司开发的最早的反病毒软件之一。

1992 年：黑夜复仇者转化引擎（DAME）被创造，它是一个能把普通病毒转变成多态性病毒的工具包。病毒创造实验室（VCL）也面世了，它是首个病毒制作工具包。

1998 年：Strange Brew 是首个破坏 Java 文件的病毒，虽然现在已经无害但仍能在市面上找得到。这种病毒修改 CLASS 文件使得文件代码的中间段含有一个文件自身的副本，并且从病毒修改区域开始执行。

1999 年：梅丽莎病毒，W97M/Melissa 通过执行邮件附件文档中的宏，转发病毒文档到受感染用户的 Outlook 地址簿中的 50 个人。病毒也感染其他 Word 文档，并随后作为附件发出。梅丽莎传播比之前任何一个病毒都快，大约感染了 100 万台计算机。

2000 年："爱虫"病毒，也被称为"我爱你"病毒，像梅丽莎病毒一样，通过 Outlook 邮件软件传播。病毒本身是一个 VBS 附件，并会删除包括 MP3 和 JPG 在内的文件。它还会发送用户名和密码资料给病毒作者。

2002 年：LFM-926 病毒于 1 月初出现，当它感染 Shockwave Flash（.swf）文件时会显示文字信息"登录 Flash 电影"。以名人名字命名的病毒继续发展，其中较知名的有"Shakira""Britney Spears"和"Jennifer Lopez"。Klez 蠕虫病毒是一个例证，表明了一类增长趋势，即通过电子邮件传播后使文件超过有效负载，复制原作的隐性副本并试图关闭常见的反病毒产品。Bugbear 蠕虫病毒也于 9 月首次出现。它是一个包含多种感染机制的复杂蠕虫。

2003 年：1 月，相对良性的蠕虫病毒"Slammer"成为当时传播最快的蠕虫病毒，在大约 10 分钟之内感染了 75 000 台计算机，在被感染的头一分钟内每 8.5 秒它的数量就翻一倍。"Sobig"蠕虫病毒成为第一个加入垃圾邮件大军的病毒。受感染的计算机系统有成为垃圾邮件转发点的可能。

2004 年：5 月，大约有一百万台运行 Windows 的计算机受到传播速度极快的 Sasser 病毒感染。受害者包括商业用户，如英国航空、银行及政府办公室，也包括英国海岸防卫队。蠕虫并没有对计算机或数据造成不可挽回的损害，但它使得计算机的运行速度减慢并造成无端的退出或重启。

第 18 单元　办公自动化

办公自动化（OA）是应用计算机和通信技术去改善办公人员和管理人员工作效率的技术。

在 20 世纪 50 年代中期，这一术语就被用作各种数据处理，主要是文书工作自动化的同义词。经若干年搁置后，20 世纪 70 年代中期该词再次被用来描述字和文本处理系统的交互使用，这种系统后来又与强有力的计算机结合导致所谓的"未来综合电子办公室"的出现。

基于个人计算机的办公自动化软件在许多国家已经成为电子管理不可缺少的组成部分。文字处理程序取代了打字机；电子表格取代了账簿；数据库程序取代了传统的纸选票、库存品和职员列表；个人管理程序取代了纸日记簿等。

办公自动化包括以下 6 种主要技术。

- 数据处理——通常指由计算机计算的数值型信息。
- 字处理——文本型信息，即文字与数字。
- 图形——可以是数字和文字形式信息，将其从键盘输入到计算机，再以图形、图表、表格或更易于理解的视觉直观形式显示在屏幕上。
- 图像——画面形式的信息。取实际画面或照片输入到计算机并在屏幕上显示。
- 声音——语音形式的信息处理。
- 网络——用电子装置把计算机和其他用于处理数据、文字、图形、图像和声音的办公室设备连接在一起。

主要生产厂家最初出售的系统是针对办公人员和秘书的，这些设备主要用作字处理和记录处理（保存像名称和地址那样的小型顺序文件，最后经分类和合并形成信函）。现在，注意力已集中在直接为经理和业务人员提供办公自动化系统上了，这些系统强调管理通信功能。

当今的机构配置了各种各样的办公自动化硬件和软件，包括电话及计算机系统、电子邮件、字处理器、桌面印刷系统、数据库管理系统、双向电缆电视、办公室对办公室的卫星广播、在线数据库服务、声音识别及合成系统。这些设备都力图使目前手工完成的任务和功能自动化。专家认为达到办公自动化的关键在于综合性——将各硬软件紧密结合成一个完整的系统，使得信息处理和通信技术应用最充分，人的干预最少。

现代商业活动的节奏急切地需求关键性的信息。与此同时，官方和商家也需要大量的文档。因此，现代商业办公室都在对传统的办公方法进行研究与调整，以找到可以随时随地获取信息和传递信息的较好途径。这些部门正在寻求用来产生、记录、处理、归档、交流或发布信息的最有效方法。现代技术把办公自动化作为一个经济解决方案的基础。现在和将来的办公室系统都将致力于开发集成信息处理网络，这种网络能把一个公司日常处理的业务有效地汇集在一起。

（1）文字处理。文字处理是指使用计算机创建、编辑、打印文档的方法和程序。大多数标准的字处理程序具有这些特点：支持脚注和邮件合并功能，但不支持表格或列。界面采用可定制的工具条，并且编辑屏采用可缩放的草稿模式，可选择性地显示标题、脚注和页脚。不可编辑的打印预览方式显示整个页面或页的外观。当选择一种新语言时，字体、键盘布局和输入方向均可变化，并且键盘布局可以定制。

文字处理软件各不相同，不过，所有的文字处理软件都支持一些基本功能。只支持这些基本功能（可能还有其他一些功能）的文字处理软件称为文本编辑程序。不过，大多数文字处理软件还有另外一些功能，以便于用户对文档进行更复杂的操作及格式化。这些高档的文字处理软件有时也称为全功能文字处理器，它们通常具有以下功能。

- 文本插入与删除
- 剪切、粘贴和复制
- 页面尺寸和边距
- 查找和替换
- 自动换行
- 打印
- 文件管理
- 字体说明
- 脚注和交叉引用
- 图形
- 页眉、页脚及页码
- 版式
- 宏
- 合并
- 拼写检查
- 内容及索引表
- 词库
- 多窗口
- 所见即所得

（2）桌面出版系统。桌面出版系统是一种重要的应用软件，它表明了个人计算机与办公自动化之间的关系。该术语是 1985 年由 ALDUS 公司创始人保罗•布赖纳德提出的。桌面出版系统最简单的形式是使用个人计算机制定和打印各式各样的排版和接近排版质量的文件。此工序包括排字、绘图、形成文件，最后在激光打印机或排字机上出版。

（3）视频会议。视频会议是将远程地点通过单向或双向电视相连接。如果会议室内能安装所需的声像设备，通过举行远距离会议可代替面对面的会议，可节约旅行所花费的时间和经费。许多商业公司正试验通过视频会议召开销售和部门会议。这种花费仍然很高，尤其是当视频会议使用直接双向卫星信道连接时。

（4）可视图文。可视图文是电子出版的一种形式，它由与终端和个人计算机相连的数据库组成。可视图文按帧或页的信息块来存储和显示信息。可视图文提供一个简单的用户接口，它可显示多页包含文本和图形的信息，并向未经训练的用户提供访问一个或多个数据库的简单途径。

使用可视图文的办公室能用电子方式颁布标准操作过程指南、销售文献、室内通信稿、公司范围内的电话号码表及其他定期变化的信息。有些可视图文系统允许用户处理电子事务，例如，办公室职员可以检查办公室用品数据库的清单，然后将各种用品排序存放。

阅读材料 1：Excel

Excel 是一个电子表格程序，用于组织数据、完成计算、做出决策、将数据图表化、生成专业水准的报告、在 Web 上发布组织好的数据及在 Web 站点上存取实时数据。使用 Excel，可以在工作表上创建一个超链接用在本地网络上，即本单位的企业网或互联网上访问其他 Office 的文件。也可以把工作表存储为静态或动态网页，这些网页可以从浏览器浏览。静态网页在用户查看时不能修改。动态网页则在浏览器中给浏览者提供了许多 Excel 功能。另外，可以生成并使用查询，从一个 Web 页上检索信息并将它直接放在工作表中。

电子表格可以用于各行各业。例如，会计可以用电子表格核对财务报表和编制工资表；在商业上，电子表格可用来编制预算和对报价进行比较；教师用电子表格记录学生的考试成绩；科学工作者利用电子表格分析试验数据；家庭主妇利用电子表格记录家庭开支。因此，可以看到电子表格应用的广泛性和灵活性。更为重要的是，电子表格可以使我们摆脱重复和费力的计算，使我们集中精力对计算结果做分析和评价，充分发挥我们的专业潜力。

一个 Excel 文件是一个工作簿，它可以包含几个工作表。在工作表中，在每行与每列的交叉点，有一方形区域，称为"单元格"。单击单元格，在它周围出现黑框，该单元块即被激活。每一单元块可以由行数字和列字母表示。在 Excel 中，整数数据包括字母型数据、数值型数据、日期型数据和逻辑型数据，此外，注释和公式也可以输入到单元格中。通过选择操作数，可以进行修改和编辑，被选定的操作数被加重显示。

♦ 选择单元格中的内容

在要选定的内容的始端按下鼠标的左键并保持住，然后把它拖到要选定的内容的末端。

♦ 选择单元格

单击单元格，此单元格即被激活（被选定）；单击行首或列首，该行或该列即被分别地选定。

♦ 选择连续的单元格

把鼠标指针放在单元格的左上角，然后把它拖到某一区域的右下角，就可以选定这一区域中的连续的单元格。

♦ 选择非连续单元格

按下 Ctrl 键的同时单击那些单元格。

♦ 选择当前工作表的全部内容

按下 Ctrl+A 组合键或单击位于该工作表左上角的"全选"按钮。

在 Excel 中，公式和函数用来对电子表格中的数据进行统计和计算，当数据改变时，计算的结果会自动地更新。所有的公式都以"="开始，它包含算术操作符、文字操作符、比较操作符和引用操作符——总共 4 种操作符。函数是在 Excel 中预定义的内置功能，包括 SUM、AVERAGE、COUNT 和 MAX 等。

电子表格中的各种统计图表示单元块中的数据，使这些数据简单且直观、易于理解；同时，当数据改变时，这些图自动地跟着变。

在 Excel 中，图包括"嵌入图"和"独立图"。前者与工作表一起放置、显示和打印，后者是独立产生的工作表，它可以独立于原来的数据表，分开打印。在 Excel 中有二维图和三维图。利用工具栏中的"图引导"或"插入"菜单中的"图画"命令按钮，它们可以逐渐地产生。使用者可以编辑这些图，有折线图、条形图、饼图和其他的图供选用。

Excel 不仅能做简单的数据管理和比较，还可以建立数据库，能利用数据表、增加或删除记录，并做排序、筛选、分类和汇总数据的工作。

第 19 单元　虚拟现实

虚拟现实（VR）是一项 20 世纪 80 年代后出现的新技术。然而仅仅几年，就已经渗透到各个领域——科学、技术、工程、医学、文化、娱乐，并且它的应用潜力的确引人注目。虚拟现实的定义之一：虚拟现实是对由计算机生成的、能够实时追踪用户的多感官信息的一种笼统表述。芝加哥电子可视化实验室（EVL）采用的就是这种定义。换句话说，虚拟现实意味着用户被计算机生成的图景所包围，这些图景要依据用户的运动（如视角的转换）而产生变化。

借助现有的成果，利用计算机硬件和软件及先进的传感器，研究人员能够创造三维人工虚拟环境，你可在其中漫步，四下观望和触摸环境中的每个物体。这种环境中的一切事物均与其他物体如此真实和谐地结合在一起，以致你可能觉得自己处在一个真实的物理环境中，但实际上你却是在一个虚拟世界中漫游。

追踪是任何一种浸入式系统的一个重要方面。为了创造一种真实环境的幻象，而不仅仅是计算机图景，这些图景必须随着虚拟世界中的用户位置的变化（由此产生的视角变化）而变化。为达到这一目的，系统必须明确用户眼睛的位置及它们的观察方向。这样就可以使图景所呈现的视角随着用户的运动而变化，这就使用户对目标产生的印象与在现实环境中相同。这是十分困难的，在六面体系统的完全浸入式环境中则更加困难。

HMD 和 BOOM

头盔式显示器（HMD）典型地也包括听觉通道用的耳机，以及测量用户的位置与方向的一些设备，在 20 世纪 90 年代的大多数时间里，它们一直是主要的 VR 视觉设备。

在 HMD 中，投影机把实时图景投射到附着于用户所戴头盔内的小屏幕上。最初，HMD 显示平面图片，追踪用户的头部运动。但在现代 HMD 中，由于这项技术变得更加便宜，立体图片则成为标准。通常说来，HMD 或分辨率过低或过于笨重。由于一个 HMD 一次只能由一个人使用，因此对图景的讨论就非常困难。

为了提高图景的质量，Fakespace 发明了 BOOM（双目全方位监视器）。机械臂上固定着非常小的监视器，就像使用双目望远镜一样，用户使用监视器观察。用户移动机械臂时改变了视角，因此实现了追踪。当一个用户不用这个 BOOM 后，另一个用户可以从相同的视角观察到相同的事物，这一点 BOOM 要优于 HMD。因为使用了真正的监视器，所以分辨率很好。

HMD 和 BOOM 是类似的设备，其类似性在于用户完全沉浸在虚拟环境中而不查看他/她的实际周围环境。BOOM 解决了 HMD 的几个局限（如分辨率、重量、视野），但其代价是要求用户站或坐在固定的位置而减少了沉浸的感觉。

CAVE

具有墙后投影生成图像的房间的概念是 1992 年在 EVL 发明的。墙上的图景通常是立体的，以便能表达出深度。与普通图形系统相比，它的主要优势在于用户被投影图景所包围，这意味着这些图景是用户的主要视野。它通常被称为 CAVE，即洞穴式自动虚拟环境。第一台 CAVE

是由 EVL 的全体教职员和学生共同创造的。

从那时起,这种虚拟现实的背投方法赢得了强烈的追从。它具有很多明显的好处:可以轻易地让共处一室的数人同时看到图景;它可以在同一环境中很容易地混合真实物体和虚拟物体。而且由于用户将自己的手脚看成虚拟世界的一部分,它们就会产生一种强烈的身临其境的感觉。

当今各种各样的类 CAVE 环境存在于世界各地。它们中的大部分已达到四投映面,图景通常会被投影到三面墙壁和地板上。添加天花板上的投影会使人产生一种完全处于封闭的虚拟世界中的感觉。投影到房间的全部六个面就可以允许用户向各个方向转身并观察。于是,他们的感觉和体验就不会受到限制,这是实现完全浸入所必需的。浸入式可视化环境可被用于从科学到艺术、从工业模拟到教育等各个领域。

当 HMD 需要用户在虚拟空间交互作用时(他们不能在他们的"实际"环境中彼此看到),沉浸室提供的重要好处是允许用户在实际世界中进行交互、讨论和分析。然而,在沉浸室内生成情景的计算费用是十分高的,必须以高刷新率为沉浸室中的每面墙生成两个图像。另外,每面墙需要高质量的投影仪,并且因为使用了背投,为了投影长度需要配给大的空间。因为成本超过 50 万美元,沉浸室仅存在于少数大的研究机构和公司。

VR 响应工作台

VR 响应工作台的操作方法是把一个计算机生成的立体图像投影到一面镜子上反射出来,然后投射到一个台子表面上,围绕此台子的一群用户观察此台子表面。使用有快门的立体双筒镜,用户观察显示在桌面上方的一个三维图像。通过使用磁传感器跟踪小组长头与手的移动,工作台允许改变视角并且与三维影像进行交互。小组其他成员随着小组长的操纵而观察影像,这样便于在观察者之间就影像进行交流,并由小组长确定进一步的一些行动。交互作用是使用语言识别、姿势识别用的连指手套及模拟的激光指示器来完成的。

VRML

虚拟现实建模语言(VRML)是一种描述交互式三维对象及客观世界的文件格式。VRML 用于 Internet、Intranet 和本地客户系统。VRML 也是组合三维图形和多媒体的通用转换格式。VRML 的应用领域很广,包括工程、科学可视化、多媒体演示、娱乐教育标题、Web 页面和共享虚拟世界等方面。

VRML 能表示静态及动态的三维和多媒体对象,并通过超链接到诸如文本、声音、电影及图像等其他媒体。VRML 浏览器,同制作 VRML 文件的制作工具一样,广泛应用于许多不同的平台上。VRML 支持扩展模式,即允许定义新的三维动态对象,并允许应用程序组开发在基本标准上加以扩展的可互操作的记录过程。在 VRML 对象和通常使用的 3D 应用程序接口特性间有相应的映射。

VRML 规范定义了一种可将 3D 图形和多媒体结合起来的文件格式。总体上,每个 VRML 文件是三维的基于时间的空间,包含能够通过不同机制动态修改的图像和音频对象。VRML 定义了对象及支持合成、封装和扩充机制的初始集。

阅读材料 1:计算机与新闻业

技术的发展也惠及了报刊业新闻工作。数十年来,排字工人将记者手写的报道转变成一栏栏整洁排列的铅字。今天，记者使用计算机和文字处理软件，打印出他们的报道，并进行初步的拼写与语法检查。

记者的报道通过计算机网络送交编辑，而编辑则使用相同的软件对报道进行编辑，使其适应时间与版面限制。排字过程已经被桌面排版软件和计算机直接制版技术所取代。用桌面排版软件生成的数字版面发送到光栅图像处理器，由其转换为组成文字和图像的点。在版面经过光栅图像处理器处理之后，版上成像机使用激光将点蚀刻到实际的印版上，印版随后被装到印刷机上来印刷那些版面。计算机直接制版远比排字要快且灵活，出版者因此可以为适应新近发生的情况而做出最后一分钟的改变。

个人计算机也为新闻采集过程增加了一个新的方面。记者曾局限于进行个人采访、观察和在图书馆里搜集事实，但现在可以广泛地利用互联网资源和电子邮件。网站和在线数据库可提供关于各种话题的背景信息。其他的资源包括新闻组和聊天室。记者通过新闻组和聊天室可密切关注对时事的舆论，确定潜在的消息源。

大型网络大多保持交互式网站，用以提供在线民意测验和旨在收集浏览者看法的公告板。虽然在线民意测验的调查对象并非具有代表性的人口样本，但他们可以帮助新闻机构估计浏览者的看法，并确定新闻报道是否全面、有效。

电子邮件改变了记者与同事及消息人士的交流方式。它常常是与遥远地点或遥远时区的人们进行联系的唯一切实可行的方法，而且对不愿暴露身份的消息人士也行之有效。然而，"核实"电子邮件消息来源可能比较困难，因此记者在没有大量佐证的情况下往往不会依靠这些消息来源。

对广播业新闻工作来说，数字通信在今天的"现场直播"电视报道中发挥着重要作用。大多数新闻机构都保持有远程制作车，有时称为"卫星新闻采集车"。它们行驶到突发新闻的现场，竖起天线，然后开始播报。这些成套完备的移动制作设施包括摄像控制设备、录音与录像设备、卫星或微波发射机等。

然而，现场报道不再需要装满设备的卡车了。视听编辑设备与摄像机已经数字化，使它们比以前易于使用，而且大小适合装入小提箱，新型的"背囊新闻记者"携带有小型数码摄像机、笔记本电脑和卫星电话。

背囊新闻记者可用"火线"电缆将其小型摄像机连接到笔记本电脑上，把拍摄的视频镜头传送到硬盘，然后使用普通用户级视频编辑软件对视频镜头进行编辑。由此产生的视频文件经压缩后通过卫星电话传送给新闻编辑室的技师，新闻编辑室的技师解压缩后即可播出这些视频文件——这一切不过是数秒钟的事。

背囊新闻记者使用小型摄像机和压缩手段的一个不利方面是，视频质量通常不如使用专业摄像机拍摄的图像清晰。标准要求高的新闻机构对于使用这种质量较差的视频图像开始犹豫不决，但他们发现观众宁可实时观看低质量的图像，也不愿事后观看高质量的图像。对许多观众来说，图像边缘有点粗糙正好使镜头显得更具说服力，更有"亲临其境"的感觉。

计算机、互联网及通信技术使即时播出全球现场报道成为可能。但是，对于现场报道不是没有争议的。手拿麦克风赶到灾难现场的记者，没有什么时间深思、核实和各方查证，因此极其严重的错误、诽谤性图像或令人反感的视频镜头有时便会出现在新闻报道中。技术为新闻记者提供了采集与报道新闻的一系列功能强大的工具，但也使他们更负有责任进行准确的、对社会负责的报道。

第20单元 人工智能

人工智能的定义在两个主要方向上有所不同，一个关心的是思维过程和推理，另一个则强调行为。此外，有些定义用人类表现来度量成功，而另一些则用智能的一个理想概念来度量，我们把它称为合理性。如果一个系统能做正确的事情，那么它就是合理的。所有这些给出了我们在人工智能中要寻求的4个可能的目标：像人类那样思维的系统、合理思维的系统、有人类那样行为的系统及有合理行为的系统。

人工智能的目标

- 像人类一样的行为

由阿伦图灵提出的图灵试验是要对智能给出一个满意的、可操作的定义。图灵把智能行为定义为在所有认知工作中能达到人类水平的行为能力，这种能力足以能戏弄询问者。粗略地说，他提出的试验是计算机要由一个人通过电传打字机来向其提出问题，如果这个人不知道另一端是人还是计算机的话，那么该试验就算通过了。为计算机编写程序让其通过这种试验有许多工作要做，计算机必须具备这些能力：自然语言处理——以使其能用英语（或某种其他人类语言）成功地交互；知识表达——以存储询问之前和询问过程中的信息；自动推理——以利用所存储的信息回答提问并推断出新的结论；机器学习——以适应新的环境并发现和推断模式。

- 像人类一样的思维

如果我们说一个给定的程序能像人一样思维的话，我们必须有一些方法能确定人是怎样思维的。我们需要知道人脑内部的实际工作情况。要做到这一点有两种途径：通过自省——在我们自己的思维流逝的瞬间努力抓住它，或者通过心理实验。一旦我们有了一个足够准确的智能理论，就有可能把该理论表示为计算机程序。如果程序的输入/输出和时机选择的行为与人类的行为相当，那就证明某些程序的机制对人类也可能有效。认知科学的交叉学科领域把人工智能中的计算机模型和心理学中的实验技术结合在了一起，以试图建立起准确的、可以检测的人类智能活动理论。

- 合理的思维

19世纪后期和20世纪初期，正规逻辑的发展为世界上所有事物及彼此之间关系的陈述提供了一套精确的符号。到1965年，出现了这样的程序：如果给定足够的时间和内存，它们就能用逻辑符号来描述一个问题并能求出问题的解（如果解存在的话）。

这种方法有两个主要的障碍。首先，对非正规知识用逻辑符号所要求的正规术语进行表达是不容易的，尤其是当知识不是100%确定的时候。其次，在能够用"准则"求解问题和实际求解之间有很大的区别。即使只有几十个论据的问题也能耗尽任何计算机的计算资源，除非它有关于先进行哪一步推理的某种指导。

- 合理的行为

合理行为是指在给定的信念下能达成目标的行为。智能体具有意识和行为。在这种方法中，人工智能可以被认为是对理性智能体的研究和构造。在关于人工智能"思维法则"的方法中，所强调的全是正确性的推理。做出正确的推理有时可以作为一个理性智能体的一部分，这是因为一种方法能够合理地实施也就能够对给定行为达到某人目标的结论进行逻辑推理，然后再按照该结论去行动。另一方面，正确的推理并不是合理性的全部，因为经常有这样的情况存在：在没有证明要做的事情是正确时，有些事情还必须得做。

人工智能的研究课题

人工智能的基本课题包括取得知识、表达知识和运用知识。在应用领域，目前的研究主要集中在这些课题上。

◆ 解决难题

这里的难题是指那些根本没有算法或算法在计算机程序中无法执行的难题。例如，路径计划、电力调度、股票市场分析、机器人行动计划等，以及在游戏中有一些难题，如汉诺塔问题、农夫过河问题、八个数问题、八皇后问题和旅行商问题。

◆ 自动翻译

自动翻译是指进行两种不同语言之间翻译的计算机程序。机器翻译不仅仅是"查字典"和"逐字"翻译。真正的翻译应建立在理解语义学和语法规则的基础上。

◆ 智能控制和智能管理

智能控制意味着将 AI 技术用于控制系统来解决诸如错综复杂、不完全、不明确或不确定的问题。随着将 AI 引进管理系统、信息管理系统、办公自动化系统和决策支持系统，其功能和技术能够基于专家系统、知识工程、模式识别和人工神经元网络进行整合，从而形成新一代的计算机化的管理系统。

◆ 智能决策

随着 AI 与决策过程相结合，专家系统作为一个智能成分，与模型库、方法库、数据库和知识库一道，逐步形成最具智能化的决策系统。

◆ 智能模拟

智能模拟是指动态模型试验。根据 3 种不同形式的知识——描述性知识、目标知识和处理知识，产生另一种形式的知识——结论性知识。AI 技术用于整个模拟过程，包括建立模型、实际操作和结果分析，以指导和改进模拟模型。

◆ 机器学习

将人工智能技术运用到软件工程已经取得了一些令人振奋的成果。在众多的人工智能技术中，机器学习方法在过去的 20 年里已被用于软件开发。机器学习解决这样一个问题，即如何借助经验来编程从而提高完成一项任务的绩效。它致力于创造和编译与设计和构造非自然信号相关的可证实的知识。机器学习算法已被用于许多不同的问题域。一些典型的应用是数据挖掘问题，在这些问题中，大型数据库包含有价值的能被自动地发现的隐含的规律性；那些缺乏用来开发有效算法所需要知识是无法被充分理解的问题域；或者是程序必须动态地适应不断改变的环境的域。

阅读材料 1：专家系统

依赖人类领域专家的知识作为系统的问题求解策略是专家系统的主要特征。专家系统是使用经过编码的知识来解决专门领域中的问题的一组程序，这个专门领域通常需要人的专门知识。专家系统的知识从各种专家源获得，并且以适合于系统在其推理过程中使用的形式编码。

专家知识必须从专家或其他专门知识源，如教科书、杂志文章和数据库中获得。这类知识通常需要在诸如医学、地质、系统配置或工程设计等一些特定领域中的许多培训和经验。一旦已收集了足够多的一批专家知识，它必须以某种形式进行编码，放进知识库，然后进行测试，并且在系统的整个生存期间不断改进。

专家系统在以下几个重要方面与常规计算机系统不同。

（1）专家系统使用知识而不是数据来控制求解过程。大部分使用的知识实际上是启发式的，而不是算法型的。

（2）知识作为一个与控制程序分开的实体进行编码和维护。因此，它不与控制程序一起编译，这就允许增量式地增加和修改知识库，而不需要重新编译控制程序。

（3）专家系统能够解释一个特定的结论是怎么得到的，以及在咨询过程中为什么需要所要求的信息。

（4）专家系统对知识使用符号表示法，并且通过符号计算进行推理，这非常类似于自然语言处理。

（5）专家系统经常用元知识进行推理，即它们用有关它们自身的知识及它们自己的知识范围和能力来推理。

专家系统的推理应受到公开检验，提供问题求解状态的相关信息及程序做出选择和决定的解释。对医生或工程师等人类专家来说，如果要他或她接受计算机的建议，合理的解释很重要。实际上，很少有人类专家愿意听取他人的建议，更何况是一台不能说明理由的机器。

人工智能和专家系统设计的探索性本质要求程序便于原型化、测试和修改。人工智能的编程语言和环境应支持这种反复迭代的开发方法。例如，在完全的产生式系统中，单一规则的改动不会影响全局的句法。规则可以添加，也可以删除，无须过多更改整个程序。专家系统的设计者们通常认为知识库是否便于修改是衡量系统是否成功的一个主要因素。

专家系统的另一特性是启发式问题求解方法的运用。专家系统的设计者发现，非正规的"窍门"和"经验法则"是教科书和课堂正统理论的必要的补充。有时，这些规则是以可理解的方式扩充了理论知识，更多的时候仅仅是工作中经验性的捷径而已。

有趣的是大多数专家系统针对的都是相对专业性较强的、专家级的领域。这些领域通常已认真研究过，问题求解策略有明确的定义。而那些依赖于模糊定义的常识性概念的问题则很难用这些方法解决。无论专家系统的前景如何，过高估计这项技术的能力是一个错误。其局限性表现在以下几方面。

- 获取问题领域"深层"知识的困难性。
- 缺乏强健性和灵活性。
- 无法提供深入的解释。
- 验证的困难性。
- 不能从经历中学习知识。

尽管有这些局限性，专家系统仍在大量的应用中证明了它们的价值。

第 21 单元　电子商务

电子商务概要

电子商务这个术语一般是指以电子方式通过一个计算机网络进行的商业交易。它包括商业和销售过程依靠互联网和万维网技术而得以实现的所有方面。电子商务的商品包括许多种物质产品、数字产品和服务。电子商务网站提供的物质产品包括衣服、鞋子、滑板、汽车等商品。这些产品大多可以通过邮政部门、包裹运送服务公司或货车运输公司运送给购买者。

随着计算机的广泛使用、互联网的完备和普及、信用卡的普及、安全交易协议的建立和政府的支持和鼓励，以及人们开始以电子手段作为商业媒介，电子商务发展得越来越繁荣。

电子商务本身不是一种纯技术，而是互联网与信息技术在商务中的运用过程。在这个运用过程中，以纸质为基础的用于信息交流、传递、处理及存储的工具被以电子媒体为基础的工具所取代。

电子商务商品越来越多地包括数字产品，如新闻、音乐、视频、数据库、软件，以及一切类型的基于知识的产品。这些产品的独特之处在于它们可转换成比特位，并通过万维网交付。消费者一旦填完订单即可得到它们，而且无人支付运费。电子商务商家还兜售服务，如在线医疗咨询、远程教育或定制缝纫。在这些服务中，有些可由计算机完成，有些则需要人工完成。服务可能以电子方式交付，如远程教育课程。

许多电子商务活动被归类为 B2C（企业对消费者），在这种电子商务中，单个消费者从在线商家购买商品和服务。C2C（消费者对消费者）是另外一种流行的电子商务模式，在这种电子商务模式中，消费者在流行的在线拍卖中相互买卖。B2B（企业对企业）电子商务涉及一家企业从另一家企业购买商品或服务。B2G（企业对政府）电子商务的目的在于帮助企业向政府售货。

电子商务网站

从一个购物者的角度来看，电子商务似乎很简单：连接到一个在线商店，浏览电子商品目录，选择商品，然后付款。然而，在后台一个电子商务网站使用数种技术来展示商品，记录购物者的选择，收集支付数据，试图保护客户的隐私，并防止信用卡号落入不该得到的人的手中。

一个电子商务网站的统一资源定位符，如 www.amazon.com，充当在线商店的入口。此处的一个网页——有时称为电子店面——欢迎客户的到来。并提供网站各部分的链接。出售的商品和服务出现在客户的浏览器窗口中。

一个电子商务网站通常包括某种机制供客户用于选择商品和付款。在一个小型企业中，客户订单可能手工处理。然而，大多数大容量电子商务企业尽量利用自动化；它们的订单处理系统自动更新库存清单，然后打印装箱单和邮件标签。

如果你使用过在线购物，就很可能用过在线购物车。大多数购物车起作用，是因为它们使用"cookies"存储你在网站上的活动信息。"cookies"以两种方式之一与购物车协作，使用哪种方式要取决于电子商务网站。一个电子商务网站可能将"cookies"作为你装入购物车的一切物品的存储箱。有些电子商务网站只是将"cookies"用作一种独一无二地标识每个购物者的方法。这些网站生成一个独特的标识号，与你所选的物品一起存储在一个服务器端数据库中。

在线拍卖

在线拍卖是老式庭院旧货出售、捐赠品义卖及拍卖的电子对应物。在一个诸如 eBay 之类的互联网拍卖网站的主持下,卖家公布要卖的商品或服务,买家竞买公布的物品。拍卖网站不拥有或持有商品,确切来说,它只是为在线出价过程和买卖双方的交易提供便利。

在线拍卖中,你可对新的、旧的、清仓销售的、库存过多的或翻新的物品出价竞买。也有不同种类的服务可供竞买。商品包括的类型很宽泛,如古董、收藏品、书与连环漫画、硬币、邮票、娱乐纪念品、艺术品、住宅与花园、照相机与便携式摄像机、计算机、车与船、体育用品、衣服、珠宝、乐器、玩具,甚至戏票、特惠旅行、不动产。

可供拍卖的服务也像商品一样,有各种各样的。例如,你可出价竞买辅导课、打印机修理、室内装饰、定制的服装、购物帮助、印刷与个性化、图形设计及万维网与计算机服务。

在线拍卖网站上,专门设计的计算机软件取代拍卖商。服务器数据库存储拍卖品信息和客户对卖家的评分。服务器端脚本接受出价并通知中标者。不仅拍卖网站实现了自动化,而且供出价人使用的软件工具也很热门。例如,狙击机器人可在最后一分钟出价,而拍卖提醒工具可在你的关注物品清单上的物品可供竞买时通知你。

电子商务的益处

计算机网络使信息交换更快更廉价,互联网也几乎渗透到了世界的每一个角落。中小型企业(SME)也能与世界任何地方的商业伙伴建立一个全球性的关系。高速的网络解决了地理上的问题。买卖能在传统意义上的市场之外进行,还能挖掘新市场,更轻易地找到商业机会。中小企业无力建立海外办事处和中心,但现在可以在世界的任何一个角落提高知名度。

以纸张为基础的购买和出售经常需要大量的文秘工作人员,信封需要准备,并且差错需要纠正。组织机构能够通过自动计费装置和账单付款实现文秘职员的减少。通过利用电子方法以更多的方式可以实现保留重要职员。例如,当客户在线获得联机的信息而不是通过电话询问时就可以减少销售和维护人员。当纸目录变成在线版本时就可以减少用于打印和邮寄目录的人员。在线输入订单要比通过电话会话需要更少的职员时间,同时产生较少的差错。

互联网给公司提供了很大的市场和大量的产品促销机会。除此之外,与买家间的关系也能得到增强。使用多媒体设备,互联网能很有效地建立起公司形象、产品和服务品牌。具体和精确的销售数据有助于减少备货量和运作资金。具体的顾客信息,如消费的形式、个人喜好和购买能力等,能帮助商家更有效地进行市场策划。

许多客户转向在线式金融市场交易,因为在这里他们对自己的资金有更大的控制权,还因为这些系统与电话订单相比更加容易使用。目前,其他许多公司通过采用直接拨号到他们的计算机的方式来支持股票、债券、期货和买卖的特权交易。这种方式为那些愿意选择电子交易的客户提供大的折扣率,因为在一天 24 小时内的任意时间里,订单会随着"确认"几乎同时被接收。投资者可以在任何时间检查他们的账户状态,而不必等待每个月的结算单。

阅读材料 1:电子商务简介

电子商务定义为:运用计算机应用软件在网络上通信,使买卖双方完成整个交易或部分交易。电子商务就是通过电子媒介来进行商务活动的,也就是意味着用简单的、快速的和低廉的电子通信来交易，而无须交易双方面对面地碰头讨论。现在，它主要通过互联网和电子数据交换来完成。电子商务是从20世纪60年代开始发展的。

商家可以通过互联网收集关于产品、消费者、竞争对手的信息资料，从而提高自身竞争力。商家可以在任何时间通过互联网与顾客、消费者建立密切的联系，持续地提供最新的产品和服务信息，这样，他们就可以保持自己的竞争优势。另一方面，数据能随时更新，并清除了信息过时的问题。

传统商务与电子商务的区别

在传统经营环境中，一个公司要花相当多的时间和气力来调查和决定生产与销售什么或多少，然后，在媒体上，如电视、广播或路边广告牌做广告，与买家通过电话、传真、邮件或直接见面方式接触。看过货后，买家再讨价还价，然后下订单或与卖方签订协议，卖家准备好货物后将货物送给买方，买方以现金或支票付款。所以，在传统经营环境里，人们需要花更多的时间和精力进行信息交流，有时还会出现信息错误和延误。

在电子商务环境中，人们利用互联网做生意，买卖双方的接触不再需要面对面，而是通过网络。在网上，买卖双方都有更多的机会了解对方，不仅可以一对一，而且可以多对多。既无地域限制，又无时间限制，双方都能找到称心如意的对家。买家利用浏览器根据想要的域名的网站或在线商店的首页寻找产品目录，而且可在那些网页上看到产品的包装、外形、款式、色泽和价格。有时，甚至还可在网上议价。如果满意这些货物，可以单击选择键的方式下订单。卖方立即就可以收到信息，如果他们接受订单，他们用互联网去安排货物生产、配送和货款支付，所有数据被及时传递以使企业经营变得更有效率。

根据交易的性质对电子商务分类

电子商务一般按照其交易性质分类，可分为以下几个不同类型。

企业对企业的电子商务（B2B）。今天大多数电子商务都属于这一类，它是组织间的电子市场交易。

企业对个人的电子商务（B2C）。它主要是与个体顾客进行的零售交易。企业通过互联网为顾客提供在线购物，允许顾客在线选购并支付账单。这种方式为零售商和消费者节省了时间。

消费者对消费者的电子商务（C2C）。这类电子商务是指消费者直接销售给消费者。例如，个人在分类广告网站发布广告及销售个人房屋资产、轿车等。个人对个人的电子商务另一个例子就是在互联网上发布个人服务广告和将知识及专家经验作为商品出售。许多拍卖网站也允许个体提供项目来进行拍卖。于是，许多人利用企业内部互联网和其他组织的内部网络来发布销售或服务的广告。

个人对企业的电子商务（C2B）。这一类包括个人向组织推销产品或服务，以及个人寻找卖家、与卖家互动，并最终达成交易。

政府对公民电子商务（G2C）。在政府提供的各种服务中，有许多是可以通过电子媒介完成的。公共服务电子化不仅能提供给公民省时和优质的服务，还能提高效率和成本效益。

政府对企业电子商务（G2B）。这种交易模式经常描绘政府能通过像互联网那样的电子媒介向企业购买产品和服务的方式。

第22单元 电子支付系统

网上支付

在电子商务过程中,支付是非常重要的一个环节。网上支付意味着以电子方式处理资金转移、收入或付出。网上支付,有时称为电子支付,是应电子商务应用的要求而发展起来的。

网上支付的前身是银行经营的电子汇款。首先,客户在银行A存入钱,当需要付款给在某个城市的某人时,客户委托银行A从他的银行账户转一笔钱到某人所在城市的银行B去,某人在银行B有账户,当银行B证明了某人身份后,把钱转入他的账户。早期银行的数据转换是以电话或电报方式进行的。最近几年,他们用EDI系统替代了电话与电报。这个金融EDI系统被称为电子资金转账系统EFT。EFT系统节约了数据处理时间并大大降低了经营开支。

现在,许多银行和软件开发公司正致力于开发一种新电子付款方式——数字资金系统。表面上,数字资金仅仅是一组数字,但这些数字不是随意或任意产生的。在数字钱币背后,有真正的钱币在支持它,否则,将发生混乱。作为流通媒质,数字资金具有普遍使用的能力,即它必须被许多银行、商家或消费者接受。否则,它只能在小范围内使用,不适应电子商务应用的特色。

还有许多种数字资金或电子货币,它们主要可以分两类:一类是实时付款式的,如电子现金、电子钱包和智能卡等;另一类是事后付款机制,如电子支票(网上支票)或电子信用卡(万维网信用卡)。

先进的加密与确认系统能使数字现金使用起来比真纸币更安全和隐私,与此同时,还保留了纸币的"匿名支付"的特点。数字现金系统是建立在数字签名和加密技术之上的,这个系统常利用公钥/私钥密钥对方法。

首先,用户存钱到银行,开个账户。银行将给客户可用于客户端的数字现金软件和一个银行的公共密钥,银行用服务器端数字钱币软件和私人密钥对信息加密,用户可用客户端数字钱币软件和银行的公共密钥将信息解密。当客户需要一些数字现金支付某人时,就用数字现金软件生成他所需要的数字现金,给它加密并传给某人。某人将这个信息再加上他自己的账户信息和存款指令一起加密并且发送给银行。银行用私人密钥解密信息并证明客户与某人的身份,钱从客户账户上转至某人账户。

由于数字现金只是一组数字,因此需要确认其真实性。银行必须具备能追踪电子现金是否被复制或重复使用的能力,而又不能将个人购买行为与从银行购现金的人联系起来,以保持真钱币的"匿名"特色。在整个过程中,商家也不必知道任何客户的个人资料,它可用银行的公共密钥检验客户送来的数字现金。

另一种网上支付方法是电子支票,它的前身是用在增值网中的财务EDI系统。有些电子支票软件可装置到智能卡上,这样,客户能非常方便地携带它而且可用在任何计算机上写电子支票。一些电子支票软件可为客户提供电子钱包,使他们在没有银行账户的情况下也可以写支票。

电子银行

电子银行不是真正的银行,它是一种虚拟银行,在这个虚拟银行中,银行家、软件设计者、硬件生产者和ISP们共同合作,完成各自领域的工作并很好地与各方协调。在电子银行,银行服务内容放在服务器的主页上供客户选择。客户无须自备金融软件,他们用浏览器链接到银行的网站并索取服务。所有服务都在银行服务器上被处理，银行通过网站与客户互动。客户也可以用便携式智能卡在任意一台联网计算机上来进行银行支付，如，付账、购买电子货币及注册网上银行等。他们不需要买 PC，可以用任何已联网的计算机来完成银行支付工作。

这些新型银行服务软件采用了现代信息技术，能提供方便、实时、安全和可靠的银行服务。此外，它们还可以提供其他相关服务，如股票信息、税收方法及政策咨询、外汇汇率和利息率信息。软件越方便，则顾客越多，利润也更多，服务则更好。这样，银行和客户为双赢关系，客户获得了便宜而方便的服务，银行极大地降低了运作成本。因为在网上，经营一个网站的开支远远比建一个银行分支便宜。而且，在网上服务一个人的成本与服务一万个人的成本没有什么不同。随着网上银行的飞速发展，软件公司和网络提供商得到更多业务，它们都与银行处于双赢关系。

电子资金转账

电子资金转账（EFT）是从一个机构的计算机把"资金""汇"到另一个机构的计算机的自动处理过程。电子资金转账是电子数据交换的一个子集，是指用电子方法实现买家向卖家的价值转移。

电子资金转账主要由银行实施，以清除书面处理工作。在对付大量书面工作中列出的 EFT 方法有基于显示屏的现金管理系统、Swift 系统（用于国际）、Bits 系统（用于国内）、直接输入（磁介质）、ATM（自动出纳机）、EFTPOS（基于信用卡，适用于国际和国内）、家庭银行业务、电话银行业务和票据支付。

信用卡

信用卡（借记卡共享其网络）是在线消费者支付的首选方案。信用卡成为一种支付方法而流行的原因有：易于使用，没有技术障碍，几乎每个人都有一张；消费者信赖发卡公司能够提供安全交易；大多数商人接受信用卡。在信用卡交易中的参与者有顾客、顾客的信用卡发行人（发行的银行）、商人、商人的银行、在 Internet 上处理信用卡事务的公司。

信用卡支付的步骤如下。

（1）顾客通过网站上的购物车或目录选择产品或服务。

（2）合计订单，发出一个购买请求。

（3）以规定的形式，顾客输入一些敏感的数据，如信用卡号、发货地址等。

（4）这些数据通过网络服务器被安全地提交、收集和传递。

（5）Web 服务器的支付系统与商家的卖方账户有关的处理器安全地进行联系，证实该信用卡号是有效的。

（6）如果上述步骤正确通过，那么购买价格将从顾客信用卡的"发行银行"划拨到商家的卖方账户。

阅读材料 1：电子商务安全

电子商务的安全是建立一个网上商店最应该关注的问题。SSL 证书、加密、公钥和私钥等，这些都很容易混淆。如果你根本不为安全问题担忧或不是很清楚采取必要措施保护你的网站的重要性，你的网站和生意将会很有风险。

◆ 建立安全控制机制

安全等级也是一个需要考虑的问题。你需要保护你的系统免受黑客和病毒的破坏。需要用到防火墙、入侵检测系统和杀毒软件。另外，也需要用到一些常用的安全措施，如保护用户的 ID 和密码、经常更换密码。支付交易需要用到更高等级的安全措施。如果想从网上获得客户的个人资料，就必须保证可靠的转账和数据的储存。一个资料保密的声明也是有必要的。

◆ SSL——安全套接层

SSL（安全套接层）技术使两个设备之间连接起来，例如，一台计算机和网络服务器能够安全地传输数据。对于一个电子商务网站或任何接收敏感信息的网站来说，一个 SSL 证书是绝对有必要的。你不仅需要 SSL 证书，还要真正使用它。

◆ 保护电子邮件的安全

如果你通过电子邮件的方式从你的网上商店收到了订单，那仅仅拥有 SSL 证书是没有用的。电子邮件是以纯文本的方式传送的，它有可能在传输和阅读的时候被截取。

有几种方法保护电子邮件。PGP 是一种加密技术，可以在邮件送出的时候将它加密，然后在接收的时候解密。如果要运用这种技术，发送者和接收者都需要在他们的计算机上安装可以鉴别和解密文件的认证证书。这种技术虽然已经发展了好几年，可是它依然不是很容易安装也没有发挥它最大的作用。

通过你的网站接收到安全的电子邮件的另一种方法是运用一个与安全命令在同一个服务器的电子邮件账户进行邮件传输。正因为它们不会通过公众网络被发送，所以在你试图下载它们之前是安全的。

◆ 经常升级软件

经常升级软件看起来不值一提，但是修补和软件升级往往可以修补那些使黑客进入你的系统的安全漏洞。经常修补你的系统。

◆ 注意服务器上安装的脚本和文件

免费的脚本和软件听起来很好，但是它们有时编写得非常不好，或者会存在能使黑客侵入你的系统的安全漏洞。要保证你安装的软件经过了彻底的测试并在出问题的时候能够获得有效的支持形式。

◆ 使用安全的 FTP 传输文件

FTP 和远程登录都以纯文本形式传输密码。黑客们可以看到这些信息并可进入整个系统。使用安全的 FTP 和 SSH 进入网站，这整个过程全部被加密就好像用 SSL 那样，保证没有人可以截获和读取你的密码或其他敏感信息。

网上购物开始了全球商品与服务。万维网已经以前所未有的方式扩大了国际市场，给消费者提供了无限的选择。然而电子购物带来了世界性的问题，特别是当与其他国家的卖主进行交易时。

当在网上购物时，掌握一些技巧来帮助你是必要的。

◆ 了解交易的对象

在与公司作交易时做些额外的工作以确定公司是合法的。确认公司的名称、物理地址，包括所属国家、E-mail 地址或电话号码，这样可以针对疑惑和问题与公司联系。并且考虑只与那些有明确政策的卖主进行交易。

♦ 了解条款、条件和成本

预先找出什么是自己需要的,什么是自己不需要的。获取一份完整的、详细的成本清单,包括销路、一份清楚的货币名称、交付或性能的条款,以及付款的条款、条件和方法。查找购买中的有关限制、局限性或条件的资料。

♦ 在线支付时保护自己

查找描述公司的安全公示信息。核对浏览器是否安全,并且在在线交易期间加密个人和账户资料。这样会减少资料受到黑客攻击的可能性。

♦ 小心隐私

所有的交易都需要用户的资料来处理订单。有些是用来通知消费者有关产品、服务和升级的,有些是用来为其他卖主共享或出售资料的。只从那些尊重个人隐私的网上卖主处购货,查找卖主在网站上的保密政策,政策声明应该显示哪些个人识别信息是被收集的,它如何被使用,卖主应该给用户机会拒绝把资料销售或共享给其他卖主。还应该告诉用户如何改正或删除公司已有的有关自己的资料。

第23单元 电子营销

通过互联网,厂商可以不利用中介,直接接触客户。只要制造商销售的是已经确立的品牌而且他们的主页已经有了很好的知名度,制造商的直销就可以实现。如果一个制造商的网站知名度不高,仅仅开办一个主页,被动地等待顾客点击,这个也许对销售贡献不大。因此,公司有必要大张旗鼓地宣传他们的网站地址。任何经济上划算的广告都可以被用来实现这一目的。其中一个例子是,把自己的网站链接到知名的电子目录网站,而且大多数厂商都使用中介机构的目录服务。这些中介网站被称为电子购物中心。

像百货公司和折扣店这样的既有经销商至少在电子商务的初级阶段不会在电子零售中扮演主要角色。传统的分销商主要利用他们的主页和电子网页目录来吸引顾客去实体商店购物。因此,我们需要研究电子商务经销商、经纪商和网上百货商店的竞争结构。

起初,电子营销主要关心的是安全技术,这些技术对于基于互联网的营销是必要的,如强大的搜索能力和安全电子支付。然而,在今天,管理的重点正转移到如何利用网络营销的机会和现有销售渠道一起来提升竞争力。

越来越多的公司将其客户分成不同的客户群,并为每个客户群提供不同的针对性的信息。在有些情况下,当公司采用网络时,这些目标群体的规模可能更小,有时候甚至可能一个客户就是一个目标群。网站访客行为的最新研究甚至标明,网站可以采取一些方法来响应不同的时间、有不同需求的访问者。

绝大多数的公司使用术语"营销组合"来描述他们用以实现其销售、促进其产品和服务目标的元素的组合。当一家公司决定使用某些元素的时候,它便把那个特定的营销组合称为它的营销战略。即使那些在同一行业的公司,他们都努力地在市场中表现出与众不同。公司的营销策略是一个与现有客户和潜在客户能获得公司信息的网上展示相结合的重要工具。

大多数的营销课程把营销的基本问题归纳成4P,即产品、价格、促销和场所。产品是一家公司正在出售的实际的项目或服务。产品的本质特性很重要,但是客户对产品的感知和产品的实际特性一样重要，客户对产品的认知度被称为"品牌"。

"营销组合"的价格元素是客户为获得产品所付出的资金数量。近几年来，不少营销专家认为公司应该在一个较宽泛的意义中来考虑价格因素；也就是说，价格应该是客户为获得产品所付出所有费用的总和。从该总成本中减去客户从所获得的产品中得到的收益就能产生客户从交易中获得的客户价值的估计值。Web 为创造性定价提供了新机遇，也为在线拍卖、反向拍卖及群体购买战略的价格协商创造了新的机遇。这些基于网络的机会正在帮助公司发现新的增加客户价值的方法。

促销包括发布有关产品消息的任何方法。在互联网上，极大的推广可能性存在于与现有客户和潜在客户的交流之中。公司已经使用互联网，通过电子邮件及其他的方法与他们的客户进行拓展业务。

多年来，营销经理都梦想着能存在这样的一个世界：当客户有所需求时，公司能够在最短的时间内满足他们的需求。场所是指在各个不同地点获得产品或服务的一种需要。自商业存在开始，公司一直在苦恼如何在最好的时间把正确的产品送到正确的地方的销售问题。虽然互联网尚未解决所有的这些物流管理和分配问题，但总有一天它能够做到。

◆ 以产品为基础的营销策略

许多公司的经理总是依照其公司所销售的产品或服务来考虑他们的生意。因为公司花费了很大的精力、时间和费用来设计并制造那些产品和服务，所以这是一个合乎逻辑的考虑生意的方法。如果你要求经理描述一下他们公司正在卖的产品，通常会提供给你一本关于他们正在销售或用之制造服务的实物的详细目录。当客户可能根据产品分类目录来购买产品，或者是依据产品分类目录来考虑他们需求的时候，这种以产品为基础的组织类型就显得颇具意义。

◆ 以客户为基础的营销策略

与广播和印刷广告这些传统大众媒体的沟通结构相比较，网络沟通结构更加复杂。当公司通过网络开展其商业活动的时候，应该能够创造一个有足够柔性的网站来应对不同访问者的需求。公司应该建立能够满足不同类型客户特定需求的站点，而不是仅仅建一个产品展示的站点。制定一个基于客户的良好的营销策略的第一步就是要甄别拥有共同特性的客户群体。在 B2B 网站上，基于客户的营销方法被大力倡导。相对于 B2C 网站的运营者，B2B 卖家更清楚地知道，要满足他们客户的需求，就必须客户化他们的产品和服务。近几年来，B2C 站点逐渐地将以客户为基础的营销元素加入到他们的网站之中。

做广告就是为了影响买卖双方的交易而进行的信息传播的努力。互联网重新定义了广告的含义。互联网使得消费者能直接与广告商及广告互动。互联网为广告商提供了双向通信和电子邮件的能力，同时使得广告商把广告费花到他们想针对的特定群体身上，这比传统电话营销更准确。最终互联网实现了真正的一对一广告。

用于广告的主要方法有以下几种。

◆ 旗帜

旗帜广告包括一篇简短的文字或所推销的产品的图形信息。广告商尽量设计出能吸引消费者注意力的旗帜广告。利用旗帜广告的一个主要优势是能够为目标观众而定制。人们可以决定针对哪个细分市场。旗帜广告甚至可以定制成一对一的目标广告。在别人的网站上放置旗帜广告，有几种不同的放置形式。最常见的形式是交互旗帜广告、交换旗帜广告和付费广告。

- 场地租赁

搜索引擎往往在其首页提供空间做商业租赁。租期取决于网站主人和承租人的合同协议。

- 电子邮件

电子邮件作为一种营销渠道正在兴起，它提供低成本的实施方式和高于其他广告渠道的更好、更快的反应率。一个电子邮件地址列表可以是一个非常强大的工具，因为针对的是你有所了解的一群人。

- URL（统一资源定位器）

应用统一资源定位器作为广告工具的最大好处就是它的免费性，任何人都可以把他的URL提交给搜索引擎并加入列表中。因为关键词的作用，URL的应用还可以锁定目标观众，同时将那些不想要的观众过滤出去。

阅读材料1：物流与供应链管理

正如商品的流动，信息也必须是流动的。在合适的时间、合适的条件下，携正确的文件把商品送到正确的地点，这些有关"正确"的问题必须要弄清楚。卡车司机应从哪里装运？谁会接收货物？某种物品的库存是多少，将要生产多少？货运现在到哪了？信息贯穿于整个物流管理系统中。

顶级的物流效率和效果要求有一个很好的集成物流信息系统（ILIS）。没有随时能够访问的准确信息，集成的物流管理操作就会失去效率和效果，便不能维持战略竞争力。ILIS的优先应用领域包括管理库存状态、追踪货品和发货、取货和运输、订货便利化、订货准确化、内部物流和外部物流的协调，以及订单处理。通过ILIS信息流的质量是至关重要的。所谓的"垃圾进—垃圾出"可能出现在任何信息系统。在信息的质量上，有三点值得我们关注：①获得正确的信息；②保持信息的准确性；③有效地沟通信息。

一个集成的物流信息系统可以被定义为：通过人员、设备和一定流程，将所需信息加以收集、整理、分析、评估，然后将它们以及时、准确的方式发给恰当的决策人以帮助他们做出高质量的物流决策。ILIS收集来自所有可能渠道的信息，以协助集成物流经理做出决策。它也与市场、金融和制造业信息系统建立接口。所有这些信息都将被高层管理者用来制定战略决策。ILIS有4个主要的组成部分：订单处理系统、研究和智能系统、决策支持系统及报告和输出系统。

在供应链管理方面，许多全球性的制造公司都参与实施了新的信息系统和技术。财务系统、生产计划、物流及库存管理系统是最早的应用领域。大多数公司需要4~5年的时间，并花费数百万美元的直接成本支出。

供应链管理改进的目的是使供应链关系更快捷和更一致，并且往往把库存作为最后的缓冲区，而不是解决问题的首要办法来降低周转金成本。在高度竞争中，重点是创造价值，价值的创造主要办法是提高信息的使用和客户数据的质量，改善售后服务及订单履行，而为上游过程定义更一致的信息则是次要办法。

大部分公司都开始面对"超竞争"，在这种竞争中，公司把自己定位在一个彼此越来越争斗的位置。而不是在适度竞争中彼此包容相伴的位置。在适度竞争中，壁垒是用来限制新的进入者的，并且，只要产业领导者一起合作约束竞争行为，那么可持续的优势就是可能的。但是，处于超竞争状态中的公司必须不断寻机破坏产业领导者的竞争优势，并创造新的机遇。

适度竞争市场与超竞争市场中的供应链管理项目的运作重点是有争议的。在前者中，上游项目的投资（新财务系统、生产计划或库存管理系统）可能带来大量的好处，其中包括持续的信息共享和跨部门合作的改进。而在超竞争条件下，运作的重点则是有高投资回报和高客户附加价值的过程和信息系统。运作的重心将转向需求侧，并且强调客户互动、账户管理、售后服务及订单处理。为了维持超竞争市场中的竞争优势，公司可能会寻求取消详细管理报告和控制，以及市场预测和生产计划的需求。取而代之，公司可能从他的经销商和零售商那里获得实时、在线的产品流向信息。公司也可以通过派遣组织或者授权员工的形式，不断地改进过程来实现控制和管理报告的简化。

第 24 单元 移动商务

随着万维网的引入，电子商务彻底改变了传统的商务，大大地推动了销售额及商品和信息的交换。最近，无线网络和移动网络的出现使得电子商务进入了新的应用及研究主题：移动商务。移动商务可以定义为：通过移动的无线电通信网络进行的涉及货币价值的任何交易。一个不太精确的说法是把移动商务描述为人们通过连接网络的移动设备所获取一系列应用及服务的兴起。移动商务可以有效、方便、随时随地给客户传递电子商务。很多大公司已经意识到移动商务带来的好处，开始为客户提供移动商务选择。

商务是指大批货物的买卖或交换，涉及从一处到另一处的运输。移动商务技术的便利与无处不在推动了商务的发展。有很多例子表明移动商务是如何促进商务的。例如，消费者现在可以用他们的便携电话为自动售货机里的商品付款或支付停车费；移动用户不必去银行或自动取款机就可以查看他们的银行账户并进行余额转账等。

及时的投递对今天的企业取得成功至关重要。移动商务让企业可以跟踪其移动库存并适时地投递，这样可以改善客户服务、减少库存并增强公司的竞争力。大多数投递服务公司如"联合包裹服务公司"和"联邦快递"都已经把这些技术应用到其全球范围的企业运作中。

交通是指车辆或行人在某一范围或沿着某一路线的运动。车上的旅客或行人是运动的目标，是移动商务理想的客户。交通管制对于很多大城市来说是一件头疼的事。用移动商务技术可以在很多方面轻易改善交通。例如，移动手持设备预计会拥有全球定位功能，如确定司机的位置、指示方向、对该区域的交通现状进行咨询；交通控制中心会根据车里移动设备发出的信号监控并管制交通。

旅行费用对企业来说可能是非常高的。移动商务通过向商务旅行者提供移动旅行管理服务可以帮助减少经营成本。它可以给有需求的客户提供有吸引力的、难忘的体验，它利用移动频道查找附近理想的旅馆、购票、进行运输安排等。它也可以扩展那些公关公司的经营范围，增加现有的经营渠道，从而帮助移动用户识别重要的消费者，吸引他们，为他们服务并且留住他们。

从拥有智能手机或平板电脑到搜索产品或相关服务，浏览这些信息到最后下单购买，这只是几个简单的步骤。结果就是移动商务的业务一年的增长超过了 50%，远高于桌面电子商务业务一年 12%的增长率。移动商务的高增长率当然不会一直持续下去。

一份对排名前 400 的移动公司销售的研究表明，在移动商务中，零售物品占了 73%，旅游业务占了 25%，同时还有 2%是票务。越来越多的顾客正在利用其移动设备搜索人物、地点及餐馆，或者在零售店看过的商品的交易。顾客从桌面平台到移动设备的快速转换驱动着在移动营销花费上的上涨。

与电子商务系统相比，移动商务系统要复杂得多，因为它必须包含与移动计算相关的成分。下面简略地描述了一个典型的由移动用户提交请求后所引起的处理过程。

- 移动商务应用：内容提供商通过提供两套程序实施应用。客户端程序，如微浏览器上的用户界面；服务器端程序，如数据库访问及更新。
- 移动站：移动站向终端用户显示用户界面，这些用户在界面上细化他们的要求。然后，移动站把用户的要求传递到其他组件并在用户界面显示处理的结果。
- 移动中间件：移动中间件的主要目的是不留痕迹地、清晰地把互联网的内容绘制到移动站上，该站支持各种操作系统、标记语言、微浏览器及协议。大多数移动中间件也通过加密信息在一定程度上保证交易的安全性。
- 无线网络：移动商务之所以可行的很大原因是无线网络的应用。用户的要求被传递到最近的无线接入点（在无线局域网环境里）或一个基站（在蜂窝网络环境里）。
- 有线网络：对于移动商务系统来说，该部分是可选的。
- 主机：该部分同电子商务所使用的这部分一样。用户的要求通常在这里得到执行。

没有信任与安全就不会有移动商务时代。在客户连接支付业务或用其移动设备进行购物时，如果供货商或支付提供者不能给客户安全感，那怎样才能吸引客户呢？对消费者来说他们需要对以下事情感觉舒心：他们不会被要求为他们没有使用的服务付费；他们的支付细节不会落在别人手里；有现成的适当的机制来帮助解决纠纷。

当我们在评论移动安全性的各个方面时，应该牢记的是安全性总是需要全面考虑的。一个系统的安全性只相对于它最弱的部分，而保证网络传输只是安全的一部分。从技术方面来说，网络安全涉及很多不同的方面，每个方面都对应着受到的威胁或薄弱环节的一个层面。防御其中一个层面不能保证你不受其他层面的侵扰。

阅读材料 1：O2O

Groupon、Open Table、Restaurant.com 和 SpaFinder 公司都有什么共同之处呢？它们都促进了 O2O 商务的前进。Groupon 的发展可谓非同凡响，但它在一个更大的分类系统中仅仅是一个小小的分支，我们把那个大的类别称为"线上-线下"商务，或者 O2O 商务，与 B2C、B2B、C2C 一样都是商务术语。

O2O 的关键在于它在线上招揽顾客，并把顾客带到现实世界的商店中。它让顾客消费、给商家持续提供客流量，从而带来线下交易（同样它也使得顾客能够发现许多商家）。由于每笔交易都是线上的，它天生就具备可测量性。这与目录模式有很大区别，因为新增的支付款项可以确定销售数量，并且全程跟进。

从分析可知，O2O 市场很"大"，或者说 Groupon 在互联网的历史中一直能够比几乎所有其他公司有着更快的高利润收益的增长，这是一个不争的事实。电子商务中的消费者平均每年大约支出 1000 美元。假设美国人平均每年挣 40000 美元。那么其余的 39000 美元怎么办？如果你考虑到电子商务中的顾客比大多数美国人挣得更多，这一比例将会更大，但是结论是一样的。

税后的可支配收入有很大一部分是在本地消费的。你会将钱花在咖啡厅、酒吧、健身馆、餐馆、加油站、管道工、干洗店和美发沙龙。除了旅行，在线 B2C 商务即是你在线订购的大件物品，并让商家装箱送货到家。这很无聊，即使电商业界已经知道了线上销售的物品有所增长。

FedEx 公司不能快递真实社会中的体验，如餐馆、酒吧、瑜伽、航海和网球课，但是 Groupon 可以。而且，对于你本地开设的瑜伽馆，将顾客安排到一个没有满员的学习班中不会带来更多的边际成本。这意味着这种本地再次销售的业务模式通常比传统的电子商务模式更赚钱，而后者是低价买进商品以便存储，再议高价卖出，并且管理易腐烂或易贬值存货时还有所不同。

O2O 商务公司重要的一点在于业绩是随时可量化的，这也是 O2O 商务的原则之一。传统的电商可通过 cookie 和像素等来追踪兑现情况。因为 Zappos 的每一个完成订单的确认页面上都有"追踪码"，所以它可以确定其在线营销的投资回报率。线下的商务就没有这种得天独厚的条件，如酒保不会查看你的 iPhone 的浏览记录。但是 O2O 可以让这一切简单起来。因为交易是在线上进行，在线下世界中也可以使用同样的手段，整个这一切都由 Open Table 或 SpaFinder 公司等中介来安排。这已经被证实是一个比向当地的公司销售广告收益更高并且扩展性也更强的模式。这完全得益于线上中介的销售。

O2O 商务挤压了传统电子商务（打包快递商品）——仅仅是因为线下商务本身比线上商务更优越，并且只是把发现商品和支付的过程转移到了线上。风险投资商和企业家要想着怎样将发现商品、支付及业绩度量从线下商务转移到线上才是明智之举，而非仅仅抄袭"团购"的概念。在本地消费者以数万亿计美元逐渐开始在线上消费之时，这将在整个互联网包括广告、支付与商务的业界中造成连锁反应。

第 25 单元　网络安全

当各类大小组织都越来越依赖网络来进行他们的活动时，对一般民众，以及社会的基础设施的特定部分的恐怖袭击事件也增加了。像计算机、数据和信息一样，网络也已成为资产。与顾客、供应商、员工及其他组织的通信都主要通过网络来处理，网络通信渠道的缺失会大大地削弱一个组织。失去了信息的获取和一个可靠的通信系统，一个生意就不能再运转下去。网络必须像其他资产一样受到保护，并伴有适当的控制和适当的安全性。所以对网络和计算机安全这个科目有强烈的兴趣也不足为奇了。

随着互联网的兴起，以及它的商业管理用途的增加，大多数组织的部分网络更加开放，也更容易受到未授权访问、计算机病毒及比以往更富有侵略性的攻击。运行在计算机网络上的应用系统的潜在破坏或基本数据的讹误是组织对网络安全十分重视并采取行动的很好理由。通常情况下，存储在联网计算机上数据的价值已远远超过网络本身的花费。

安全威胁

对一个网络的安全威胁可分为那些含有某种形式的未经授权访问和所有的其他情况。一旦有人得到了对网络的未经授权的访问，那么他们可以做事的范围就很大了。一些人只是对突破安全的挑战很感兴趣，而没有采取进一步行动的兴趣。他们可能选择监控网络流量，即所谓的窃听，其目的是了解一些特别的事情或将其泄露给其他人，或者为了分析通信量曲线，通信量分析可以得到观测数据。我们能经常想起的这类未授权访问都是主动安全攻击，在一些人得到了对网络的未授权访问后，他们就采取一些明显动作了。主动攻击包括修改消息内容、冒充他人、拒绝服务和种植病毒。

网络安全建议

◆ 使计算机保持在最新状态

为了帮助网络上的计算机更加安全，请打开每台计算机上的自动更新，Windows 可以自动安装重要更新和推荐更新，或者只安装重要更新。重要更新可以让用户受益匪浅，如获得更高的安全性和可靠性。推荐的更新可以处理非关键的问题，并帮助增强用户的计算体验。可选更新不会自动下载或安装。

◆ 使用防火墙

防火墙不仅有助于防止黑客或恶意软件通过网络访问计算机，而且还有助于阻止计算机向其他计算机发送恶意软件。

◆ 在每台计算机上运行防病毒软件

防火墙有助于阻挡蠕虫和黑客，但是并不能保护计算机免受病毒的侵害，所以应该安装并使用防病毒软件。病毒可能来自电子邮件的附件、CD 或 DVD 上的文件或从 Internet 下载的文件。请确保防病毒软件是最新的，并将其设置为定期扫描。

防病毒程序扫描计算机上的电子邮件和其他文件，以查看是否存在病毒、蠕虫和特洛伊木马程序。如果找到，防病毒程序就会在其破坏计算机和文件之前将它隔离，或者将其完全删除。Windows 没有内置的防病毒程序，不过计算机制造商可能已安装了某种防病毒程序。检查安全中心以查看自己计算机是否具有病毒防护。如果没有安装，可转到 Microsoft 防病毒合作伙伴网页，查找防病毒程序。

因为每天都会识别出新病毒，所以选择一个具有自动更新功能的防病毒程序非常重要。防病毒软件更新时，将向其病毒列表添加新识别出的病毒以备检测时使用，这样可保护计算机免受新病毒攻击。如果病毒列表过期，则计算机容易受到新病毒的威胁。更新通常需要缴纳年订阅费，让订阅保持最新以便接收定期更新。如果不使用防病毒软件，则会将计算机暴露于恶意软件的破坏之下。并且还存在将病毒传播给其他计算机的危险。

◆ 使用路由器共享一个 Internet 连接

可以考虑使用路由器来共享一个 Internet 连接。这些设备通常具有内置防火墙和其他功能，有助于更好地保护网络以免受黑客的攻击。

◆ 不要总是以管理员的身份登录

使用需要 Internet 访问权限的程序时，如 Web 浏览器或电子邮件程序，建议以标准用户身份登录，而不是以管理员身份登录。这是因为只有当用户以管理员身份登录时，许多病毒和蠕虫才能在计算机上存储并运行。

◆ 使用间谍软件防护

间谍软件是一种可显示广告、搜集用户的信息或更改计算机上设置的软件，并且通常是在未征得同意的情况下进行的。例如，间谍软件可能会在 Web 浏览器上安装不需要的工具栏、链接或收藏夹，更改默认主页，或者频繁地显示弹出广告。

有些间谍软件没有任何可检测的征兆，但却在秘密地收集敏感信息，如访问了哪些网站或输入了哪些文本。大多数间谍软件是通过下载的免费软件安装的，但有些情况下，只要访问某个网站也可感染间谍软件。为保护计算机免受间谍软件的侵扰，可使用反间谍软件程序。

管理者的责任

网络安全政策是管理部门对网络安全的重要性的声明，也是他们对网络安全所承担的义务。该政策需要用一般条款来描述应该怎样做，但并不涉及达到保护目的的方法。编写这个政策是非常复杂的，因为在现实中，它通常是一个更宽泛的文件，即该组织的信息安全政策的组成部分。网络安全政策需要明确申明管理层对网络安全和需要保护内容非常重要的立场。管理层必须明白一个绝对安全的网络是绝对不可能存在的。

此外，由于科技的进步和那些想入侵网络和计算机的人的创造性，网络安全是一个不断变化的目标。今天，为最大限度地减少安全风险而落实的措施在将来还需要升级，通常升级需要附加资金，管理者需要为之支付。

简单写写政策并不能将实践、过程或软件顺利实施来改善安全状况。这需要后续通过与所有员工的沟通，使他们认识到高层管理人员在安全上布置的重点和重要性。在各个层面，管理层都需要去支持这项政策，并定期与员工用各种方式强化它。IT和网络的工作人员可能需要安装额外的硬件、软件和程序来自动进行安全检查工作。

阅读材料1：网络防火墙

当你把局域网连接到互联网后，你的用户就能够与外部世界进行通信联系。然而，同时你也让外界能进入局域网并与之交互。

网络防火墙的目的是在网络周围设置一层外壳，用于防止连入网络的系统受到各种威胁。防火墙可以防止的威胁类型包括以下几种。

- 非授权的对网络资源的访问

入侵者可能渗入网上的主机，并对文件进行非授权访问。

- 拒绝服务

例如，网络以外的某个人可能向该网上的主机发送成千上万个邮件信息，企图填满可用的磁盘空间，或者使网络链路满负荷。

- 冒充

某个人发出的电子邮件可能被别有用心的人篡改，结果使原发件人感到难堪或受到伤害。

防火墙基本上是一个独立的进程或一组紧密结合的进程，运行在路由器或服务器上，控制经过防火墙的网络应用程序的信息流。一般来说，防火墙置于公共网络（如互联网）入口处。它们可以被看作交通警察。防火墙的作用是确保一个组织的网络与互联网之间所有的通信均符合该组织的安全策略。这些系统基本上是基于 TCP/IP 协议的，并且取决于实现它们能实施安全路障并为管理人员提供下列问题的答案。

- 谁一直在使用网络？
- 谁要上网但没有成功？
- 他们什么时间使用网络？
- 他们在网络上去了何处？
- 他们在网上做什么？

防火墙可以通过滤掉某些原有的不安全的网络业务而降低网络系统的风险。例如，网络文件系统（NFS）可以通过封锁进出网络的所有 NFS 业务而防止为网络外部人员所利用。这就保护了各个主机，同时使其一直能在内部网络中服务，这在局域网环境中很有用。一种回避网络计算有关问题的方法是把组织的内部网与其他外部系统完全断开。当然这不是一个好办法，其实需要的是对访问网络进行过滤，同时仍允许用户访问"外部世界"。

在这种配置中，用一个防火墙网关把内部网和外部网分开。网关一般用于实现两个网络之间的中继业务。防火墙网关还提供过滤业务，它可以限制进出内部网络主机的信息类型。

通常有3种类型的防火墙实现方案，其中某些可以一起使用，以建立更安全的环境。这些实现方案是包过滤、应用程序代理和电路级或通用应用程序代理。

包过滤

把你的网络数据看成一个你必须送到某个地方的干净的小数据包。该数据可能是电子邮件、文件传输等的一部分。使用包过滤时，你自己来传送此数据包。包过滤器起交通警察的作用，它分析你想到哪儿去，你随身携带了什么。但包过滤不打开数据包，如果允许，你仍然要把它送到目的地。

多数商品化的路由器都有某种内建的包过滤功能。然而，有些由ISP控制的路由器不可能给管理人员提供控制路由器配置的能力。在这些情况下，管理人员可能选择使用接在路由器后面的独立包过滤器。

应用程序代理

为理解应用程序代理，来看一看这样的情况，你需要递交一个干净的小网络数据包。用应用程序级代理，情况是相似的，但现在你需要依靠另外一个人来为你传递此数据包。因此，术语"代理"说明这个新情况。包过滤适用的规则也适用于应用程序代理，有一点不同，即你不能越过应用程序代理递交包。有人会为你做此事，但此代理人首先要看一下包的内部，确认它的内容。如果代理已有递交该包的内容的许可，他就会为你递交。

当然，由于代理必须有能力打开"包"来看看或对其内容进行译码，因此安全和加密问题也随之而来。

电路级或通用应用程序代理

与应用程序级代理一样，你需要依靠某个人来为你传递数据包。差别是，如果这些电路级代理要把数据包递交到你要求的目的地时，它们就会这么做。他们不需要知道内容是什么。电路级代理工作在协议的应用层的外面。这些服务器允许客户机通过此集中式服务，并接到客户机指定的TCP端口。

附录 A 计算机专业英语缩写词表

A

AAT（average access time） 平均存取时间
ACL（access control lists） 访问控制表
ACK（acknowledgement character） 确认字符
ACPI（advanced configuration and power interface） 高级配置和电源接口
ADC（analogue to digital converter） 模数转换器
ADSL（asymmetric digital subscriber line） 非对称用户数字线路
ADT（abstract data type） 抽象数据类型
AGP（accelerated graphics port） 图形加速端口
AI（artificial intelligence） 人工智能
AIFF（audio image file format） 声音图像文件格式
ALU（arithmetic logic unit） 算术逻辑部件
AM（amplitude modulation） 调幅
ANN（artificial neural network） 人工神经网络
ANSI（american national standard institute） 美国国家标准协会
API（application programming interface） 应用程序设计接口
APPN（advanced peer-to-peer network） 高级对等网络
ARP（address resolution protocol） 地址解析协议
ARPG（action role play game） 动作角色扮演游戏
ASCII（american standard code for information interchange） 美国信息交换标准代码
ASP（active server page） 活动服务器网页
ASP（application service provider） 应用服务提供商
AST（average seek time） 平均访问时间
ATM（asynchronous transfer mode） 异步传输模式
ATR（automatic target recognition） 自动目标识别
AVI（audio video interface） 声音视频接口

B

B2B（business to business） 商业机构对商业机构的电子商务
B2C（business to consumer） 商业机构对消费者的电子商务
BBS（bulletin board system） 电子公告牌系统
BFS（breadth first search） 广度优先搜索法

BGP (border gateway protocol) 边界网关协议
BIOS (basic input/output system) 基本输入输出系统
BISDN (broadband-integrated services digital network) 宽带综合业务数字网
BLU (basic link unit) 基本链路单元
BOF (beginning of file) 文件开头
BPS (bits per second) 每秒比特数
BRI (basic rate interface) 基本速率接口
BSP (byte stream protocol) 字节流协议
BSS (broadband switching system) 宽带交换系统

C

CAD (computer aided design) 计算机辅助设计
CAE (computer aided engineering) 计算机辅助工程
CAI (computer aided instruction) 计算机辅助教学
CAM (computer aided manufacturing) 计算机辅助制造
CASE (computer assisted software engineering) 计算机辅助软件工程
CAT (computer aided test) 计算机辅助测试
CATV (community antenna television) 有线电视
CB (control bus) 控制总线
CCP (communication control processor) 通信控制处理机
CD-R (compact disc recordable) 可录光盘，只写一次的光盘
CDFS (compact disk file system) 密集磁盘文件系统
CDMA (code division multiple access) 码分多路访问
CD-MO (compact disc magneto optical) 磁光式光盘
CD-ROM (compact disc read-only memory) 只读光盘
CD-RW (compact disc rewritable) 可读写光盘
CGA (color graphics adapter) 彩色图形适配器
CGI (common gateway interface) 通用网关接口
CI (computational intelligence) 计算智能
CISC (complex instruction set computer) 复杂指令集计算机
CMOS (complementary metal oxide semiconductor) 互补金属氧化物半导体存储器
COM (component object model) 组件对象模型
CORBA (common object request broker architecture) 公共对象请求代理结构
CPU (central processing unit) 中央处理单元
CRC (cyclical redundancy check) 循环冗余校验码
CRM (client relation management) 客户关系管理
CRT (cathode-ray tube) 阴极射线管
CSMA (carrier sense multi-access) 载波侦听多路访问
CSU (channel service unit) 通道服务单元
CU (control unit) 控制单元

D

DAC (digital to analogue converter) 数模转换器
DAO (data access object) 数据访问对象

DAP（directory access protocol） 目录访问协议
DBMS（database management system） 数据库管理系统
DCE（data communication equipment） 数据通信设备
DCE（distributed computing environment） 分布式计算环境
DCOM（distributed COM） 分布式组件对象模型
DDB（distributed database） 分布式数据库
DDE（dynamic data exchange） 动态数据交换
DDI（device driver interface） 设备驱动程序接口
DDK（driver development kit） 驱动程序开发工具包
DDN（data digital network） 数据数字网
DES（data encryption standard） 数据加密标准
DFS（depth first search） 深度优先搜索法
DFS（distributed file system） 分布式文件系统
DHCP（dynamic host configuration protocol） 动态主机配置协议
DIB（dual independent bus） 双独立总线
DIC（digital image control） 数字图像控制
DLC（data link control） 数据链路控制
DLL（dynamic link library） 动态链接库
DLT（data link terminal） 数据链路终端
DMA（direct memory access） 直接内存访问
DMSP（distributed mail system protocol） 分布式电子邮件系统协议
DNS（domain name system） 域名系统
DOM（document object mode） 文档对象模型
DOS（disk operation system） 磁盘操作系统
DQDB（distributed queue dual bus） 分布队列双总线
DRAM（dynamic random access memory） 动态随机存储器
DSD（direct stream digital） 直接数字信号流
DSM（distributed shared memory） 分布式共享内存
DSP（digital signal processing） 数字信号处理
DTE（data terminal equipment） 数据终端设备
DVD（digital versatile disc） 数字多功能盘
DVD-ROM（DVD-read only memory） 计算机用只读光盘
DVI（digital video interactive） 数字视频交互

E

EC（embedded controller） 嵌入式控制器
EDIF（electronic data interchange format） 电子数据交换格式
EEPROM（erasable and electrically programmable ROM） 电擦除可编程只读存储器
EGA（enhanced graphics adapter） 增强型图形适配器
EGP（external gateway protocol） 外部网关协议
EISA（extended industry standard architecture） 增强工业标准结构
EMS（expanded memory specification） 扩充存储器规范
EPH（electronic payment handler） 电子支付处理系统
EPROM（erasable programmable ROM） 可擦除可编程只读存储器

ERP（enterprise resource planning） 企业资源计划

F

FAT（file allocation table） 文件分配表
FCB（file control block） 文件控制块
FCFS（first come first service） 先到先服务
FCS（frame check sequence） 帧校验序列
FDD（floppy disk device） 软盘驱动器
FDDI（fiber-optic data distribution interface） 光纤数据分布接口
FDM（frequency-division multiplexing） 频分多路复用
FDMA（frequency division multiple address） 频分多址
FEC（forward error correction） 向前纠错
FEK（file encryption key） 文件密钥
FEP（front effect processor） 前端处理机
FET（field effect transistor） 场效应晶体管
FIFO（first in first out） 先进先出
FM（frequency modulation） 频率调制
FPU（float point unit） 浮点部件
FRC（frame rate control） 帧频控制
FTAM（file transfer access and management） 文件传输访问和管理
FTP（file transfer protocol） 文件传输协议

G

GAL（general array logic） 通用逻辑阵列
GCR（group-coded recording） 成组编码记录
GDI（graphics device interface） 图形设备接口
GIF（graphics interchange format） 图形转换格式
GIS（geographic information system） 地理信息系统
GPI（graphical programming interface） 图形编程接口
GPIB（general purpose interface bus） 通用接口总线
GPS（global positioning system） 全球定位系统
GSX（graphics system extension） 图形系统扩展
GUI（graphical user interface） 图形用户接口

H

HDC（hard disk control） 硬盘控制器
HDD（hard disk drive） 硬盘驱动器
HDLC（high-level data link control） 高级数据链路控制
HDTV（high-defination television） 高清晰度电视
HPFS（high performance file system） 高性能文件系统
HPSB（high performance serial bus） 高性能串行总线
HTML（hyper text markup language） 超文本标记语言

HTTP（hyper text transport protocol） 超文本传输协议

I

IC（integrated circuit） 集成电路
ICMP（Internet control message protocol） 互联网控制消息协议
IDC（international development center） 国际开发中心
IDE（integrated development environment） 集成开发环境
IDL（interface definition language） 接口定义语言
IEEE（institute of electrical and electronics engineers） 电气和电子工程师协会
IGP（interior gateway protocol） 内部网关协议
IIS（Internet information server） 互联网信息服务器
IP（Internet Protocol） 互联网协议
IPC（inter-process communication） 进程间通信
IPSE（integrated project support environments） 集成工程支持环境
IPX（Internet packet exchange） 互联网报文分组交换
ISA（industry standard architecture） 工业标准结构
ISDN（integrated service digital network） 综合业务数字网
ISO（international standard organization） 国际标准化组织
ISP（Internet service provider） 互联网服务提供者
IT（information technology） 信息技术
ITU（International Telecom Union） 国际电信联盟

J

JDBC（Java database connectivity） Java 数据库互连
JPEG（Joint Photographic Experts Group） 联合图片专家组
JSP（Java server page） Java 服务器页面技术
JVM（Java virtual machine） Java 虚拟机

K

KB（kilobyte） 千字节
KBPS（kilobits per second） 每秒千比特
KMS（knowledge management system） 知识管理系统

L

LAN（local area network） 局域网
LBA（logical block addressing） 逻辑块寻址
LCD（liquid crystal display） 液晶显示器
LDT（logic design translator） 逻辑设计翻译语言
LED（light emitting diode） 发光二极管
LIFO（last in first out） 后进先出
LP（linear programming） 线性规划
LPC（local procedure call） 局部过程调用
LSIC（large scale integration circuit） 大规模集成电路

M

MAC（medium access control） 介质访问控制
MAN（metropolitan area network） 城域网
MCA（micro channel architecture） 微通道结构
MDA（monochrome display adaptor） 单色显示适配器
MFM（modified frequency modulation） 改进调频制
MIB（management information bass） 管理信息库
MIDI（musical instrument digital interface） 乐器数字接口
MIMD（multiple instruction stream，multiple data stream） 多指令流，多数据流
MIPS（million instruction per second） 每秒百万条指令
MIS（management information system） 管理信息系统
MISD（multiple instruction stream，single data stream） 多指令流，单数据流
MMDS（multi-channel multipoint distribution service） 多波段多点分发服务
MMU（memory management unit） 内存管理单元
MMX（multi media extensions） 多媒体增强指令集
MPC（multimedia PC） 多媒体计算机
MPEG（moving picture experts group） 运动图像专家组
MPLS（multi-protocol label switching） 多协议标记交换
MPS（micro processor series） 微处理器系列
MTBF（mean time between failure） 平均故障间隔时间
MUD（multiple user dimension） 多用户空间

N

NAOC（no-account over clock） 无效超频
NAT（network address translation） 网络地址转换
NC（network computer） 网络计算机
NDIS（network device interface specification） 网络设备接口规范
NCM（neural cognitive maps） 神经元认知图
NFS（network file system） 网络文件系统
NIS（network information services） 网络信息服务
NNTP（network news transfer protocol） 网络新闻传送协议
NOC（network operation center） 网络操作中心
NSP（name server protocol） 名字服务器协议
NTP（network time protocol） 网络时间协议
NUI（network user identifier） 网络用户标识符

O

OA（office automation） 办公自动化
OCR（optical character recognition） 光学字符识别
ODBC（open database connectivity） 开放式数据库互联
ODI（open data-link interface） 开放式数据链路接口
OEM（original equipment manufactures） 原始设备制造厂家
OLE（object linking and embedding） 对象链接与嵌入

OMG (object management group) 对象管理组织
OMR (optical-mark recognition) 光标阅读器
OOM (object oriented method) 面向对象方法
OOP (object oriented programming) 面向对象程序设计
ORG (object request broker) 对象请求代理
OS (operating system) 操作系统
OSI (open system interconnect reference model) 开放式系统互联参考模型
OSPF (open shortest path first) 开发最短路径优先

P

PBX (private branch exchange) 用户级交换机
PCB (process control block) 进程控制块
PCI (peripheral component interconnect) 外部部件互连
PCM (pulse code modulation) 脉冲编码调制
PCS (personal communication services) 个人通信业务
PDA (personal digital assistant) 个人数字助理
PDF (portable document format) 便携式文档格式
PDN (public data network) 公共数据网
PHP (personal home page) 个人主页
PIB (programmable input buffer) 可编程输入缓冲区
PMMU (paged memory management unit) 页面存储管理单元
POP (post office protocol) 邮局协议
POST (power-on self-test) 加电自检
PPP (peer-peer protocol) 端对端协议
PPP (point to point protocol) 点到点协议
PPSN (public packed-switched network) 公用分组交换网
PR (performance rate) 性能比率
PROM (programmable ROM) 可编程只读存储器
PSN (processor serial number) 处理器序列号

Q

QC (quality control) 质量控制
QLP (query language processor) 查询语言处理器
QoS (quality of service) 服务质量

R

RAI (remote application interface) 远程应用程序界面
RAID (redundant array independent disk) 独立冗余磁盘阵列
RARP (reverse address resolution protocol) 反向地址解析协议
RAM (random access memory) 随机存储器
RAM (real address mode) 实地址模式
RAS (remote access system) 远程访问服务
RDA (remote data access) 远程数据访问
RDO (remote data object) 远程数据对象

RFID（radio frequency identification） 无线射频识别技术
RIP（raster image processor） 光栅图像处理器
RIP（routing information protocol） 路由选择信息协议
RISC（reduced instruction set computer） 精简指令集计算机
ROM（read only memory） 只读存储器
RPC（remote procedure call） 远程过程调用
RPG（role play game） 角色扮演游戏
RTS（request to send） 请求发送
RTSP（real time streaming protocol） 实时流协议

S

SACL（system access control list） 系统访问控制表
SAF（store and forward） 存储转发
SAP（service access point） 服务访问点
SCSI（small computer system interface） 小型计算机系统接口
SDLC（synchronous data link control） 同步数据链路控制
SDK（software development kit） 软件开发工具箱
SGML（standard generalized markup language） 标准通用标记语言
SHTTP（secure hyper text transfer protocol） 安全超文本传递协议
SIMD（single instruction stream, multiple data stream） 单指令流，多数据流
SISD（single instruction stream, single data stream） 单指令流，单数据流
SLIP（serial line Internet protocol） 串行线路接口协议
SMDS（switch multi-megabit data services） 交换多兆位数据服务
SMP（symmetric multi-processor） 对称式多处理器
SMTP（simple mail transport protocol） 简单邮件传输协议
SNA（system network architecture） 系统网络结构
SNMP（simple network management protocol） 简单网络管理协议
SNTP（simple network time protocol） 简单网络时间协议
SONC（system on a chip） 系统集成芯片
SONET（synchronous optic network） 同步光纤网
SPC（stored-program control） 存储程序控制
SQL（structured query language） 结构化查询语言
SRAM（static random access memory） 静态随机存储器
SRPG（strategies role play game） 战略角色扮演游戏
SSL（secure socket layer） 安全套接层
STDM（synchronous time division multiplexing） 同步时分复用
STP（shielded twisted-pair） 屏蔽双绞线
SVGA（supper video graphics array） 超级视频图形阵列

T

TCB（transmission control block） 传输控制块
TCP（transmission control protocol） 传输控制协议
TCP/IP（Transmission Control Protocol /Internet Protocol） 传输控制协议/网间协议

TDM（time division multiplexing） 时分多路复用
TDMA（time division multiplexing address） 时分多址技术
TDR（time-domain reflectometer） 时间域反射测试仪
TFT（thin film transistor monitor） 薄膜晶体管显示器
TFTP（trivial file transfer protocol） 简单文件传送协议
TIFF（tag image file format） 标记图形文件格式
TIG（task interaction graph） 任务交互图
TLI（transport layer interface） 传输层接口
TM（traffic management） 业务量管理，流量管理
TPS（transactions per second） （系统）每秒可处理的交易数
TTL（transistor-transistor logic） 晶体管-晶体管逻辑电路
TWX（teletypewriter exchange） 电传电报交换机

U

UART（universal asynchronous receiver transmitter） 通用异步收发器
UDP（user datagram protocol） 用户数据报协议
UHF（ultra high frequency） 超高频
UIMS（user interface management system） 用户接口管理程序
UNI（user network interface） 用户网络接口
UPA（ultra port architecture） 超级端口结构
UPS（uninterruptible power supply） 不间断电源
URI（uniform resource identifier） 环球资源标识符
URL（uniform resource locator） 统一资源定位器
USB（universal serial bus） 通用串行总线
UTP（unshielded twisted-pair） 非屏蔽双绞线
UXGA（ultra extended graphics array） 超强图形阵列

V

VAD（virtual address descriptors） 虚拟地址描述符
VAGP（variable aperture grille pitch） 可变间距光栅
VAN（value added network） 增值网络
VAP（value-added process） 增值处理
VAS（value-added server） 增值服务
VAX（virtual address extension） 虚拟地址扩充
VBR（variable bit rate） 可变比特率
VC（virtual circuit） 虚拟线路
VCPI（virtual control program interface） 虚拟控制程序接口
VDD（virtual device drivers） 虚拟设备驱动程序
VDR（video disc recorder） 光盘录像机
VDT（video display terminals） 视频显示终端
VDU（visual display unit） 视频显示单元
VFS（virtual file system） 虚拟文件系统
VGA（video graphics adapter） 视频图形适配器
VHF（very high frequency） 甚高频

VIS（video information system） 视频信息系统
VLAN（virtual LAN） 虚拟局域网
VLIW（very long instruction word） 超长指令字
VLSI（very large scale integration） 超大规模集成
VMS（virtual memory system） 虚拟存储系统
VOD（video on demand） 视频点播系统
VPN（virtual private network） 虚拟专用网
VR（virtual reality） 虚拟现实
VRML（virtual reality modeling language） 虚拟现实建模语言
VRR（vertical refresh rate） 垂直刷新率
VTP（virtual terminal protocol） 虚拟终端协议

W

WAN（wide area network） 广域网
WAE（wireless application environment） 无线应用环境
WAIS（wide area information service） 广义信息服务
WAP（wireless application protocol） 无线应用协议
WAV（wave audio format） 非压缩的音频格式文件
WDM（wavelength division multiplexing） 波分多路复用
WDP（wireless datagram protocol） 无线数据包协议
WFW（windows for workgroups） 工作组窗口
WML（wireless markup language） 无线标记语言
WORM（write once, read many time） 写一次读多次光盘
WWW（world wide web） 万维网
WYSIWYG（what you see is what you get） 所见即所得

X

XGA（extended graphics array） 增强型图形阵列
XML（extensible markup language） 可扩展标记语言
XMS（extended memory specification） 扩展存储器规范
XQL（extensible query language） 可扩展查询语言
XSL（extensible stylesheet language） 可扩展样式表语言

Z

ZA（zero and add） 清零与加指令
ZBR（zone bit recording） 零位记录制

附录 B 计算机专业英语词汇表

A

abnormal end 异常终止
abstract data type 抽象数据类型
acceleration card 加速卡
access control 访问控制
access list 访问控制表
access permission 访问许可
accessory program 附件程序
active desktop 活动桌面
active window 活动窗口
acyclic directory structure 非循环目录结构
adapter card 适配卡
adaptive scheduler 自适应调度
addressability 寻址能力
addressing mechanism 寻址机制
address space 地址空间
affordable 付得起的
algorithm 算法，规则系统
ambient intelligence 情景智能
analog signal 模拟信号
animation 动画
anomalies 异常，反常
anti-static coating 防静电涂层
anti-virus program 防病毒程序
application integration 应用程序集成
application layer 应用层
aspect ratio 纵横比
assembler 汇编程序
asymmetric encryption 非对称加密
asynchronous 异步的
asynchronous primitive 异步原语
atomic action 原子操作

atomicity property　原子属性
audio-output device　音频输出设备
authentication　鉴别，证实
authorization　授权，认可
avalanche　雪崩，崩塌

B

banner advertising　旗帜广告
bar code reader　条形码读卡器
baud　波特
baud rate　波特率
big data　大数据
binary digit　二进制数字
binary tree　二叉树
bind　绑定
bitable　双稳态的
bitmap graphic file format　位图图形文件格式
black box　黑盒子
block diagram　框图
block structure　模块化结构
Boolean logic　布尔逻辑
boot block　引导块
boot sector　引导扇区
boot strap　引导程序
boot up　启动
broadband modulation　宽带调制
broadcast address　广播地址
browser　浏览器
bubble jet printer　喷墨打印机
built-in　内装的，固有的
bulk storage　大容量存储器
bulletin board　告示板，公告板
bus-contention　总线争用

C

carriage return　回车
cartography　绘图法
certificate authority　证书认证
chain reaction　链式反应
check box　复选框
chaotic　混乱的
child node　子节点
child window　子窗口
chipset　芯片组

circuit switching   电路交换
classification   分级，分类
client program   客户程序
clipboard   剪贴板
client/server   客户机/服务器
close box   关闭框
cloud computing   云计算
coaxial cable   同轴电缆
coincided   相符，极为类似
command button   命令按钮
communication deadlock   通信死锁
communication satellite   通信卫星
community cloud   社区云
compression ratio   压缩率
connectionless service   无连接服务
connection-oriented service   面向连接的服务
content provider   内容提供商
context   上下文
control box   控制框
control panel   控制面板
critical region   临界区
cross platform   跨平台的
cryptography   密码学
cybercash   电子现金
cyberspace   计算机空间

D

data flow diagram   数据流程图
data link layer   数据链路层
data model   数据模式
debit-card   借记卡
data stream   数据流
data structure   数据结构
database engine   数据库引擎
database interface   数据库接口
database server   数据库服务器
data source   数据源
data transfer rate   数据传输速度
data window object   数据窗口对象
debugger   调试程序
decoder   译码器
decryption   解密
dedicated server   专用服务器
descrambler   伪随机序列译码器
digital camera   数码相机

digital cash　数字现金
digital wallet　数字钱包
demodulator　解调器
device contention　设备竞争
device dependent　设备相关的
device driver　设备驱动程序
device independent　设备无关的
device object　设备对象
dialog box　对话框
digital certificate　数字证书
digital signature　数字签名
discrete mathematics　离散数学
disk drive　磁盘驱动器
diskless workstation　无盘工作站
display adapter　显示适配器
distraction　娱乐，消遣；心烦，分心
distributed database　分布式数据库
distributed processing　分布式处理
distributed system　分布式系统
dot matrix printer　点阵打印机
dot pitch　点距
downlink　下行链路
download　下载
drop-down listbox　下拉式列表框
drop-down menu　下拉式菜单
drum　硒鼓
dynamic binding　动态绑定

E

elasticity　灵活性，伸缩性
electricity grid　电网
electronic cash　电子现金
electronic check　电子支票
electronic commerce　电子商务
electronic wallet　电子钱包
electronic meetings　电子会议
electronic banking　电子银行
electronic money　电子货币
electronic payments　电子支付
embedded computer　嵌入式计算机
embedded real-time system　嵌入式实时系统
encapsulation　封装，将相关的数据和过程打包在一个对象中
encryption key　加密密钥
end user　终端用户

entrepreneur 企业家，创业人，主办人
enumerate 枚举，列举
environment variable 环境变量
epidemics 流行病，蔓延
etched circuit 蚀刻电路
etching technology 蚀刻技术
Ethernet 以太网
expansion slot 扩展插槽
explosion 扩张，爆发
extraordinary 非常的，异常的，特别的

F

fatal error 致命错误
fault tolerance 容错
fiber-optic table 光纤
file handle 文件句柄
file server 文件服务器
file system 文件系统
firewall 防火墙
firmware 固件
flash memory 闪存
flexibility 弹性，适应性
flowchart 流程图，框图
flow control 流量控制
folder 文件夹
font format 字样格式
foreign key 外键，数据库中用以建立同其他表间的关联
fragmentation 存储碎片
frame 帧
frame rate 帧速
front-end 前端，前台
full-duplex 全双工
fuzzy set 模糊集

G

game theory 博弈论
gateway 网关
gigabit network 千兆网
global scheduler 全局调度
graphics tablet 图形输入板
grayscale 灰色标度
grid computing 网格计算
group editor 群编辑器

groupware 群件

H

hacker 黑客
hanging indent 悬挂式缩进
hardcopy 硬复制
hashed file 散列文件
headline 大字标题
head pointer 头指针
head node 头结点
header and footer 页眉和页脚
heap sort 堆排序
heterogeneity 异质，异种
hexadecimal system 十六进制
hierarchical directory structure 层次目录结构
high-level language 高级语言
home page 主页
host 主机
hub 集线器
Huffman codes 哈夫曼编码
Huffman tree 哈夫曼树
hyperlink 超链接
hypermedia 超媒体
hypertext 超文本

I

icon 图标
image map 图像映射
index 索引
inductance 电感
industrial sensors 工业传感器
information superhighway 信息高速公路
inheritance 继承
instant messaging 即时通信
instruction 指令
integrated package 集成软件包
intelligent bridge 智能网桥
interactive around-sound 交互式环绕声
interception 侦听
interface 接口，界面
interlacing 隔行扫描
internal memory 内存储器
Internet of things 物联网
Internet telephone 网络电话

interoperability 互用性，协同工作能力
interpreter 解释器，解释程序
int-jet printer 喷墨打印机
iteration 迭代

J

job scheduler 作业调度程序
joypad 游戏手柄
joystick 操纵杆
junk e-mail 垃圾邮件
jurisdiction 权限
justification 对齐，版面调整

K

kernel 核心
keypad 辅助小键盘
keystroke 键击
kilobit 千比特
kilobyte 千字节

L

laser-etched 激光蚀刻的
laser printer 激光打印机
layout 规划，布局
leased line 租用线路
legacy application 遗留应用
letter quality 字符模式
license 许可证
life cycle 生命周期
light-pen 光笔
linear linked lists 线性链表
linear list 线性表
local scheduler 本地调度
local variable 局部变量
location 定位，位置，配置
log file 日志文件
log on 登录
log out 注销登录
logical link control 逻辑链路更新
logical schema 逻辑模式
logic circuit 逻辑电路
logic complementation 逻辑补码法
low-level language 低级语言

M

machine code 机器码
machine language 机器语言
magnetic pot 磁场
magnetic tape 磁带
mailbomb 邮件炸弹
mail merging 邮件合并
mainframe 大型机，主机
main memory 主存储器
main window 主窗口
manufacturer 制造者，厂商
marginal cost 边际成本
media player 媒体播放器
megabit 兆位，百万位
megabyte 兆字节
megahertz 兆赫兹
mega-pixels 百万像素
mention 说到，提起，提及
menu bar 菜单栏
microcontroller 微控制器
microelectronic 微电子的
middleware 中间件
miniaturize 使小型化
minicomputer 小型计算机
mobile apps 移动应用程序
mobile commerce 移动商务
mobile communication 移动通信
mobile middleware 移动中间件
mobile phone 移动电话
mobile user 移动用户
mother board 系统板，主板
multidocument interface 多文档界面
multiline edit box 多行编辑框
multiple inheritance 多重继承
multi-threaded 多线程
multi-processor 多处理器
multitasking 多任务
mutual exclusion 互斥

N

nanosecond 纳秒
narrowband 窄带
natural language 自然语言

netiquette 网络礼仪
network layer 网络层
network administer 网络管理员
neuron 神经元
newsgroups 新闻讨论组
noise figure 噪声指数
non-blocking primitive 非阻塞原语
non-impact 非击打式
normalization 标准化，正常化
null string 空串

O

object-oriented system 面向对象系统
online business applications 在线商业应用
online file storage 在线文件存储
online-to-offline commerce 线上-线下电子商务
open system 开放系统
optical fiber cable 光导纤维电缆
optimal scheduling algorithm 最优调度算法
optimal tree 最优树
optimization 最优化，最佳化
ordered tree 有序树
orthogonal list 十字链表
outline view 大纲视图
outsource （将业务、工程等）外包
overfrequency 超频
overlapped 重叠
overloading 重载

P

page description language 页面描述语言
parallel port 并行接口
passive component 无源元件
peer entities 对等实体
pervasive computing 普适计算
photo-diode 光敏二极管
photo-resistor 光敏电阻
photo-sensitive drum 感光鼓
pie chart 饼图
pin printer 针式打印机
pirate 盗版者，盗版
plug and play（or PnP, P&P） 即插即用
pointing device 定位设备
polling task 轮流查询任务

polymorphism 多态
portal 门户网站，入口，正门
pop-up menu 弹出式菜单
pop-up window 弹出式窗口
precision transforms 精确度变换
preemptive multitasking 抢先式多任务
prevalent 流行的，盛行的，优势的，普遍的
primary key 主键，唯一标识数据库表中每条记录的一个或多个列
printed circuit board 印制电路板
privacy setting 隐私设置
private cloud 私有云
private key cryptography 私钥加密
privileged instruction 特权指令
procedural programming 面向过程程序设计
proliferate 扩散，激增
proprietary 专有的，独占的，专利的，有财产的
proxy server 代理服务器
public key 公开密钥
public key cryptography 公钥加密
pull mode 拉模式
push mode 推模式

Q

quad speed 四倍速
quadratic probing 二次探测
quantizer 数字转换器，编码器
quantometer 光谱分析仪
query 查询
queue 队列

R

radio button 单选按钮
real time system 实时系统
recursive function 递归函数
refresh time 刷新率，更新率
register contention 抢占寄存器
register pressure 寄存器不足
register renaming 寄存器重命名
relational model 关系模型
remote terminal 远程终端
resident 驻留的
resolution 分辨率
resource contention 资源冲突
response window 响应式窗口

ribbon cartridge 色带盒
right-click 右击
ring network 环形网络
router 路由器

S

safe mode 安全模式
screen saver 屏幕保护程序
screen-sharing technology 共享屏幕技术
seamless interconnection 无缝连接
search engine 搜索引擎
security certificate 安全认证
serial port 串行接口
security certificate 安全认证
sequential access 顺序存取
shared variable 共享变量
shareware 共享软件
shortcut key 快捷键
social media 社交媒体
solid ink 固体油墨
sound card 声卡
source code 源代码
spam 电子垃圾
speech generator 译音发生器
speech recognition 语音识别
speech synthesizer 语音合成器
stem from 起源于，出自，由……造成
storage class specifier 存储类标识符
streaming video 流体视频
structure chart 结构图
supercomputer 超级计算机
surround sound 环绕立体声
surveillance 监督，监视
syntax 句法
synthesizer 综合者，合成器

T

tangible 实在的，确实的，有实质的
taskbar 任务栏
telecommunicating 电子通信
tether 拘束，束缚，限度，范围
thin client 瘦客户端
thread 线程
three-dimensions 三维

throughput 吞吐量
time-sharing 分时
time slicing 时间分片
time-varying 时间变换的
title bar 标题条
touch-sensitive display 触控式显示器
traffic monitoring systems 交通监控系统
transparency 透明，透明度
trunk cable 中继电缆
trunk line 干线，中继线

U

ubiquitous computing 普适计算
ubiquity 泛在的，无所不在的
unauthorized access 未授权访问
undirected graph 无向图
uni-programming 单道程序设计
unordered tree 无序树
update 更新，修正
upgrade 升级
upload 上传
urban 城市的，在城市里的
usability 可用，有用，适用
user account 用户账号
user-defined 用户自定义
user ID 用户标识符

V

vacuum tube 真空管
vector graphics 向量图形
vendor 自动售货机
video bandwidth 视频带宽
video camera 视频相机
video capture card 视频采集卡
video clips 视频片段
video conferencing 视频会议
video phone 可视电话
videotext 可视图文，可视数据
virtual address space 虚拟地址空间
virtual block caching 虚拟块高速缓存
virtual memory technology 虚拟存储器技术
virus checker 病毒检查程序
voice mail 语音邮件
voice synthesis 语音合成

volume label　卷标
voice control　语音控制
vulnerability　脆弱性，攻击，弱点
vulnerable　易受攻击的，易受伤的

W

wave band　波段
wave form　波形
wave length　波长
web page　网页
wireless service provider　无线服务供应商
wiretapping　搭线窃听
workgroup hub　工作组集线器
workstation　工作站
worm　蠕虫
write protect notch　写保护口

Z

zero access　立即存取
zero complement　补码
zero suppression　消零
zoom in/out　移近/拉远镜头

参 考 文 献

[1] 汪海涛,郑鹤彬,陈永亮. 计算机实用英语[M]. 北京:电子工业出版社,2015.
[2] 吕云翔. 计算机英语实用教程[M]. 北京:清华大学出版社,2015.
[3] 王小刚. 计算机专业英语[M]. 4版. 北京:机械工业出版社,2015.
[4] 金志权,张幸儿. 计算机专业英语教程[M]. 6版. 北京:电子工业出版社,2015.
[5] 张芳芳. 计算机网络专业英语[M]. 北京:电子工业出版社,2014.
[6] 苏兵,张淑荣. 计算机英语[M]. 2版. 北京:化学工业出版社,2014.
[7] 刘纪红,邓庆绪,李曼宁,等. 物联网专业英语[M]. 北京:清华大学出版社,2014.
[8] 王春生,刘艺,等. 新编计算机英语[M]. 2版. 北京:机械工业出版社,2013.
[9] 卜艳萍,周伟. 计算机专业英语[M]. 北京:清华大学出版社,2010.
[10] 甘艳平. 信息技术专业英语[M]. 北京:清华大学出版社,2009.
[11] 张勇. 计算机专业英语[M]. 北京:北京大学出版社,2008.
[12] 温丹丽. 计算机类专业英语[M]. 北京:中国电力出版社,2008.
[13] 吴强. 计算机专业英语[M]. 北京:电子工业出版社,2008.
[14] 许春勤. 计算机专业英语[M]. 北京:电子工业出版社,2008.
[15] Shelly G B, Cashman T J, Vermaat M E. Discovering Computers 2007: A Gateway to Information[C]. Boston, MA: Thomson Course Technology, 2007.
[16] Joyce J, Moon M. Windows Vista Plain & Simple[M]. Indiana: Wiley Publishing, Inc., 2007.
[17] 王晔. 电子商务专业英语教程[M]. 北京:电子工业出版社,2007.
[18] 顾大权. 实用计算机专业英语[M]. 北京:国防工业出版社,2007.